MERIWETHER LEWIS

Of courage undaunted,

possessing a firmness

& perserverance of purpose

which nothing but impossibilities

could divert from its direction.

—Thomas Jefferson

MERIWETHER LEWIS

The Assassination of an American Hero and the Silver Mines of Mexico

Kira Gale

RIVER JUNCTION PRESS Omaha NE

River Junction Press LLC
Omaha, Nebraska
www.riverjunctionpress.com

Distributed in U.S. and Canada by IPG
www.ipgbook.com
Printed in the United States of America
First edition, 2015

ISBN: 978-09911409310

1 2 3 4 5 6 7 8 9 10

www.lewisassassination.com

Publisher's Cataloging-In-Publication Data
(Prepared by The Donohue Group, Inc.)

Gale, Kira.
Meriwether Lewis : the assassination of an American hero and the silver mines of Mexico / Kira Gale. -- First edition.

pages : illustrations, maps ; cm

Issued also in various ebook formats.
Includes bibliographical references and index.
ISBN: 978-0-9914093-1-0 (hardcover)
ISBN: 978-0-9914093-0-3 (paperback)

1. Lewis, Meriwether, 1774-1809. 2. Explorers--West (U.S.)--Biography. 3. United States--Politics and government--1797-1801. 4. United States--Politics and government--1801-1809. 5. Mexico--History--Wars of Independence, 1810-1821. 6. Lewis and Clark Expedition (1804-1806) 7. Wilkinson, James, 1757-1825. 8. Biography. I. Title.

F592.7.L42 G354 2015
917.804/2/092

FRONT COVER: Meriwether Lewis and James Wilkinson, by Charles Willson Peale (public domain), 1804 North American map by John Pinkerton. FRONTISPIECE: Meriwether Lewis, 1807 portrait by Charles Balthazar Julen Févret de Saint-Mémin (Missouri History Museum, St. Louis)

Dedicated to my husband, Henry Hanna Gale, who has always supported me in my efforts to learn more, and who has been my companion on countless road trips across America exploring our country and its history.

Contents

CHAPTER SUMMARIES

son—Lewis is determined to "get a wife"—Lewis, his brother Reuben, John Pernier and Seaman set off for St. Louis in November—Lewis surveys family land holdings in Kentucky—Lewis hires printer, Joseph Charless in Louisville, to relocate to St. Louis—Lewis arranges for Laws of the Territory of Louisiana to be printed by Charless when they arrive in St. Louis—Charless returns to Louisville to settle affairs—Lewis becomes both publisher and editor of the Missouri Gazette, first newspaper west of Mississippi River—Lewis rents house and shares it with newlyweds William and Julia Clark—Clark whipping his slaves who are unhappy about moving to St. Louis—Clark threatens to sell York to a severe master—York wants to stay in Louisville where his wife is enslaved—York's eventual fate—Lewis buys "Great Mound" of St. Louis—ancient Indian mounds in Cahokia—Lewis buys property for his mother in St. Louis, plans to sell property in Albemarle—explanation of drafts and bills of exchange, shortage of money on frontier—due to embargo, government is broke—Northern Confederacy wants to secede from union—Lewis founds first Masonic Lodge in St. Louis, serves as Grand Master—Lewis presides over territorial legislature—two forts established in 1808, Fort Madison for Sac and Fox, and Fort Clark (Osage) for Osage—Lewis writes three long letters regarding Indian affairs to Washington—British inspired Indian attacks, Shawnee chief Tecumseh recruiting allies for war against U. S.—Clark builds fort and makes a treaty with the Osage—Manuel Lisa returns to St. Louis from Montana—St. Louis Missouri Fur Company organized with 11 partners, including William Clark and Reuben Lewis—another expedition is planned to return Big White—Lewis becomes guardian of 13 year old Toussaint Jessaume, agrees to pay schooling and expenses in St. Louis for five years—Lewis has multiple enemies—Burrites and land claimants with fraudulent claims—Judge Lucas receives death threats—Smith T. organizes petition drive to not reappoint Lucas to Land Commission—Frederick Bates an enemy of Lewis—Postmaster John Hay an enemy of both Lewis and Clark—mail with bills or drafts takes 3 or 4 months to reach Washington—historian Donald Jackson's unfair treatment of Lewis—George Prevost plots secession with northern states—President Madison chooses cabinet based on sectional politics—Secretary of War Dr. William Eustis is Burr's best friend—Secretary of State Robert Smith and Eustis refused to reimburse Lewis for money he has personally spent to print Territorial Laws and return Big White—Missouri Fur Company under leadership of Manuel Lisa returns Big White to his village—Sacagawea and Char-

List of Maps

Maps by Kira Gale

Preface

Someday people will no longer remember what it was like to do research and write a book in the days before the internet. I started researching local history in the Omaha and Council Bluffs area in the early 1990's, after I got my first computer, a 128 megabyte Macintosh. It was before there was an internet, but I was sold on the delete, copy and paste, and print functions of a personal computer, rather than typing on a typewriter with carbon copies. If I needed information, I asked a reference librarian. I put coins in the copy machine at the library. Interlibrary loan books took months to arrive.

I filled about thirty fat notebooks with local history research in those years. My basic interest is in history connected to place. At first I didn't realize that early Omaha and Council Bluffs history was connected to the early American West. The original Council Bluffs got its name from Lewis and Clark's first meeting with Indians living in Louisiana Territory in 1804. The oldest American community west of the Missouri River is Bellevue, Nebraska, established in 1822. Bellevue is located in the old Council Bluffs area south of Omaha, and many residents in old Council Bluffs were connected to St. Louis.

Gradually I began building a library by going to second hand book stores. Visiting a used book store and finding a book was just a matter of luck in those days. I did research at the University of Nebraska-Omaha, which has a fine collection of early American history books and academic journals. In the 21st century, Google, Wikipedia, JSTOR, reprint books, internet archives and used bookstores on the web have developed into a truly golden age of research. Only a few books in the bibliography are not in my personal library. I have over fifty running feet of file folders on research topics, which I shuffle around in open topped file boxes. I bought collections of research notes by early researchers, and used the resources of the National Archives. It is a great time to do research.

I begin writing my first book, *Lewis and Clark Road Trips: Exploring the Trail Across America,* in 2001. It was supposed to come out during the bicentennial years of 2003–2006. However, it took years to complete, and was published at the end of the bicentennial in 2006. It contained over 800 tourist destinations, on facing pages of attractions and maps with driving directions. My husband and I traveled over 8,000 miles on the Lewis and Clark Trail. (He estimates we were lost 20% of the time.) I got to see the country, from Washington D.C. to the Pacific Coast, and the Montana border to New Orleans. It was a wonderful experience. My next project is to revisit the *Road Trips* book and bring it up to date in the digital age.

The Death of Meriwether Lewis: A Historic Crime Scene Investigation, was written with the cooperation of co-author, James E. Starrs, and published in 2009, the 200th anniversary of Lewis's death. Professor Starrs taught forensic law at George Washington University. In 1996 he organized an official coroner's inquest into the death of Lewis which was held in Lewis County, Tennessee. The coroner's jury asked the Department of Interior to allow the exhumation of Meriwether Lewis's remains in order to determine, if possible, the cause of his death.

It was national news, but permission for exhumation was denied. The Meriwether Lewis Gravesite & National Monument is located on the Natchez Trace near Hohenwald, Tennessee. Over 200 descendants of the Lewis family signed a petition in support of the request for exhumation. For a brief few months in 2008, the Department of Interior allowed the permit process to proceed, but then it reversed its decision. In 2009, I appeared on C-Span Book TV at the Southern Book Festival in Nashville, and on Brad Meltzer's *History Decoded* series on the History Channel.

The book contains the transcript of testimony at the coroner's inquest. Some of the most famous experts in the forensic world volunteered their time and expertise to testify as to what an exhumation might reveal. Jim Starrs asked me to supply additional material and I added twenty documents pertaining to the death of Lewis with my commentary.

I also included a narrative, "The Case for Murder," in which I proposed Lewis was assassinated by General James Wilkinson. I had read Vardis Fisher's book, *Suicide or Murder?: The Strange Death of Meriwether Lewis* years ago when it first came out, which convinced me it was murder. Another book, *The Jefferson Conspiracies: A President's Role in the Assassination of Meriwether Lewis* by David Leon Chandler, suggested Wilkinson was behind the murder, but I didn't agree with his main thesis, that Jefferson was involved.

Then I was asked to do unpaid research on the life of James Wilkinson for a proposed television mini-series. Wilkinson is, without a doubt, the most complicated person in the early American Republic to research. His career covered approximately 50 years, from the early days of the Revolutionary War to the independence of Mexico and his death in Mexico City in 1825. When he titled his massive three volume publication, *Memoirs of My Own Times,* he meant it.

This extended my research into the international geo-politics of the early American Republic, and this book is the result of these experiences. I have provided summaries for individual chapters in the Table of Contents, because I know it is a complex subject. Wilkinson did the same in his *Memoirs.* Readers may find it difficult to accept one of my main conclusions, that Wilkinson was a patriot as well as a double agent and a sociopathic killer. It is the key, however, to understanding his role in the tumultuous years following the American War of Independence from Great Britain and the democratic revolutions sweeping the world.

Although Wilkinson is often portrayed as a villain, this is the first attempt to examine his role in assassinating, or attempting to assassinate, people he considered his enemies.

Please visit www.lewisassassination.com to stay current. It is my intention to create video blogs on the topics presented in this book.

Kira Gale

Acknowledgements

Thanks to Earl Weidner, Tony Turnbow, and Jerry Gandolfo for sharing their research with me. Thanks for editing are due to Jack Young and Rick Blond. Rick shared his photos of the Lewis and Clark Trail with me. Rick took these photos while using the *Lewis and Clark Road Trips* book on his travels, which makes them very special.

Thanks to librarians and archivists: Dennis Northcott, Missouri History Museum; Jennifer Clark, Jefferson National Expansion Memorial; Monica Smith, Cairo Public Library; Anne Shepherd, Cincinnati Museum Center; Penny White, University of Virginia Special Collections; Mark Walters, Interlibrary Loan Director, University of Nebraska at Omaha; and Jack Robertson, Jefferson Library at Monticello. I have been helped by staff at: the Tennessee State Library and Archives; Albemarle-Charlottesville Historical Society; Boutetourt County Historical Society; Natchez Historical Society; Filson Historical Society; Chicago Historical Society; Newberry Library of Chicago; Archives of the University Missouri-Columbia; Special Collections of the University of Mississippi; the National Archives in Washington, Chicago and Kansas City; Beinecke Library of Yale University; Massachusetts Historical Society;University of Southern Illinois Library; and the University of Texas at Austin Archives. Librarians and archivists are a special group of people and I thank them all for their helpfulness.

The Lewis and Clark Study Group at the Western Historic Trails Center in Council Bluffs has been my support group for fifteen years. My husband Henry has always encouraged and facilitated my work. I am fortunate to have his support and companionship.

The interpretation of events is entirely my own.

Kira Gale

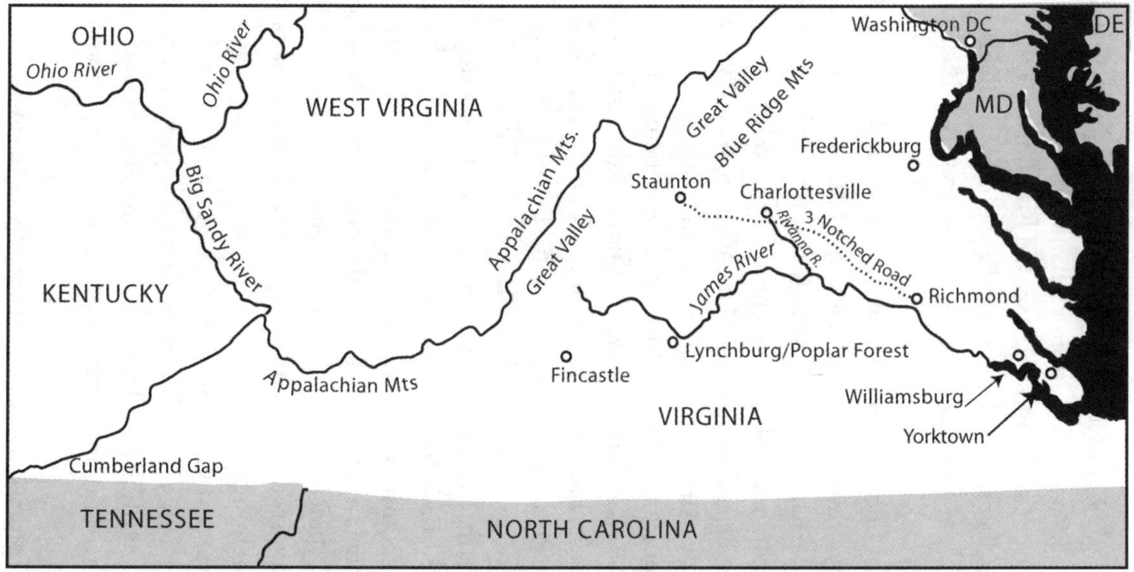

Virginia, the "Old Dominion," was the oldest and largest of the original thirteen colonies. West Virginia was part of Virginia and Kentucky was a county.

Chapter 1

Albemarle County

Meriwether Lewis was born at the dawn of the American Revolution, on August 18th, 1774 in Albemarle County, Virginia. It was one of the epicenters of the revolution that would go onto to shake the world. Thirty one year old Thomas Jefferson, married and with a newborn baby of his own, lived eleven miles away at his mountain top home, Monticello.

Jefferson had just completed writing *A Summary View of the Rights of British America,* a radical attack on the British Parliament. It would serve as talking points for Virginia's delegates attending the First Continental Congress in Philadelphia in September. Jefferson became ill with dysentery while traveling to the meeting and missed attending the congress.

The United Colonies had called the congress in response to the British Parliament closing the ports of Boston in retaliation for the Boston Tea Party. An open rupture and civil war was threatening, but it was not yet a reality. Virginia's delegates issued Jefferson's *Summary View* as a pamphlet which was circulated throughout the colonies and Europe. Two years later Jefferson attended the Second Continental Congress at Philadelphia, where he was chosen to write the Declaration of Independence, issued on July 4th, 1776.

The First Continental Congress was presided over by Jefferson's cousin, Peyton Randolph, Speaker of the Virginia House of Burgesses. The Randolphs, Jeffersons, Meriwethers and Lewises were among the leading families of Virginia's western frontier, bound together by ties of kinship and intermarriage. They were of English and Welsh descent. For generations, their families had migrated west along the James River, seeking new lands for their tobacco plantations. By the 1770's they had settled in the low rolling foothills on the east side of the Blue Ridge Mountains called the Virginia Piedmont.[1]

The Lewis family lived at Locust Hill, near Three Notched Road, the main east-west road of central Virginia. Marked by trees with three notches and mileage numbers cut in them, the road went from Richmond near the Atlantic Coast to Staunton in the Great Valley on the other side of the Blue Ridge Mountains. The road had been surveyed and laid out by Peter Jefferson, Thomas Jefferson's father. Charlottesville, established in 1762 on the Rivanna River, served as the county seat.

Locust Hill sketch from "Lucy Meriwether Lewis Marks: Her Life and Her World" (www.monticello.org)

William Lewis, Meriwether's father, inherited 1,900 acres of land from his father, Robert Lewis, and built a large log home called Locust Hill, a few miles northwest of Charlottesville. It was located on an elevated piece of ground near Ivy Creek with a grove of locust trees and a beautiful view of the surrounding mountains. The Meriwether and Lewis families were the first settlers in Albemarle County and its largest landholders.[2] The Lewis's were related to George Washington by marriage. Robert Lewis's brother, Fielding, was married to Betty, George Washington's sister.

Meriwether's mother, Lucy Meriwether, who was born in 1752, was the eighth of eleven children. Her father, Thomas Meriwether, a man of great wealth, was known as a "healer"—

someone who used medicinal plants to cure illnesses. Lucy would grow up to become a noted healer herself. After the death of Lucy's father in 1756, her mother married Robert Lewis, and the family moved to Lewis's nearby Belvoir plantation.

Lucy's stepfather became her father-in-law, because when Lucy was 17 years old she married 37 year old William, his son by his first wife.[3] Like many marriages in Virginia there was a great age discrepancy between husbands and wives, with men marrying girls in their teens. Marriage between cousins was also common in Virginia, as it kept the land in the family.[4]

The men of Virginia's western frontier became the elite ranger corps of the American revolutionary army due to their skill with hunting rifles and tomahawks. They were called "shirtmen" because they wore fringed hunting shirts made from heavy cotton fabric over their clothes. At first their enemies called them the "damn'd shirtmen," but it was soon adopted as a patriotic name. In June, 1775, members of the Virginia House of Burgesses wore hunting shirts with tomahawks strapped to their sides to honor the shirtmen.[5] A year later the burgesses would declare themselves to be the Virginia General Assembly, independent of British rule.

Meriwether's father, William Lewis, volunteered to serve without pay in the Virginia militia in 1775. But before we continue with the story of Lewis's childhood, another native son of Albemarle County needs to be introduced—George Rogers Clark, the older brother of William Clark, who became one of the greatest heroes of the Revolutionary War and the most famous of all the Virginia shirtmen.

In 1778–1779, George Rogers Clark's campaign against the British gained Northwest Territory for the United States.

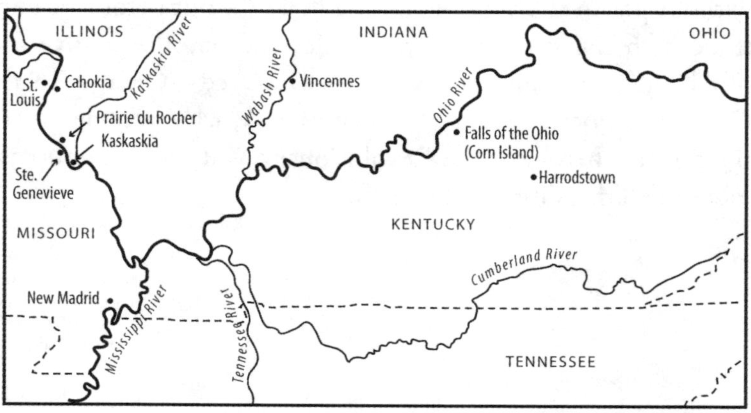

The French towns of St. Louis, Ste. Genevieve and New Madrid were under Spanish rule on the west side of the Mississippi. Kaskaskia, Prairie du Rocher, Cahokia and Vincennes were under British rule on the east side.

Chapter 2

George Rogers Clark

George Rogers Clark—who was called the "Conqueror of the Northwest"—was born in Albemarle County in 1752. He was an older brother of William Clark, the youngest son of the Clark family's ten children. William Clark's five older brothers fought in the Revolutionary War, and two became generals: George Rogers and Jonathan. The Clark family lived near the Rivanna River in Albemarle County until they inherited property in 1756 and moved back east to Caroline County, where William was born in 1770.

In the 20th Century, William Clark would become famous as Meriwether Lewis's partner in the expedition to the Pacific Coast, which was then called the "New Northwest." But during the 19th Century, his older brother, George Rogers, was much more famous for winning the "Old Northwest" for the United States during the Revolutionary War.

Meriwether Lewis and William Clark were children during the Revolutionary War. Their families were "Patriots" or revolutionaries, rather than "Loyalists." The country was about evenly split on this—about 1/3 Patriots, 1/3 Loyalists, and 1/3 undecided. It was natural for those who lived on Virginia's frontier in the foothills of the Blue Ridge Mountains to be Patriots. They wanted to cross over the Appalachian Mountains and settle in the lands claimed by Virginia, which extended all the way to the Mississippi River on both sides of the Ohio River.

The British had reserved the land west of the mountains for Indians at the end of the Seven Years War (1756-1763). Colonial militias, supported by troops from Great Britain, had won control of "New France," in what was called the "French and Indian War" by Americans, but was actually an international war fought for control of the North American continent. New France stretched from Hudson Bay in Canada to the gulf waters of Louisiana.

The struggle for control of the lands west of the Appalachians is the story of the early American Republic. It starts with the French and Indian Wars between Britain and France and their colonies in the new world; and includes Spain as the third great power in the conflict. Indians fought on all sides—where ever they thought they had the best chance of holding onto their lands. Everyone believed they had a legitimate claim to the land. They were uneasy partners in shifting alliances, conspiracies, land speculation, wars, and filibuster expeditions during the American Revolution and for many years afterwards.

George Rogers Clark, computer generated image. George Rogers Clark National Historical Park (www.nps.gov/gero)

In 1772, 20 year old George Rogers moved to the Ohio River Valley, where he worked as a surveyor for the Ohio Company. Two years later, Clark served in the Virginia militia during Lord Dunmore's War against the Shawnee and Mingo Indians. In 1776, when he was living at Harrodstown, Kentucky, he was elected as a representative of Kentucky County to the first Virginia General Assembly, the newly created legislative body replacing the House of Burgesses. After a journey of 700 miles to Williamsburg, he discovered the legislature had met and gone home. Clark then sought out the new Governor of Virginia, Patrick Henry, and obtained 500 pounds of gun powder for the defense of Kentucky against Indian attacks.

The British were supplying arms and ammunition to the Indians of the Ohio River Valley. The tribes had a combined force of about 8,000 warriors.[6] Clark returned to Kentucky with the gunpowder in the early spring of 1777. Henry Hamilton, Lieutenant Governor of Canada and Commander of Fort Detroit—who was called the "Hair-buyer General"—was paying Indians for the scalps of Americans. In the summer of 1777 Hamilton employed fifteen bands of Indians to raid the isolated settlements of Kentucky, killing or capturing men, women and children, taking scalps, torturing, and burning their homes.[7]

Clark decided the best defense against the British and their Indian allies would be to capture the British forts at Kaskaskia and Vincennes in Illinois country, the land between the Mississippi, Wabash, Ohio and Illinois Rivers. Four French villages were located here, with a population of about 900–1,000 whites and over 600 black slaves.[8]

Though French residents had lived in the Mississippi Valley for generations, they were now under foreign rule. When France lost the Seven Years War in 1763, the vast lands of "New France," were acquired by the British. The British then ceded Louisiana, the land west of the Mississippi River, to Spain in exchange for East and West Florida. French residents in the Mississippi Valley were governed by the British on the east bank, and by the Spanish on the west bank.

Clark wrote to Governor Patrick Henry:

> We must either take the town of Kuskuskies or in less than twelve month send an army against the Indians on Wabash, which will cost ten times as much and not be of half the service.[9]

Kaskaskia was located on the east bank of the Mississippi near its junction with the Ohio. Clark sent spies, disguised as hunters, to the Illinois country, who reported the British did not suspect an attack by Kentuckians and the French were not happy with British rule.

Clark once again journeyed to Williamsburg where he received permission from Governor Patrick Henry to mount a secret expedition to capture Kaskaskia.[10] He was appointed a lieutenant colonel with instructions to raise seven companies (50 men each) of militia for the defense of Kentucky. In May, 1778 Clark set out from Pittsburgh with about 150 militia volunteers, some private adventurers, and twenty families in a flotilla of flat boats bound for the Falls of the Ohio near present day Louisville, Kentucky.

The Falls of the Ohio were a strategic location, as boats were unloaded there to pass through the 20 foot drop of the rapids. Clark established a settlement on Corn Island, a 70 acre island in the middle of the rapids, where he revealed his secret plan to capture Kaskaskia. The island was to serve as a communications post during their expedition into Illinois country. The families, consisting of 60 civilians, were to raise corn on the island. A blockhouse was built for their defense. Corn Island was the founding settlement of Louisville, Kentucky.

In June, 1778 a messenger from Pittsburgh arrived with the news that France had joined in an alliance with the United States, and Clark set off to capture Kaskaskia with 175 men, chosen for their hardiness.

When they reached the Tennessee River they left the Ohio River and traveled overland, thus escaping discovery by the British and their Indian allies. Marching in single file like Indians and dressed in buckskins and hunting shirts, they arrived at the Kaskaskia River opposite the village of Kaskaskia on the evening of July 4th. Crossing the river in boats, they captured the British commander at his home without firing a shot. The fort had been destroyed by fire some years earlier.

The town of Kaskaskia, founded by Jesuits in 1703, was the capitol of Upper Louisiana during the French colonial administration. Its residents were mixed blood descendants of French traders who married women of the Illinois and other tribes. Black slaves, originally from Saint Domingue (Haiti), worked in the lead mines across the river. The men were voyageurs who traded for furs in Indian country, and farmers who grew corn and wheat for Lower Louisiana, shipping tons of flour to New Orleans annually. In 1763, at the close of Seven Years War between Britain and France, the residents of Kaskaskia became the subjects of Great Britain. Many had moved across the river to live in French towns under Spanish rule.

Clark met with the village leaders and explained they would be free under American rule, telling them the news that France had become an ally of America in the Revolutionary War. He assured Father Gibrault, the village priest, that the Catholic Church would be protected under Virginia's laws of religious freedom. Clark wrote:

> In a few Minutes the scean of mourning and distress was turned to an excess of Joy, nothing else seen or heard—Adorning the Streets with flowers and Pavilians of different colours, compleating their happiness by singing etc.[11]

The King of France had given them a bronze bell for their church in 1743. The bell was rung on July 4th to celebrate their liberation from British rule, and it became famous as the "Liberty Bell of the West."[12] The residents of Cahokia and Prairie de Roche, the French towns north of Kaskaskia, also took oaths of loyalty to the United States.

Pierre Gibault, S. J., the "Patriot Priest," and the Liberty Bell of the West at Kaskaskia, Illinois. (Indiana.gov and photo by Rick Blond).

Father Gibault went east to Vincennes on the Wabash River to persuade its residents to join the American cause. The local militia—Vincennes residents who were guarding the fort in the absence of British troops—switched sides and hoisted a new, homemade American flag over Fort Sackville.

Clark visited Fernando de Leyba, the Spanish Governor in St. Louis. The governor, newly arrived from Spain, agreed to support the Americans and French in the event of Indian attacks.

Spain joined the war against Britain in June, 1779, in order to regain its possession of the two Floridas.

Without money to pay for his expedition, Clark was signing bills of credit, payable by the state of Virginia. These bills were honored by Oliver Pollock in New Orleans, who paid $8,500 in silver to Illinois merchants. By January, 1779 Clark had received an additional $30,000 worth of supplies from Pollock, who was instructed by the Commercial Committee of Congress to give all possible support to Clark. At the end of the Revolutionary War, these bills would be disastrous for both men.[13]

For five weeks, in August and September, 1778, Clark held an Indian council at Cahokia, the French town across the river from St. Louis. Ten or twelve tribes from 500 miles around attended the gathering. During the council, two Winnebago Indians attempted to kidnap Clark, but instead they were seized and imprisoned. The French rallied to support the Americans—which impressed the Indians—and while the tribes were deciding how to proceed, Clark wrote that he:

> . . . assembled a number of Gentlemen and Ladies and danced nearly the whole night.

After that, he declared to the Indians that he came as a warrior not as a counselor, and peace or war would be their choice. They chose peace, and Clark allowed the two prisoners to go free. For years afterwards the tribes preferred to deal with Clark.[14]

In December, 1778, the British "Hair-buyer General" Henry Hamilton—upon learning of the surrender of the fort at Vincennes—marched from Detroit with a force of over 200 British regulars and French militia, and about 300 Indians to retake the fort.[15] When they arrived, the residents of Vincennes switched sides again. Four Americans at the fort were captured. Most of Hamilton's French militia returned to their homes in Detroit and Hamilton sent the Indians out to raid Kentucky settlements. He planned to winter at the fort and renew his campaign against the Americans in the spring.

Francis Vigo, a St. Louis merchant sent out by Clark to provision the fort, was at Vincennes when it was recaptured by the British. Not suspecting that Vigo was a friend of the Americans, Hamilton allowed him to return to St. Louis.

On January 29th, Vigo met with Clark at Kaskaskia and gave him detailed intelligence. Clark decided to risk everything and retake Fort Sackville in the dead of winter. If he didn't, Hamilton's spring campaign, utilizing large numbers of Indians, would destroy the Kentucky settlements.

On February 5th, Clark sent a gunboat with 46 men down the Mississippi and up the Ohio and Wabash Rivers with orders to wait for them near Vincennes. The boat carried heavy artillery and armaments.

March to Vincennes by F. C. Yohn (Wikimedia Commons)

On February 6th, Clark set out with 130 men—sixty of them French volunteers—from Kaskaskia to march to Vincennes.[16] The men traveled 180 miles across icy flood plains, often in rain, through mud and water several inches deep. What they endured is the stuff of legends. Clark wrote about their journey through the "drowned lands"—

> It was difficult and very fatiguing marching. My object now was to keep the men in spirits. I suffered them to shoot game on all occasions and feast on it like Indian war-dancers—each company, by turns, inviting the others to the feasts. . . . .myself

> and principal officers . . . running as much through the mud and water as any of them. [17]

As they neared Vincennes, they went two days without food. Game had fled the flooded area of the Wabash River and its streams. On February 23rd, six miles from Vincennes, Clark led the way into a shallow lake, four miles wide. He placed an officer with 25 men at the rear, with orders to shoot anyone who refused to march. In a wooded area, the water reached the shoulders of the tallest men, and they clung to trees and logs waiting to be rescued by two canoes.

After reaching dry land, large fires were built and soup was made from buffalo meat, corn, and tallow—the food and kettles appropriated from Indian women and children passing by in canoes. Camped within two miles of the fort, they took prisoners of five French hunters. The hunters were friendly and reported their presence was unsuspected.

Clark sent the hunters—who were unaware of how few men he had—back to the village with a letter. Clark's letter, read in the public square, warned the villagers to keep off the streets that night as the fort was going to be attacked. No one told Hamilton, but the American prisoners at the fort were alerted.

At sunset, Clark marched his men back and forth behind small hills with flags flying on tall poles, to persuade the villagers that he had a force of 1,000 men. When it was dark they approached the fort in two divisions, guided by the hunters. A few shots were fired, wounding one British soldier. It was only then that Hamilton realized he was under attack. The surprise was complete.

Sharpshooters with long rifles killed and wounded several men inside the fort during the night. At daybreak, Clark's men had their first real meal in six days. In the morning, Hamilton sent word he wanted a three day truce. Clark demanded unconditional surrender. The British had been tricked into thinking their attackers were more numerous, by the firing, noise and shouts coming from different directions. There were about 40 British soldiers and 40 French militia from Detroit inside the fort, with provisions enough for six months and more men on their way as reinforcements. But the bluff worked.[18]

While Clark and Hamilton were holding talks at a nearby church, a war party of about 20 Indians and two Frenchmen, returned to the village bringing scalps and two prisoners. They had come from raiding settlements in Kentucky. The British flag was flying over the fort. Firing a salute, they shot their rifles in the air and were defenseless when Clark's men opened fire, wounding and killing most of them. The two prisoners were released. One of the Indians, the 18 year old son of Pontiac, Chief of the Ottawa, was recognized and not harmed. The two Frenchmen were saved because the father of one of them was serving in Clark's army, and the other had a sister living near Vincennes who begged for his life. Clark allowed a man—who had lost his family in a brutal Indian attack—to scalp and tomahawk the remaining four Indians in full sight of the men in the fort.[19]

General Hamilton signed an unconditional surrender agreement, effective the next day, February 25th, 1779. British supply boats approaching the fort were captured and six tons of supplies were divided among the troops. The French militia at the fort were sent back to Detroit, and General Hamilton, his hated Indian Agent Jehu Hay and 25 other British prisoners were escorted across the Kentucky settlements to a prison at Williamsburg, where the general was put in irons and treated as a common prisoner. Thomas Jefferson, now the Governor of Virginia, refused to see him.[20]

George Rogers Clark's daring expedition to capture the British forts in Illinois country is credited with doubling the size of the United States, as the British ceded all their land claims west of Pittsburgh to the Mississipi River at the Treaty of Paris in 1783. After the claims of the states were settled, this land became the Northwest Territory in 1796.

Fall of Fort Sackville, 1779, by Frederick Yohn. (Wikimedia Commons)

Wabash River near George Rogers Clark Memorial in Vincennes, Indiana. (Photo by Rick Blond)

George Rogers Clark Memorial in Vincennes, Indiana.

George Rogers Clark statue.

Francis Vigo statue.

For more about the George Rogers Clark National Historical Park see www.nps.gov/gero. (Photos by Rick Blond)

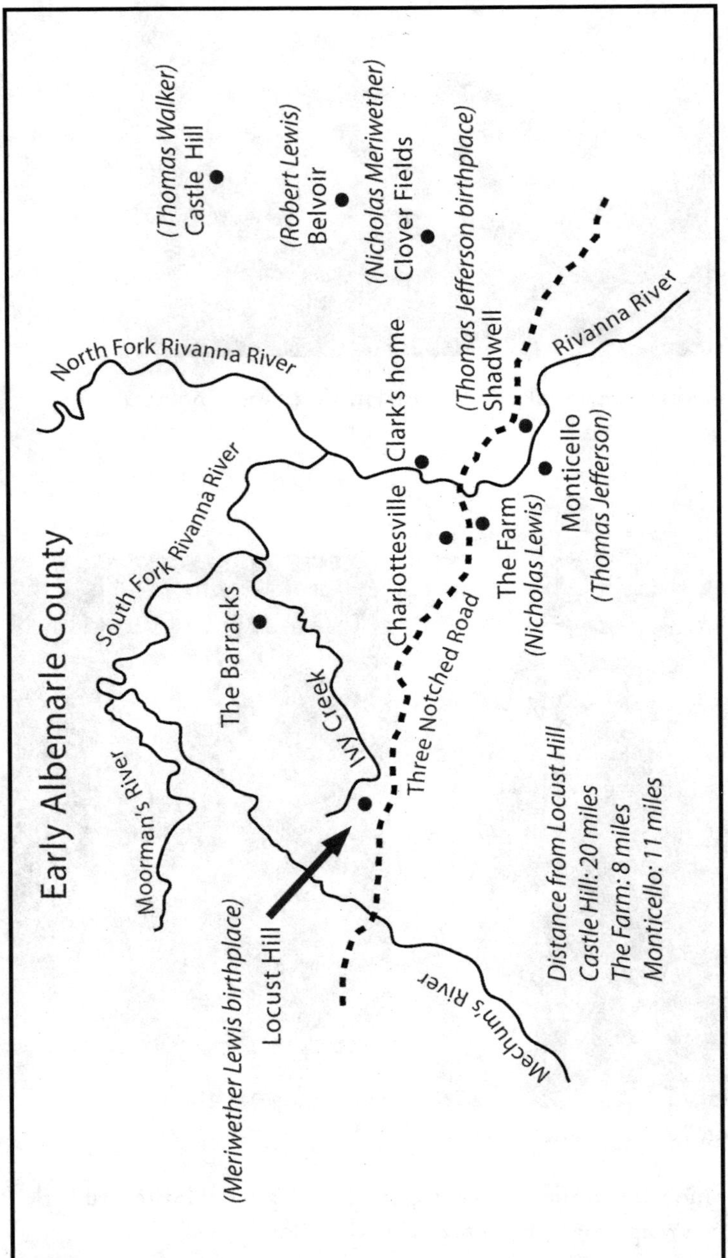

Early Albemarle County: map based on map on the "Lucy Meriwether Lewis Marks: Her Life and Her World" section of www.monticello.org website.

Chapter 3

Growing Up in Albemarle

When Meriwether Lewis was an infant, his father was away from home fighting as a Patriot. His father served as an unpaid volunteer in the Revolutionary War. Because he received no pay and paid his own expenses, there are few military records for William Lewis. He died in 1779, while on military leave in Albemarle, when Meriwether was only five years old.

William Lewis joined the First Independent Company of Albemarle County in April, 1775. The "gentlemen soldiers" pledged to muster four times a year—or oftener if necessary; to provide their own gun, shot-pouch and powder horn; and to appear on duty in a hunting shirt. Lewis, and 26 others of the Albemarle volunteers rode to Williamsburg in May, in response to the "Gun Powder Affair." [21]

Lord Dunmore, the Royal Governor of Virginia—reacting to Patrick Henry's "Give me liberty or give me death" speech—seized the gun powder stored at Williamsburg. It marked the beginning of the end of British rule over Virginia. In June, Lord Dunmore and his family fled to the safety of a British naval ship; and George Washington was appointed Commander-in-Chief of the Continental Army by the United Colonies.

In September, William Lewis became a lieutenant in the Albemarle Minutemen under the command of his older brother, Captain Nicholas Lewis, who was a close friend of Thomas Jefferson. The Minutemen were an elite force, ready to go into battle in a "minute." [22] Lewis may have remained a lieutenant in the Albemarle Minutemen, and never officially joined the First Virginia Regiment of the Continental Line because he was serving at his own expense.

The Albemarle Company fought Dunmore's loyalist troops in December, and became part of the First Virginia Regiment of the Continental Line in February, 1776.[23] They fought the Loyalists until the British abandoned Virginia in August. They were at Morristown in the winter of 1776-77; fought at Brandywine and Germantown in 1777; wintered at Valley Forge in 1777-78; and fought at Monmouth in 1778.

Meriwether's mother, Lucy Meriwether Lewis, was a well-known "yarb doctor." She was famous for her herb-doctoring, tending to the sick, including the slaves, of the Albemarle community. While her husband was away at war, she managed their tobacco plantation and its 24 slaves. George Gilmer, a neighbor both in Virginia and Georgia, wrote:

> Her person was perfect and her activity beyond her sex. She was sincere, truthful, industrious, and kind without limit.[24]

Lucy Meriwether Lewis Marks in her old age. (Wikimedia Commons)

Gilmer wrote about the Meriwether clan, who had emigrated from Wales:

> They brought more wealth with them to Virginia than was usual for emigrants in the seventeenth century. Most of them were peculiar in manners and habits; low and stout in stature, with round heads, dark complexion, and bright hazel eyes. They were very industrious and economical, and yet ever ready to serve the sick, and those who needed their assistance. They were too proud to be vain. They looked to their own thoughts and conduct rather than to what others might be thinking of them. . . No one ever looked at, or talked with one of them, but he heard or saw something which made him listen, or look

> again. They were slow in forming opinions, and obstinate in adhering to them. They were very knowing. But their investigations were minute and accurate rather than speculative and profound. Mr. Jefferson said of Col. Nicholas Meriwether that he was most sensible man he ever knew.[25]
>
> Nicholas, the oldest son of Thomas Meriwether, was brave in danger and self-possessed in the most difficult situations. He was one of four Americans who bore the wounded Braddock from the field of battle at his disastrous defeat. A gold-laced, embroidered coat sent him from Ireland, by General Braddock's sister, remained for a long time a curiosity in his plain household.[26]

In January, 1779, there was more than a British general's coat hanging on the wall at Clover Fields, there were 4,200 British and German prisoners of war living in Albemarle County. They were called the "Convention Army" because General John Burgoyne, their commander, had been allowed to sign a "treaty of convention" rather than a treaty of unconditional surrender. The terms of the treaty were that officers would sign an agreement "on the condition of not serving again in North America during the present contest." The troops were to exit America from Boston. However, George Washington opposed these terms, objecting that they would ship right back to America again. The Continental Congress decided they would remain in America, and they were moved to a series of different POW camps.

Almost 4,000 POW's arrived at Albemarle in January, 1779. Their arrival was unexpected; but by March, Jefferson was writing their living conditions were much improved—their camp was on a high hill, with convenient access to spring waters and several nearby flour mills. The officers were renting and renovating homes in the area, and the soldiers had laid out hundreds of fenced garden areas at the Barracks. They had chickens and pigeons. The Germans were provided with 200 pounds of vegetable seeds by General Riedesel, their commanding officer, who had rented a vinyard estate for his family near Monticello.[27]

In April, Albemarle citizens signed a "Declaration of Independence," renouncing King George III and pledging their allegiance to the Commonwealth of Virginia. William Lewis was the third signer on the document. Albemarle's population had almost doubled with the arrival of the Convention Army, and—with their officers and families living in the community—they felt a need to reaffirm their loyalty to the Patriots.

In May, the First Virginia Regiment was consolidated with other Virginia regiments, and in the summer they fought under General Anthony Wayne at the Battle of Stony Point. In November, Lieutenant Lewis was in Albemarle on military leave.

Lewis was returning to his unit based in Yorktown when his horse drowned in the flooded Rivanna River. He was able to swim across the river, but caught a chill. He went to his wife's old family home, Clover Fields, where he died of pneumonia on November 14th, 1779. Because of the flooded river he was buried at Clover Fields. On his death bed, he advised his wife that if Captain John Marks should come courting, she should marry him.[28]

Captain Marks, who had commanded a company at Valley Forge,[29] did come courting and on May 13, 1780, he married Lucy and became stepfather to Meriwether, age five; Jane age nine; and Reuben, age two.[30]

A few months later, the Convention Army was forced to leave Albemarle. The American military command was worried that the British army was going to invade Virginia. The coastal town of Yorktown—where Lieutenant Lewis had been based—guarded the entrance to Chesapeake Bay, the key to Baltimore and Washington D.C. as well as Virginia. In 1779 there were several British raids at Yorktown.[31]

Major Thomas Asbury, a British POW, wrote to friends on November 20, 1780:

> About six weeks ago we marched from Charlottesville barracks, Congress being apprehensive that Cornwallis in overrunning the Carolinas might by forced marches retake the prisoners. The officers murmured greatly at the step, having been

The Surrender of General Burgoyne at Saratoga by John Trumbull. A Virginia "shirt man," Colonel Daniel Morgan is given the place of honor in the painting. Morgan's Riflemen were crucial in winning the battles of Freeman's Farm and Bemis Heights, a turning point in the war. Burgoyne is shown surrendering his sword to General Horatio Gates. A young James Wilkinson may be seen behind Gates's outstretched hand. Wilkinson, age 20, actually conducted the negotiations with Burgoyne. (Wikimedia Commons)

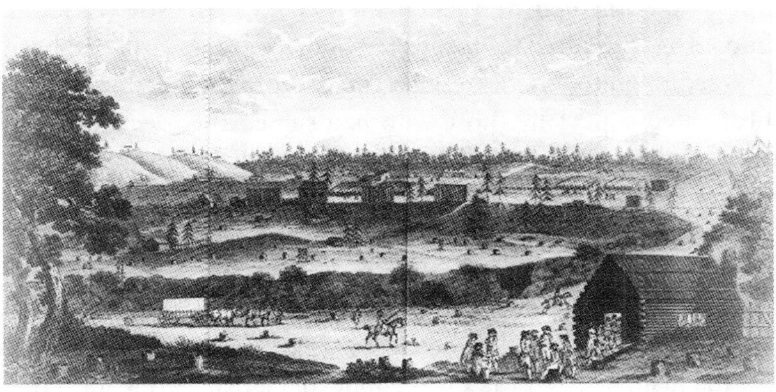

The Barracks: The Convention Army encampment at Albemarle. (Wikimedia Commons)

given to understand they were to remain until exchanged. Many had laid out considerable sums to render their huts comfortable, particularly by replacing the wood chimneys with stone, and to promote association, they had erected a coffee house, a theatre, a cold bath, &c. [32]

A few weeks later, on January 5, 1781, Brigadier General Benedict Arnold—who had switched sides and joined the British Army—led 1,600 American Loyalists in a raid on the new state capitol at Richmond. They burned buildings and destroyed government papers. For many years, it was believed George Rogers Clark's expense records from the Illinois campaign were destroyed in this attack.[33]

Arnold continued to lead raids in the Richmond area over the next months, joined by troops under the command of Major General Phillips, who had recently been a prisoner of war at Albemarle. In late May, Lord Cornwallis, commander of British forces in the South, arrived with his troops. There were now 7,200 British and Loyalist soldiers in Virginia.

Jefferson—who served for two years as Governor of Virginia—and the Virginia General Assembly met for the last time on May 10th in Richmond, and then fled to the safety of Charlottesville, where they convened again on May 24th.[34] They were pursued by Lieutenant Colonel Banastre Tarleton, known as "Bloody Ban" for his terror tactics. Tartleton had been ordered to go to Charlottesville and capture the notables assembled there, especially those who had signed the Declaration of Independence—Thomas Jefferson, Benjamin Harrison, Richard Henry Lee and Thomas Nelson, Jr.[35]

Tarleton moved quickly with a force of 250 soldiers, including 180 American Loyalists wearing green coats called the "Green Dragoons." On June 3rd, when Tarleton's men stopped at Cuckoo's tavern in Louisa County, Jack Jouett, an Albemarle militia man, realized what was about to happen, and made a ride that rivals Paul Revere's ride in American history. Jouett, a giant of a man at 6'4," took the fastest horse in seven counties, and rode 40 miles through the night to warn Jefferson of Tarleton's impending arrival. He reached Monticello before dawn on June 4th and then went to Charlottesville to warn the assemblymen.[36] Charlottesville consisted of about 12 houses, a tavern and courthouse along Three Notched Road.[37] The assemblymen agreed to meet again in three days time at the town of Staunton on the other side of the Blue Ridge Mountains.

When Tarleton and his troops reached Albemarle, they demanded breakfast at Castle Hill, a plantation owned by Dr. Thomas Walker. The Walkers delayed the British as long as possible. When the dragoons got to Charlottesville, they encountered Daniel Boone, loading records from the courthouse into a wagon. Boone was serving as Kentucky's representative at the assembly. They didn't realize who he was because he was dressed in a hunting shirt—until Jouett forgot and called him "Colonel." Jouett escaped, but Boone was captured. Boone spent the night of June 4th locked in a coal shed at "The Farm," where Tarleton was staying.[38]

The Farm was the home of Meriwether's uncle, Nicholas Lewis, who was away fighting in the war. Why Boone was released the next day is a matter of speculation, but it may have been that a relative of Boone's wife, serving in Tarleton's regiment, secured his release. Tarleton's men took all of the Lewis family's flock of ducks except one. After they left, Mrs. Lewis sent a servant with the one remaining duck to Tarleton.[39]

Tarleton's men treated Monticello respectfully, perhaps in recognition of the good treatment the Convention Army had received. Jefferson sent his family and several important guests away from Monticello, but took his own time in safeguarding his papers and getting his favorite horse shod. After he spied green coats with his telescope, he left Monticello with only a few minutes to spare before they arrived.[40]

Patrick Henry was another one of the assemblymen who escaped to Staunton ahead of Tarleton. The story is told that when he and two others stopped at a house to get some food, the woman scolded them and said if Patrick Henry had been there, he never would have run away. His friends laughed, and revealed his identity.[41]

Jefferson's term of office was up in June, and he refused to serve again, saying that Thomas Nelson, Jr., a militia general from Yorktown was better suited to govern Virginia. The final, decisive battles of the Revolutionary War, took place in Virginia, ending with the surrender of Cornwallis at Yorktown on October 19, 1781.

How did all of this affect Meriwether Lewis? He grew up in wartime and lived at one of its centers, due to the presence of Thomas Jefferson. His father had died and was buried at Clover Fields; there was a prisoner of war camp at the other end of Ivy Creek, and he had a stepfather serving in the war. But he was—after all—a Meriwether and a Lewis, and he was surrounded by a close knit family of relatives.

There are several stories about his mother that have come down to us. One story is that Lucy used a gun to scare off some drunken British officers from the POW camp when they came to Locust Hill. Another story is that a party of deer hunters went out from Locust Hill to get some deer, but had no luck. When they got back, they found Lucy had shot a deer in her back yard, skinned and dressed it, and had it cooked and waiting for them. Her hams were famous, and Jefferson bought several a year from her, although he smoked his own hams at Monticello.[42]

Jefferson wrote about Meriwether Lewis, as a young boy:

> At first [he] went to common day schools, learning to read, to write & Arithmetic with ordinary facility, he was early remarkable for intrepidity, liberality & hardihood, at eight years of ago going alone with his dogs at midnight in the depth of winter, hunting wading creeks when the banks were covered with ice & snow. He might be tracked through the snow to his traps by the blood which trickled from his bare feet.[43]

Meriwether would have been eight years old in the winter of 1782-83 when he was hunting raccoons at night. The peace treaty with Great Britain was signed in 1783, and war veterans began to make plans for the future.

John Marks, Lucy, and Lucy's youngest son Reuben moved to Georgia in 1785. The year before some Albemarle men had gone to Georgia to look at land in the Piedmont area of North Georgia. Marks was elected Sheriff of Albemarle County in 1785, but he decided to move to Georgia, leaving the post vacant.[44] He purchased land near the other families from Albemarle along the Broad River in Wilkes County.

Meriwether and his older sister Jane remained behind in Albemarle. Jane had gotten married at age 15. Meriwether was 11 years old, and near the age for advanced schooling. Jefferson wrote that he was left in the care of a family guardian:

> At eleven years old he was taken from his mother and remained until thirteen with his guardian, when he was past [his 13th birthday] to Latin schools kept by Dr. Everett, Parson Maury & Parson Wardell.[45]

Meriwether's guardian was his 24 year old cousin, William Douglas Meriwether, a surveyor.[46] Meriwether lived at Clover Fields with his cousin, who had inherited the property from his father, Nicholas Meriwether, Lucy's brother. His father's brother, Nicholas Lewis, lived close by at The Farm and served as his unofficial guardian. Nicholas Lewis, whose land was adjacent to Monticello, was acting as Jefferson's business agent while Jefferson served as Minister to France in 1784–1789.

The Locust Hill plantation had been left in the care of an overseer. Due to Virginia's Law of Primogeniture, Meriwether, as the eldest son of his father, inherited the entire property when his father died. It would remain in the care of his mother until he came of age. Lucy, as his father's widow, had dower rights to the use and income from one third of her husband's estate during her lifetime.

Meriwether visited his mother and stepfather in Georgia, in 1790 when he was 16 years old. Peter McGhee, an Albemarle neighbor who moved to Georgia with the Marks family related:

> In the section of the country in which they were the Indians sometimes made incursions, committed murders and carried

> off property. An alarm of Indians on the warpath got out creating great alarm, the neighbors met and consulted as to what they should do; they determined that they could not defend themselves, could easily be burned out, they determined the best thing, the only thing would be to take to the woods, they selected a situation made a large fire while assembled around it at night, a gun was heard not far off, a cry of Indians excited great alarm, mothers caught up their children and were weeping, when a boy in the crowd threw a bucket of water on the fire put out the light, in a moment all was quiet and a sensation of safety was felt by all, that boy was Meriwether Lewis.[47]

Meriwether's half brother and sister, John Hastings and Mary, were born in Georgia. Their father, John Marks, died in 1791. After his stepfather died, Meriwether had to assume his responsibilities as head of the family. The 17 year old was in Virginia studying at a classics school run by Reverend James Waddell, a famous teacher known as the "Blind Reverend." Patrick Henry said the minister was one of two of the greatest orators he ever heard.[48] In April, 1792, Lewis quit school. The next month he went down to Georgia to escort his mother and the three children back to Locust Hill.

Meriwether had previously studied with the Reverend Matthew Maury, the son of Thomas Jefferson's teacher, in 1788–89. In 1790 he studied with an Albemarle physician, Dr. Charles Everitt. One of Lewis's classmates at Dr. Everitt's school, left this description of him:

> Meriwether Lewis, afterwards distinguished for his expedition up the Missouri was one of our classmates. He was always remarkable for perseverance which in the early period of his life seemed nothing more than obstinacy in pursuing the trifles that employ that age: a martial temper: a great steadiness of purpose self-possession, and undaunted courage. His person was stiff and without grace, bow legged, awkward, formal

> and almost entirely without flexibility: his face was comely & by many considered handsome. It bore to my vision a very Strong resemblance to Bonaparte.[49]

Meriwether managed the farming operations at Locust Hill, but adventure called to him, and the military life appealed. In 1794, when he was twenty years old, he joined the Virginia militia which had been called up by President Washington to put down the tax protest known as the "Whiskey Rebellion" in western Pennsylvania. The militia of four states had been called up in the first test of federal power to use state militia to put down internal rebellion.

Thomas Jefferson wrote his recollections of Lewis's life from 1794 to 1801:

> From eighteen to twenty he remained on his farm an affectionate son and assiduous and devoted farmer, observing with minute attention all plants and insects which he met with. In his twentieth year he joined a volunteer corps as a private under T. Walker against the insurgents. During the same year he was appointed lieutenant in the U.S. army. In his twenty third year he was promoted to a Captaincy and returned to Albemarle to recruit. He again joined the army and acted as paymaster until he was made private secretary to the President.[50]

He got his facts a little bit mixed up regarding Lewis's military service, but the observations about Lewis's dedication to farming and interest in nature ring true.

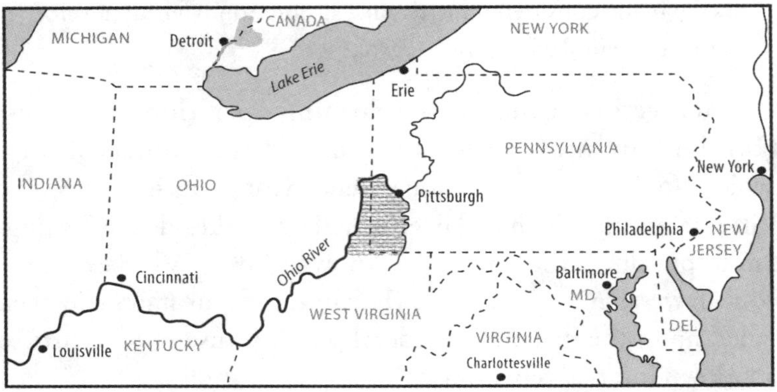

The center of the Whiskey Rebellion was south of Pittsburgh, in the area bounded by the Monongahela River on the east and the Ohio River on the west.

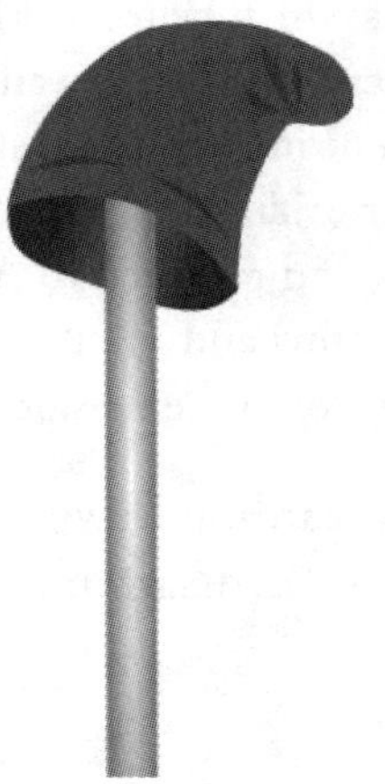

THE LIBERTY POLE

The Liberty Pole was often erected during the American Revolution. It was a pole stuck in the ground and topped with a red "liberty cap," or a red flag. The Phrygian cap dates back to the Roman Empire, and is found on the seal of the United States Senate. Liberty poles were erected along the roads and town squares of the western counties during the Whiskey Rebellion. (Wikimedia Commons)

Chapter 4

The Whiskey Rebellion: 1794–1795

In the fall of 1794, 20 year old Meriwether Lewis enlisted as an ensign in a Virginia militia unit commanded by General Daniel Morgan, the famed commander of "Morgan's Riflemen," the sharpshooters who played a prominent role in defeating General Burgoyne at Saratoga. The Virginians went to western Pennsylvania to suppress an armed uprising against the federal government called the Whiskey Rebellion. It was a popular revolt against a new excise tax which taxed whiskey at its place of origin at the manufacturer's still, rather than at the retail level. The tax had been created in 1791 to pay for the federal government assuming the states' war debts. Liberty poles were raised and the cry against tyranny was heard in the Monongahela River area south of Pittsburgh, where 1,200–1,300 rye whiskey stills were located—an estimated one quarter of all the stills in America.[51]

Large distilleries didn't mind the tax, because they could pass the cost onto customers and it might force small distilleries out of business. Distilleries in the Pittsburgh area sent an estimated 100,000 gallons of whiskey to New Orleans in 1794.[52] Small producers, however, received very little cash income from whiskey, which instead they used for barter to obtain goods from storekeepers. A farmer rarely saw $20 in cash in a year.[53] The rebellion was centered in Washington County in the southwestern corner of the state.

There was a concern as to whether Great Britain or Spain, might intervene on behalf of the rebels.[54] In addition, there was speculation that Secretary of Treasury Alexander Hamilton had deliberately provoked the crisis in order to demonstrate a need for a standing army.[55] In 1794 Hamilton and President George Washington organized a coalition of nearly13,000 state militia draftees and gentlemen volunteers to go across the Appalachian mountains to western Pennsylvania to restore order.

The 62 year old president, accompanied by Hamilton, went to war in a carriage. Washington was revisiting the scenes of his youth when he had served as an aide to British General Braddock during the French and Indian War. He had also been a surveyor in the Ohio River Valley, and was still a leading land owner. The president returned to the nation's capitol at Philadelphia after visiting Fort Cumberland, Maryland, and Hamilton proceeded on as the unofficial civilian head of the expedition. Hamilton had been Washington's most trusted staff aide during the Revolutionary War.

In the end, it proved to be a non-starter as a war. Twenty low level rebels were brought back to Philadelphia, but only two stood trial, and they were pardoned. Over 2,000 of the hard core rebels moved farther west to escape prosecution.[56] General Anthony Wayne's decisive victory at the Battle of Fallen Timbers in August, 1794—which marked the end of the Indian Wars in the Ohio River Valley—opened the region for exploration and settlement.

Fifteen hundred of Morgan's men remained in the Monongahela region over the winter of 1794–95 to maintain order.[57] Lewis's unit camped on the farm of one of the rebel leaders, Andrew McFarlane. His brother James McFarlane had been killed in an attack on tax collector's John Neville's house, an event which triggered the crisis.

Lewis wrote to his mother on November 24, 1794 from "McFarlin's Farm" that he was fifteen miles below Pittsburgh on the Monongahela River, and he intended to visit Kentucky in six months time when his enlistment was up. He wanted her to send the papers concerning land titles in Kentucky, and money to pay the taxes. He ended with the request to tell all his relations and acquaintances that: "I am quite delighted with a soldier's life."[58]

In a letter dated December 7th, he explained that the man who carried the letter to her could be entrusted with the papers and money for the Kentucky taxes and that he himself was employed in building huts. He wrote:

> Remember me to all the girls of the neighborhood as well as all my relations and acquaintances and hope you will never find me less than your affectionate son.[59]

He wrote to his mother on Christmas Eve:

> The situation of the soldairy is truly deplorable exposed to the inclemency of the weather . . . without any shelter more than what eight men can derive from a small tent. Many are sick but fortunately few have died as yet. My old Friend Capt. Thos. Walker has for a fortnight past been very much indisposed and has this day removed to the country where comfortable lodgings have been procured for him and all appearances indicate a speedy recovery. . . .[60]
>
> However to my great comfort I have this day been so fortunate as for the price of one dollar to procure a quart of Rum for a chrismas dram. [I] would thank you to send me four shirts and a pice of nankeen. . . I had the misfortune to have good part of my linin borrowed from me the other day with some other of my best clothes and am not able to repair the loss in consequence of these articles being extremely high in this country.[61]

By April 6th, 1795, the 21 year old Meriwether Lewis was writing about getting married and staying out west:

> I am sorry to find from my brother s letter that my absence is a cause of uneasiness to you . . . from the present situation of the Back country I find so great an opening for acquiring lands that I have declined returning until the fall . . . Remember me to Aunt and uncle Thomson and all the girls, and tell them that I shall bring an insurgiant Girl to see them next fall bearing the title of Mrs. Lewis.[62]

Nothing more is heard of the insurgent girl. General Morgan's volunteers were disbanded in the spring of 1795 and Lewis enlisted in the Legion of the United States on May 1, 1795. He was in Pittsburgh taking a small pox inoculation when he announced in a letter to his mother dated May 22:

> . . . your kind intreaty to return, yet notwithstanding all, so violently opposed is my governing passion for rambleing to the wishes of all my friends that I am led intentionally to err and then have vanity enough to beg for forgiveness. . . . I have mentioned to Rubin my having joined the fideral Army which puts it out of my power to leave this Country until the conclusion of this summer's campaign is over. . . . I must conclude by signing myself your ever sincair tho' wanderong son.[63]

He addressed the letter to "Cittizen Lucy Markes," following the trend of the new Democratic-Republican societies, who were calling everyone "citizen" in the spirit of sympathy with the French Revolution. James McFarlane had been the chairman of the Society of United Freemen of Mingo Creek, one of three Democratic-Republican societies in western Pennsylvania accused of inciting the Whiskey Rebellion.

In a letter dated May 23rd, Lewis wrote to the father of Thomas Walker that he had taken care of placing a tombstone on the grave of his son—his "old friend"—who had died on January 15, 1795. Lewis wrote:

> Before I left McFarlings Farm every arrangement was made that I could conceive necessary either to interment of the remains or to perpetuate the memory of your son. Sir, the willows that you sent me were carefully planted, but unfortunately few of the slips have any appearance of living. There are representative trees of that kind in this place. I'll send some of the slips by an officer belonging to the garrison who has promised me he will plant them with great care himself. There being no locusts in the neighborhood, I was obliged to make use of black walnut for both post ends and railing. His grave was handsomely set on a high knoll. We erected a well polished stone two feet and a half in breth and five feet in length and on it is engraved the following inscription: In memory of

> Capt. Thomas Walker who died January 15, 1795 in the 20th year of his age.[64]

It seems likely that Thomas Walker was a schoolmate friend of Lewis's, and that his home was near Jump Mountain, about 70 miles west of Charlottesville. It is a sad commentary on the death of Meriwether Lewis that no one—family, friends, government officials, or Masonic lodge brothers—took the same care with his grave site at the time of his death in 1809.

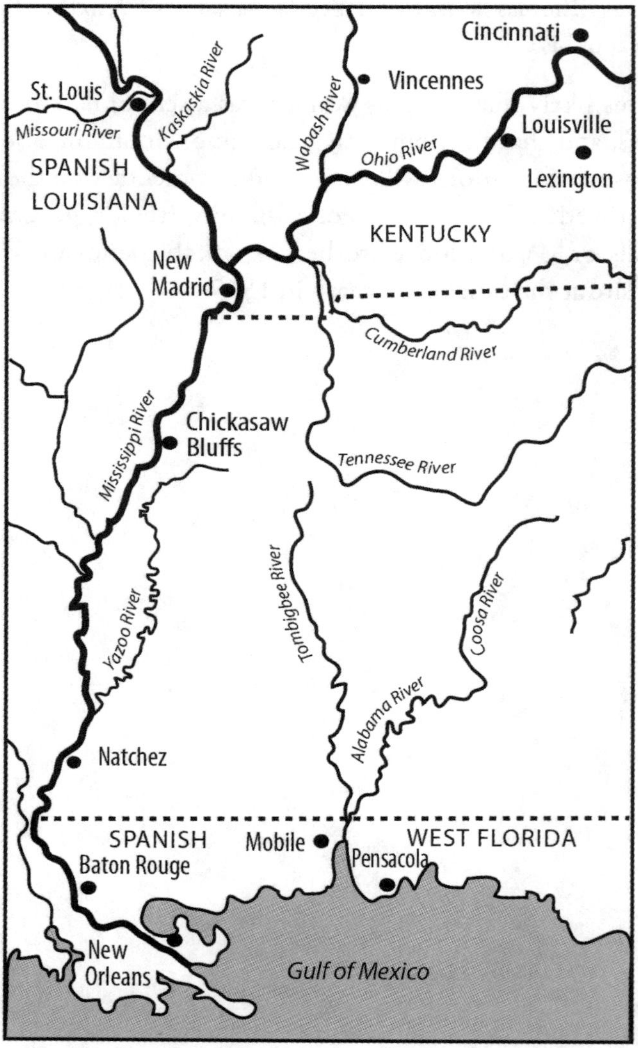

East side of Mississippi River, south of Kentucky

This map says it all—why the land south of Kentucky was so irresistible to westerners and foreign governments. It was good agricultural land, adjacent to the Mississippi River and the Gulf of Mexico and a natural location for a new independent country. The Chickasaw, Choctaw and Creek Indians still lived there. It is a central theme of the early American Republic and the story of James Wilkinson.

Chapter 5

James Wilkinson

The case will be made in this book that General James Wilkinson was responsible for the assassination of Meriwether Lewis and others. For this reason, the story of Wilkinson's activities—before he meets Meriwether Lewis in 1795—needs to be told.

Wilkinson dominated the western frontier from the time of his arrival in Kentucky in 1783 until his death in Mexico City in 1825. He is one of the most complex and intriguing figures in American history. No less a historian than Frederick Jackson Turner called Wilkinson, "the most consummate artist in treason the nation has ever produced." However, a study of his career has convinced me that he was a double agent, a scapegoat, and a patriot—as well as an assassin. I call him a "patriotic sociopath." There have always been mixed sentiments among historians about Wilkinson. With access to historic documents and the use of search functions on the internet, it is much easier to construct a nuanced version of his life.

Wilkinson wrote a massive three volume set of books called *Memoirs of My Own Times*, published in 1815. Much of it was material he assembled in his own defense during a series of congressional investigations and military courts-martial—all of which ended with his acquittal. On the title page of his books, he provided a quote from the English poet Richard Savage:

> For patriots still must fall for statesmen's safety,
>
> And perish by the country they preserve.

The implication is that he has been scapegoated to protect higher ups and feels victimized, although he is still a "patriot."

General Wilkinson entered Lewis's life after Lewis joined the Legion of the United States in 1795. (The Legion would become the United States Army the next year.) In 1795, Wilkinson was 38 years old and second in command to General Anthony Wayne, commanding general of the legion. Wilkinson was Wayne's mortal enemy and a mortal enemy to George Rogers Clark—although apparently Clark didn't know it. From this time on, he is a pivotal figure in Lewis's life and he would become his mortal enemy also.

Wilkinson was smart, manipulative, and intent on achieving power. He was vain, charming, and lied easily. He knew how to throw a good party and loved music and dancing. Although he drove his superiors to distraction with lengthy and verbose reports, he was a responsible army officer, who didn't shirk hard work and frontier assignments. He was interested in science, exploration, horticulture and maps, and was elected to the American Philosophical Society.

He had been born in 1757 on a plantation near Benedict, Maryland, forty miles southeast of Washington, D. C. When he was six years old his father died. As the second son, James was not entitled to inherit any property, and at 14 he was apprenticed to a family relative to begin the study of medicine. At 17 he went to Philadelphia to finish his medical studies, and it was there that he met his future wife, Ann Biddle, the daughter of a prominent Quaker family. After a few months of practicing medicine in Maryland, he enlisted in the American army in 1776.

Wilkinson rose to power during the Revolutionary War, serving as an aide to Generals Nathaniel Greene, Benedict Arnold, Arthur St. Clair, and Horatio Gates in 1776–1777, when he was 19 and 20 years old. After the Battle of Saratoga, General Gates allowed Wilkinson to conduct the surrender negotiations with British General John Burgoyne. Gates sent Wilkinson with a copy of the surrender terms to the Continental Congress and recommended that he be promoted from the rank of lieutenant-colonel to brevet brigadier-general for his services, calling him a "military genius." [65]

Wilkinson was accused of shamefully delaying his travels (taking 11 days to travel 285 miles) to reach Congress, which was meeting at York, Pennsylvania. News of the surrender reached York before he did. En route to York, he visited "his beloved" Ann Biddle at Easton, and met with Major-General William Alexander (Lord Stirling) and his aides Major James Monroe and Colonel William McWilliams at Reading. As he traveled, Wilkinson discussed General Gates and the "Conway Cabal" with several more high ranking officials.[66]

Thomas Conway
(Wikimedia Commons)

General Horatio Gates
(Wikimedia Commons)

Thomas Conway was a French army officer and a native of Ireland. He came over with the young Marquis de Lafayette and other volunteers to serve in the American army. Given a commission as a major general, Conway joined in a conspiracy to have General Gates replace George Washington as commanding officer of the army. There was also a plan for the army to invade Canada under French leadership. Many members of Congress supported the removal of Washington and the invasion of Canada, without realizing the French officers planned to betray the American army and take control of Canada for France.

The story was that Wilkinson had gotten drunk and told Colonel McWilliams about a letter written by Conway to Gates, in which Conway had written disparagingly of Washington:

> Heaven has been determined to save your country, or a weak General and bad counsellors would have ruined it. [67]

McWilliams was treasurer of George Washington's Masonic lodge in Fredericksburg.[68] Excerpts of the letter reached George Washington quickly, and Washington used the information to discredit his opponents.

Conway resigned from the army, and the invasion of Canada was halted, due to Lafayette's loyalty to General Washington. The young Lafayette had agreed to lead the army in the Canadian campaign, not realizing that his own French officers were part of a cabal to overthrow Washington. He insisted he would take orders only from Washington, and he appointed officers loyal to Washington, breaking up the conspiracy.[69]

Wilkinson acted as a secret agent in conferring with the Biddle family and army officers loyal to Washington as he traveled to York. Although General Gates had allowed him to conduct the surrender negotiations with Burgoyne's army, it had not blinded Wilkinson to Gates's disloyalty to Washington, and he became a pivotal figure in exposing the French cabal and Gates' involvement with it.

The idea that Wilkinson was actively supporting Washington is a controversial one, as historians have said that he unintentially revealed the plans of Conway and Gates because he was drunk. However, the clue is that he met with the Biddle family before his so-called drunken revelations. Clement Biddle, Ann's older brother, was a close friend of George Washington.

As a reward for negotiating the British surrender, the twenty year old Wilkinson received his commission as brevet brigadier-general. However, his irregular promotion made Wilkinson a target for officers and congressmen who were opposed to General Washington, and Wilkinson decided to resign from the army in March, 1778.

He and Ann Biddle were married in November, 1778. She was 36 years old and he was 21.The Biddles were Patriots. Ann's two army officer brothers, Clement and Owen, had been expelled from Quaker meeting because of violating their religion's belief in pacifism and Ann was expelled because she married outside her faith. Their marriage took place at the Anglican Christ Church in Philadelphia.[70] Fifteen signers of the Declaration of Independence worshipped at this church.

James Wilkinson
(1757-1825)

Ann Biddle Wilkinson
(1742-1807)

from *Tarnished Warrior: Major-General James Wilkinson* by James Ripley Jacobs

In July, 1779, the 22 year old Wilkinson was appointed Clothier-General of the U.S. Army. He took the job, even though it needed a drastic overhaul in its procedures and he had little interest or experience in business. The next month, he purchased Trevose Manor, an "attainted estate" confiscated from a Loyalist, located 18 miles northeast of Philadelphia.

He joined the Masons and became Master of the Dancing Assembly in Philadelphia. General Washington was not pleased that his new Clothier-General preferred the social life of Philadelphia to his army duties. Wilkinson resigned his post in March,1781 after making recommendations for improving the department. Their first child, John Biddle, was born in 1781; their second, James Biddle, in 1783. Wilkinson became a brigadier general in the Pennsylvania militia and entered politics, serving in the Pennsylvania state assembly for two years. At the end of the war in 1783, he made his first trip to Kentucky where he purchased land.[71]

The family moved to Lexington in the heart of Kentucky's blue grass country in 1784. Wilkinson entered business selling general merchandise for a large Philadelphia firm, Barclay, Moylan & Company. He speculated in land, buying land for himself and others. He traveled widely, and said he knew more about "western lands than any Christian in America," purchasing everything from town lots to tracts of 60,000 acres.[72]

They had moved to one of the most dangerous places in the country—Kentucky. More than one third of the Kentucky militia had lost their lives in the "Last Battle of the Revolution" at Blue Licks in August, 1782. Nearby Bryan's Station at Lexington had been the Indians' original target, but was so stoutly defended that the Indians retreated. The militia followed the Indians, and at least 77 militia men were killed in an ambush by 400-600 British-led Indians. It was the worst American defeat during the Revolutionary War on the frontier.

George Rogers Clark—now a brigadier-general in the Virginia militia—was in charge of defending Kentucky and the Illinois Country. He was blamed for the defeat, even though he wasn't at Blue Licks. He was at Fort Nelson at the Falls of the Ohio, which he had built as the key defensive post for the western country. Militia leaders accused Clark of being drunk and incompetent, to shift the focus from their own refusal to heed his advice and their own poor judgement in allowing themselves to be trapped in an ambush.[73]

In 1784, Jefferson succeeded in having Clark appointed as one of five Commissioners for Indian Affairs to treaty with the Indians of the Northwest.[74] Clark also supervised the surveying of bounty lands granted to veterans of the Illinois Regiment and Virginia State Line. Clark made his home at Clarksville, at the Falls of the Ohio on his own bounty lands grant.[75]

The Indian attacks continued. A report to the Secretary of War in 1790 stated that in the six years between 1784–1790, 1,500 residents of Kentucky were killed by Indians; and 20,000 horses and property worth 15,000 pounds of sterling had been stolen or destroyed by the Indians.[76]

Reenactors on the front porch of George Rogers Clark's reconstructed log cabin overlooking the Falls of the Ohio at Clark's Point in Clarksville, Indiana

In early 1786, 400 American residents at Vincennes—located on the Wabash River in Illinois Country—appealed to General Clark and Patrick Henry, the Governor of Virginia, to come to their aid in defending Vincennes against Indian attacks. Vincennes had been the scene of Clark's victory over the British in 1779. The governor authorized the counties in Kentucky to draft militia to attack the Shawnee Indian towns near Vincennes. Shawnee warriors had been terrorizing Kentucky settlements as well as Vincennes.[77]

General Clark and Colonel Benjamin Logan were each to command half the militia draftees raised in Kentucky. Clark, led an army of 1,200 men from Clarksville. As they marched towards Vincennes, 200 men mutinied over food supplies. The militia were not required to serve out of state, and there were no federal troops accompanying Clark.

The Indian villages on the Wabash River had been left unprotected when the Shawnee warriors went out to oppose Clark. During their absence, 800 militia under the command of Colonel Logan burned ten of their villages and destroyed an estimated 15,000 bushels of corn.[78] Logan's men also destroyed Louis Lorimier's trading post on the Great Miami River, which was providing arms and ammunition to the Indians. The next year, Lorimier obtained a Spanish land grant and moved west to Cape Girardeau on the Mississippi with a group of Shawnee and Delaware Indians.[79] (Lorimier's nephew was George Drouillard, who accompanied Lewis and William Clark on their expedition to the Pacific a few years later.)

Clark established a garrison of 140 men at Vincennes to protect the town. Because of low water on the Wabash, they were unable to receive food, clothing and other supplies. Clark seized $20,000 worth of goods and a boat belonging to three Spanish merchants who had come to Vincennes to trade. The merchants lacked the necessary trading permits from their own government; and American boats and goods were being seized by Spanish officials at Natchez. Clark called a court martial of his officers who decided the property of the merchants could be confiscated under military law.[80] This action by Clark was later used against him by Wilkinson in a wildly inflated charge of wanting to start a war with Spain.

Wilkinson saw George Rogers Clark as his rival for power when he arrived in Kentucky and devised an intricate plan to destroy his career. It counted on the fact that easterners,—unaware of the circumstances—would believe his lies about Clark's seizure of the Spanish boat and supplies at Vincennes and Clark's troubles with the militia.

Kentuckians held ten statehood conventions between 1784–1792, seeking independence from Virginia, free navigation on the Mississippi, and recognition as a state. Wilkinson played a leading role in the conventions, shaping policy and creating controversy.

He also bought land north of the Kentucky River and established the town of Frankfort in 1786, naming streets for himself and his friends. His wife Ann, however, after a short stay in Frankfort, decided they would move back to Lexington. In 1792, Frankfort became the capitol of the new state of Kentucky.

Before attending the statehood convention at Danville in

December, 1786, Wilkinson sent anonymous letters and forged documents to Edmund Randolph, the new governor of Virginia, and to members of the Confederation Congress. In the letters he spread lies about Clark being drunk and incompetent. They were ingeniously constructed so that the letters had to be shared among the recipients to learn the "full story." In that way, the contents seemed more believable.[81]

This set the stage for the letters which followed. On December 20th, Wilkinson wrote to Francisco Cruzat, the Spanish Governor at St. Louis, apologizing for Clark seizing the boat and supplies at Vincennes and warning the Spanish governor that an attack on Natchez by "desperate adventurers" would take place in the spring of 1787.[82]

He wrote two letters on the 22nd to Virginia Governor Randolph. The first letter, skillfully written, was signed by Wilkinson and endorsed with the signatures of high level friends attending the convention. It referred to "testimonials," which were copies of documents enclosed with the letter. His friends saw one set of "testimonials," but—before sending the letter—Wilkinson substituted another set of forged documents purporting to show that Clark was assembling a great army to invade Louisiana and seize it from the Spanish. The second letter, signed only by his friends, asked that Wilkinson be appointed Indian Commissioner in place of Clark.

This trickery was discovered by historians Temple Bodley and Lyman Draper, and an analysis of his byzantine methods was published in a lengthy appendix to Bodley's *Life of George Rogers Clark* in 1926.[83] Bodley was a Kentucky historian and descendant of the Clark family.

Wilkinson succeeded in his plan: Governor Randolph condemned Clark's actions at Vincennes and appointed Wilkinson as Indian Commissioner. However, the attorney general refused to indict Clark on technical grounds, which resulted in Clark not being able to clear himself in court, and Wilkinson's deceptions remained undiscovered.

Wilkinson was to employ these tactics over and over again. The consummate bureaucrat, he understood the value of a "paper trail" and created and destroyed documents to suit his purposes. In an era when letters took weeks to deliver, it was easy for him to remain undiscovered. Everything was written by hand and copies were routinely made by others. Wilkinson was ingenious in constructing complicated chains of false evidence—but he was not infallible, once the documents are examined carefully. In 1809-1811, he used the same methods to create a false trail of evidence concerning the death of Meriwether Lewis.

The Cabildo is now a museum located on Jackson Square in New Orleans. It was once the seat of Spanish government. This building replaced an earlier building which was destroyed in the Great Fire of 1788. (photo by Kira Gale)

Wilkinson boldly set out for New Orleans in June, 1787 to arrange a contract for himself with the Spanish government to be given the exclusive right to bring Kentucky goods down the Mississippi to New Orleans. Spain had closed navigation on the Mississippi River to Americans at the end of the Revolutionary War.

Rivers were the interstate highways of the time period. There were no modern roads. Economic development depended on the ability to ship bulky goods by water instead of hauling them across the Appalachian Mountains.

New Orleans was the hot spot at the mouth of the Mississippi River on the Gulf Coast. Whoever controlled the city controlled the region. The Spanish government had acquired New

Orleans at the end of the Seven Years War. It was a French city; its 5,000 residents were French speaking creoles, African slaves, and a free population of mixed bloods. It was the most cosmopolitan city on the continent.

Wilkinson paved the way for his adventure when he wrote to Cruzat, the Spanish Governor of St Louis, in December. His letter, warning about the attack on Natchez—which didn't happen because it was imaginary—, made his name known to Spanish officials and they allowed him to travel down the Mississippi without stopping him.

Don Estavan Miró, Governor of Louisiana and West Florida. (Wikimedia Commons)

Wilkinson arrived in New Orleans on July 2nd with flatboats of flour, butter, bacon and tobacco. He met with Don Estevan Miró, the Governor of Louisiana and West Florida, and told him Kentuckians were concerned the United States would cede the right of free navigation on the Mississippi to Spain for a period of twenty years, and because of that they were considering declaring their independence—and they were also talking about invading Louisiana by joining a British military expedition that would come down from Canada.

By these means, he persuaded Miró that the best policy would be to allow westerners to ship their goods to market through New Orleans and to grant Wilkinson the exclusive right to the shipping business. In return, he would encourage Americans to settle in Louisiana and become Spanish citizens, and Kentuckians to become independent. Miró bought the arguments, although he wrote the next year:

> I am aware that it is possible that it is his intention to enrich himself by means of inflating us with hopes and advantages, knowing that they will be in vain.[84]

On August 22nd, 1787 Wilkinson signed an oath of allegiance to the Spanish monarchy, becoming Spanish agent #13, and then presented a lengthy memorial (report) to Miró on the subject of western dissatisfaction with their federal government. He argued Spain should continue to deny free navigation rights on the Mississippi to westerners, because it would induce them to form an alliance with Spain; and that the Spanish government should offer financial benefits to men of real influence—in other words, to Wilkinson and his friends. He later wrote:

> . . . nor will it be denied, when my enemies are forgotten, that the projects for which I am now charged with traitorous designs, had a direct tendency to accelerate, the annexation of Louisiana to the United States. . . . I have ample and undeniable testimony to show that I omitted on no occasion, to employ my ascendancy over the officers of Spanish Louisiana, to render them subservient to the interest, and accommodation of the United States.[85]

In his report to Miró Wilkinson requested permission to bring fifty or sixty thousands dollars worth of Kentucky goods annually to New Orleans, duty-free. In return he would keep the Spanish government constantly informed about all matters concerning them. He was relying on their discretion to keep this matter entirely secret and —

> "to bury these communications in eternal oblivion."[86]

Calling himself "a good Spaniard," Wilkinson wrote to the Spanish diplomat Don Diego Gardoqui on January 1, 1789:

> It is not necessary to suggest to a gentleman of your knowledge and experience that the human race, in all parts of the world, is governed by its own interest, although variously modified. Some men are sordid, some vain, others ambitious. To detect the predominant passion, to lay hold of it, and to derive advantages from it, is the most profound part of the

political sciences.[87]

Wilkinson very skillfully played both ends against the middle—expansionist, intriguing, Americans on one end, foreign powers on the other end, and the future of the United States in the middle.

Wilkinson, after visiting the east coast, returned to Kentucky as a hero in February, 1788. He advertised as a wholesaler for products to sell at New Orleans and sent out buying agents. It was the first time that there was a market and cash payment for produce, and it brought prosperity to the back country. No one had ever managed to distribute so many silver dollars before. Tobacco was the main product brought to market, and it was potentially very profitable. Miró—who had been bribed by Wilkinson—said that if the business had been better conducted, or had better luck, he would have profited greatly.[88]

The opening of trade with New Orleans brought enormous benefits to farmers in Kentucky. It also hastened the admission of Kentucky to the union as it demonstrated the commercial potential of the region. It gave Kentuckians a chance to determine their own future, regardless of the wishes of easterners who opposed free navigation on the Mississippi River as injurious to their own trade.

At statehood conventions, Wilkinson advocated that Kentucky first become independent, and then seek admission to the United States. It was how he accelerated the admission of Kentucky to statehood. If Kentucky became independent, any one of the three foreign countries competing for control of the Mississippi Valley—Spain, Britain or France—could take Kentucky over, or form an alliance with it. It was just the push that was needed, and Kentucky became the 15th state in 1792.

George Rogers Clark knew that he had been used in this affair. He wrote:

> The real motive of my accusers [in Kentucky] was that wishing to trade with New Orleans, they thought that by pretending to take the side of the Spanish vassals, they might hope to

be favored in their projected commerce.

A few years later, George Roger's younger brother, William Clark, became a supporter of Wilkinson in the frontier army—obviously unaware of the role Wilkinson had played in destroying his brother's reputation.

In 1787, George Rogers Clark was asked by the governor to supply an account of his expenses in capturing Kaskaskia and Vincennes in 1778–1779. Clark had sent more than 20,000 expense vouchers to the state auditor in 1781. After Benedict Arnold burned the capitol at Richmond, Clark had supplied (as best he could) a duplicate set in 1783. Now he was being asked again. He said it was impossible unless the duplicates were returned to him. The original vouchers were eventually found in a storeroom in the state auditor's office in 1913, and revealed both his careful accounting and that his ability to reconstruct them a second time was remarkable.[89]

In 1789 a British agent, Colonel John Connolly, came to the Falls of the Ohio from Detroit with an offer from Lord Dorchester, the Governor-General of Canada, to provide pay, clothing and equipment for 10,000 men if they would join with the British army in an invasion of Louisiana. Wilkinson relayed the news to Governor Miró in New Orleans and told him that he had sent a reputed assassin after Connolly, and that Connolly had fled the territory with an armed escort—also provided by Wilkinson. He earned the gratitude of both Spain and Britain with this maneuver.[90]

Wilkinson was promised he would receive payments from Miró for "secret and indirect agencies" in 1789. While he was in New Orleans he told Miró that General Washington had spies watching him during his visit. He wrote in code to Spanish officials, using an English pocket dictionary as the key, and they wrote back using the same code.

His finances were in terrible shape—business expenses, boat accidents and the suicide of a partner had resulted in massive debt rather than profit. Wilkinson asked for a loan from Miró.[91]

On December 26, 1789, 7,000 silver dollars arrived in Frankfort in heavy leather saddle bags carried by two mules. Each silver dollar weighed one ounce; 2,000 dollars weighed 125 pounds. Wilkinson boasted that his young assistant, Philip Nolan, could carry a bag with one hand.[92]

Silver dollar or peso, depicting Charles III of Spain, 1776. (Wikimedia Commons)

The coins, minted in Mexico, were international currency. Millions of silver dollars were stored at the fortress at Vera Cruz and shipped through the Gulf of Mexico to world markets. Spanish Texas guarded the silver mines of the Sierra Madre Mountains. Spanish Government policy dictated that Texas serve as a barrier state. It was sparsely settled and commercially undeveloped in order to protect the silver mines.

The Spanish monarchy failed to become a modern state and to develop other commercial enterprises because it controlled so much silver and gold in the Americas. In the age of revolution, its colonial empire was ripe for a takeover, and everyone knew it. That is the real reason why Spanish-owned New Orleans and the Gulf Coast were such sought after prizes—the region not only would provide access to international shipping, but it could also become the site of a new country incorporating the silver mines of Mexico.

As Americans tried to force their way into new lands, Spanish authorities had to perform a delicate juggling act in the two Floridas, New Orleans, Louisiana and Texas to avoid war and hold onto their lands. The United States couldn't provide protection and services on the frontier, so Americans were told to wait—but westerners hated the Spanish dons, who supplied arms and ammunitions to the Indians, and they didn't think much of the federal government.

Thomas Jefferson wrote from Paris on January 25, 1786:

> I fear . . . that the people of Kentucké think of separating, not only from Virginia (in which they are right) but also from the confederacy. I own I should think this a calamitous event, and such as one as every good citizen on both sides should set himself against, our present federal limits are not too large for good government, nor the increase of votes in Congress produce any ill effect. On the contrary it will drown the little divisions at present existing there. our confederacy must be viewed as the nest from which all America North & South is to be peopled. we should take care to not to think it for the interest of that great continent to people it too soon on the Spaniards. these countries cannot be in better hands. my fear is that are too feeble to hold them till our population can be sufficiently advanced to gain it from them piece by piece. the navigation of the Mississippi we must have. this is all we are yet ready to receive. [93]

Wilkinson soon faced competition from rivals in the Mississippi River trade. His monopoly was short lived. He continued in the trade for two more years with disastrous results. He had to sell his lands, carriages, horses and oxen to satisfy his creditors in 1791.

In the summer of 1790, General Josiah Harmar led a campaign against the Miami Indians on the Ohio frontier with frontier militia. It was a failure. The next year Wilkinson served as second in command under General Charles Scott in raids against the Indians, and then led a raid under his own command. Young William Clark served under Wilkinson.

Wilkinson wasn't present at General Arthur St. Clair's Defeat on November 4, 1791. More than a thousand Shawnee, Miami, Delaware and Potawatomi warriors won the battle on the Wabash River under the leadership of Miami Chief Little Turtle and Shawnee Chief Blue Jacket. It was the worst defeat of the America military in the Northwest Indian Wars, with more than 600 soldiers killed and 300 wounded out of a total force of about 1,400. Indian casualties were about 61, with 21 dead.[94]

On November 5th, 1791, Wilkinson was sworn in as

Lieutenant Colonel Commandant of the 2nd Infantry in the Legion of the United States. His connection with Spanish officials may have been a factor in his appointment. James Ripley Jacobs, the leading biographer of Wilkinson, says he probably became "sort of a federal stool pigeon."[95]

George Washington appears to have trusted Wilkinson because of his actions in exposing the Conway Cabal during the Revolutionary War. When Wilkinson informed Miró that General Washington had spies watching him during his second visit to New Orleans in 1789, he may have been explaining why he was seen in the company of American secret agents. Washington later wrote to Alexander Hamilton that he had appointed Wilkinson because he felt it would:

> . . . feed his ambition, soothe his vanity and by arresting discontent produce a good effect.

But perhaps that wasn't the whole story.

Wilkinson said that he had returned to military service, because he had been:

> . . . disgusted by disappointment and misfortunes. My views in entering the Military Line are Bread & Fame—uncertain of either, I shall deserve both.[96]

On February 1, 1792, the new Spanish governor, Francisco de Carondelet, wrote to Wilkinson that he was finally going to receive the $2,000 annual pension he had been promised, and it would be retroactive to January 1,1789. He received the $4,000 that summer, supposedly as payments for tobacco sales. The Spanish government felt he was now a good investment.

The United States republic and Spanish officials in the two Floridas, Louisiana and Mexico shared a common goal—to maintain the status quo as long as possible. As Jefferson said, "these countries could not be in better hands." His only worry was whether Spain was "too feeble" to hold onto them. The real concern was whether Britain or France might manage to establish a new country in the west before the United States was ready to "receive" the territories from Spain.

The American Revolution had spread to France in 1789, and the Napoleonic Wars would begin in 1803. The United States adopted a neutral policy during the European conflicts. It was too weak to do anything else. The wars actually protected the new country, as Britain and France fought each other from 1793–1815, and had limited resources to devote to regaining their old territories in America.

Seen in this light, Wilkinson's role as a double agent can be understood. The strategies he suggested to the Spanish government were obvious. He enticed them with stories of Kentucky's independence, but it led to Kentucky's statehood. His biggest accomplishment was in opening the Mississippi River to Kentucky trade, which eliminated the primary reason why westerners would want to enter into a union with Spain.

Since it was widely rumored that Wilkinson was a Spanish agent, and he was always considered to be an intriguer, he found it easy to infiltrate American groups plotting to establish new countries with either British or French military support, or with disloyal Spanish officials. He sabotaged their plots, and they never seemed to realize he was behind their failures. He thwarted efforts by rivals to establish new colonies in Spanish territory. And he kept the lines of communication open with Spanish officials. He was busy.

Baron Joseph de Pontalba of New Orleans, who later moved to France, advised Napoleon to employ Wilkinson in his secret service. He wrote:

> Four times, from 1786 to 1792, preparations were made in Kentucky and Cumberland to attack Louisiana, and every time this same individual caused them to fail through his influence over his countrymen. I make these facts known to show that France must not neglect to enlist this individual in her service. [97]

In building his first relationship with Spanish officials and gaining his own career advancement, Wilkinson ruined George Rogers Clark's army career and took his place as Indian Commissioner. But he only attempted "character assassination" with Clark. He will have another encounter with Clark in the next few years.

The assassinations and attempted assassinations associated with Wilkinson will be discussed in the next chapter.

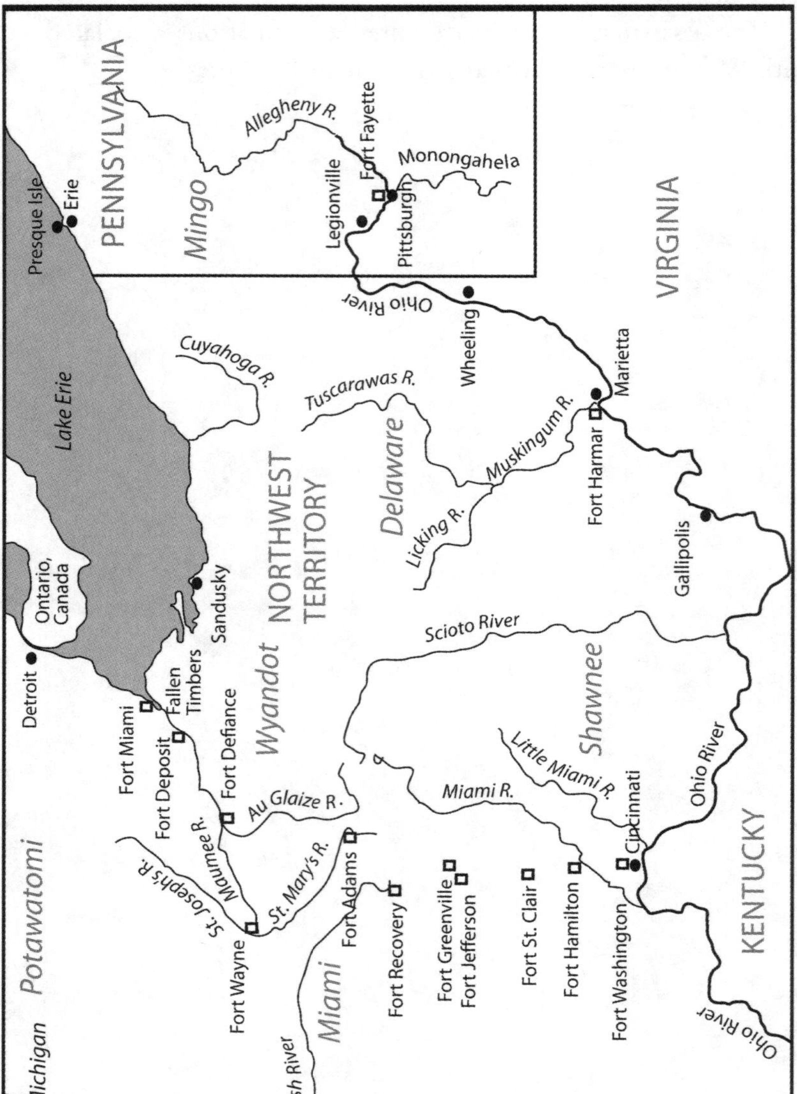

Map of the Legion of the United States, Ohio Valley1792–1796.

Chapter 6

The Legion of the United States

Major General Anthony Wayne arrived in Pittsburgh to take command of the American army in June, 1792. After St. Clair's Defeat, congress expanded and reorganized the army, and gave it a new name, the Legion of the United States. Four sub-legions were created, and Wilkinson was appointed one of the commanders.

"Mad Anthony Wayne"—a nickname gained for his bold leadership during the Revolutionary War—was a seasoned commander, who knew the value of strict discipline. He began turning new recruits into trained soldiers who would avenge the American defeats in the Indian Wars. The legion remained in Pittsburgh until November, and then moved to a new headquarters, 22 miles south of Pittsburgh on the Ohio River. Legionville would become the first basic training camp of the United States military.

At the same time, 13,000 nationalized state militia troops were arriving in western Pennsylvania to put an end to the Whiskey Rebellion. The government's show of force was meant to intimidate more than Pennsylvania rebels. The state militias got a chance to see the Ohio River Valley. The newly created Northwest Territory lay across the river—bounded by the Ohio River on the east and south, the Great Lakes on the north, and the Mississippi River on the west—the territory which would become open for settlement at the close of the Indian Wars.

Meriwether Lewis was among those who decided to stay on in the west after his militia duties were finished. He enlisted in the Legion of the United States in 1795. But before we continue with his story, the events of the next three years need to be told.

James Wilkinson was made a brigadier-general in March, 1792. He commanded 200 soldiers at Fort Washington in Cincinnati, the largest military post in Northwest Territory. The fort's roof and fences were painted red, and gardens of vegetables and flowers were planted nearby. Cincinnati consisted of about 900 residents, taverns, vice, and a surplus of merchants.[98]

Fort Washington. (Wikimedia Commons)

The Wilkinsons were now the parents of three boys, ages 12, 9 and 7. Ann left in the summer to bring the two older boys to Philadelphia to attend school, and to visit relatives. A two story frame house was built for General Wilkinson at Fort Hamilton, where he could enjoy fishing and hunting.

President Washington was determined to give peace talks with the Indians of the Northwest a chance to succeed. He sent peace commissioners throughout the region in 1792, but to no one's surprise, they were unsuccessful. The Indians, victorious in battle, were in no mood to compromise.

When General Rufus Putnam met with the Illinois and Wabash tribes at Vincennes in September, the chiefs declared the north side of the Ohio River was Indian territory and they had signed no treaties. They wanted American traders—but no American settlements and no forts. They told General Putnam:

> . . . that wherever the Virginians go, they cause trouble in some way.[99]

Under British rule, it had been Indian territory since the end of the French and Indian Wars in 1763. Despite the terms of the Treaty of Paris at the end of the Revolutionary War, the British still had forts in Northwest Territory and continued to supply arms and ammunition to the Indians.

Wilkinson chose two men to carry Washington's message of peace under white flags of truce to the Wyandot and Miami Indians. The General wrote to Colonel John Armstrong, commander of Fort Hamilton on May 24th, 1792, predicting their deaths—

> Hardin and Trueman left us day before yesterday, the former for Sandusky, the latter for the Maumee, I think it is equivocal what may be the event, but do not expect they will return.[100]

John Hardin, a Kentucky militia general, and Major Alexander Trueman of the First Infantry Regiment of the United States were highly regarded military leaders. Wilkinson had written to Hardin urging him to undertake the mission, saying:

> . . . you are better qualified for it than any man of my acquaintance, and because I think it will lead to something of advantage to you.[101]

Hardin had served under Wilkinson at the Battle of Saratoga in 1777. Often sent out on dangerous reconnaissance missions, Hardin was one of Daniel Morgan's best Virginia sharpshooters. An early biographer of Wilkinson says tradition had come down through the years that it was Hardin who actually discovered Burgoyne's army was advancing towards Saratoga, rather than Wilkinson who took credit for it. The biographer, Royal Ornan Shreve, points to a suspicious comment made by Wilkinson in his *Memoirs*:

> . . . where I found Hardin who had made no discovery . . .

and Wilkinson's fulsome praise for Hardin in a footnote—

> * Afterwards General Hardin of Kentucky, an excellent officer and most valuable citizen, who having encountered a

> thousand dangers in the service of his country, was treacherously murdered in 1791 [sic] by a party of Indians, as he approached Sandusky with a flag of truce and a talk from General Washington. A braver soldier never lived—a better man has rarely died.[102]

Humphrey Marshall, a contemporary, and an enemy of Wilkinson's, wrote in his *History of Kentucky*, published in 1812:

> Whether the general, was really the friend of Hardin, and candid in his expressions of a desire to serve him—and thought such employment conducive to that end: or viewing him as a rival in fame, who might afterwards be in his way, if not seasonably put out of it . . . as by some was strongly suspected; there is no means possessed of knowing. Certain it is that Wilkinson persuaded, and pressed Hardin, to the undertaking—as he did Major Trueman, an officer of great merit, under his command; and with whom he was known to be at variance, to undertake a like commission. . . .They were both called and they were both cut off.

Hardin and Trueman and their companions were killed by Indians who appeared friendly and who volunteered to camp with them. In each case, the Indians suddenly murdered them without warning. Marshall said that:

> . . . when the news was carried to the [Indian] town, "that a white man with a peace talk had been killed at the camp," that it excited great ferment; and that the murderers were much censured.[103]

Hardin and Trueman would never have camped with anyone they suspected of intending to harm them. These were planned and premeditated assassinations. Their deaths are also the first time that Wilkinson is accused of assassinations.

The author of these charges, Humphrey Marshall, believed he was also the target of an assassination attempt by Wilkinson. Marshall was a Federalist, a U. S. senator, and a cousin of John

Marshall, Chief Justice of the Supreme Court. Humphrey Marshall was an outspoken, relentless critic of James Wilkinson and the "Spanish Conspiracy" in Kentucky politics. In 1788, Marshall survived an assassination attempt by a man he called:

> . . . an ignorant catspaw of craftier but more prudent men.

Members of the Spanish Conspiracy included Wilkinson and his friends Harry Innes, a federal judge, John Brown, a U. S. senator, and Benjamin Sebastian, who received a pension from Spain.[104]

Fort Greene Ville (www.shelbycountyhistory.org)

General Wayne moved about 1,500 troops from Legionville to a camp one mile south of Fort Washington at Cincinnati in May, 1792. General St. Clair had built Forts Washington, Hamilton and Jefferson. Wayne would leave nothing to chance and build seven more forts before going into battle. The chain of forts would ensure that army horses would have hay, and the men would have provisions and protection as they marched north.

Peace negotiations with the Indians ended in failure in September, 1793 and the army was now authorized to prepare for war. In November, Wayne began building a fort six miles north of Fort Jefferson, capable of serving as a winter camp for 2,000 soldiers. He named the post Fort Greene Ville after his friend, Nathaniel Greene, a Revolutionary War general. Fort Greene Ville was a small city, and both Wayne and Wilkinson had houses built there. The fort, was one of the largest wooden forts ever built: 900 feet by 1,800 feet enclosing 55 acres within its walls.[105]

The reconstructed Fort Recovery is now the site of an Ohio State Museum in Fort Recovery, Ohio.

Fort Recovery was built on the site of St. Clair's battlefield near the Wabash River. Wayne arrived there on Christmas Eve with a special detachment. They found 500 tomahawked skulls on the battlefield, and buried the remains of their comrades in a common grave. They also began to locate the cannons from St. Clair's army that had been hidden under logs and buried by the Indians after the battle in 1791.[106]

Fort Adams was built in the summer of 1794 near St. Mary's River to control the river traffic. Fort Defiance was built in August on a high bank overlooking the confluence of the Maumee and Auglaize Rivers. It was reinforced to defend against artillery attacks by the regular British troops stationed at Fort Miamis. Fort Deposit was just what it name implies: a small fort where their excess baggage was deposited before the Battle of Fallen Timbers.

Throughout this all, Wayne had to deal with the disloyalty and troublemaking mischief of General Wilkinson, his second in command. Wilkinson went over his head, writing directly to Secretary of War Henry Knox, criticizing Wayne's leadership. However, Knox, a friend of Wayne's, promised to protect Wayne against Wilkinson's misrepresentations as long as he was in office.[107] The officers of the legion split into two factions, with the Wilkinson faction resenting Wayne's disciplinary measures, his abrupt manner, his restrictions on alcohol use, and the slow progress towards war. Wilkinson was doing all he could to encourage discord, hoping to take over Wayne's command. His easy manner and skill at manipulating people won him many friends.

Wayne was actually waging war by building forts and training his soldiers. It was slow and methodical, and in the end, successful. Additional forts signalled the absolute determination of the United States to take control of the Northwest. The British rebuilt an old, abandoned fort, Fort Miamis, on what was now American soil, claiming it was a "dependency" of Fort Detroit. Fort Miamis was located two miles from Colonel Alexander McKee's trading post, which was supplying arms and ammunition to the Indians.[108]

The crucial question was—would war break out between the United States and Great Britain over Northwest Territory, or would it remain a war fought between the Indians and Americans? The British promised the Indians that Great Britain would soon be going to war against America.[109]

In June, the Indian confederacy began assembling at the Auglaize River for an attack on Fort Recovery at St. Clair's old battlefield. 1,250–1,500 warriors from all the northwestern tribes, including the Lake and Canadian Indians, participated.[110] General Wayne wrote to Secretary of War Henry Knox that the Chickasaw and Choctaw Indian scouts he employed as spies had told him that:

> . . . there were a considerable Number of armed white men in the rear, who they frequently heard talking in our language & encouraging the Savages to perservere in the assault, that their faces were Generally blacked except three British Officers who were dressed in scarlet & appeared to be men of great distinction, from being surrounded by a large party of White men & Indians who were very attentive to them, these kept a distance in the rear of those who were engaged.[111]

On June 30th the Indians attacked a supply convoy leaving Fort Recovery. Because Wayne had insisted the cannons which the Indians had hidden after St. Clair's defeat be located, the Indians searched in vain for them. They had come to the fort with ammunition to use in the cannons. Undoubtedly the British officers planned to help with the artillery—but instead the cannons were used to defend the fort.[112] The battle at Fort Recovery was a significant turning point in the Indian Wars, because the Indians—who had been so confident of another victory—retreated after two days.

Lieutenant William Clark, who had fought under Wilkinson in two earlier campaigns in 1790, was strongly attached to the general. He had no idea that Wilkinson had destroyed his older brother's career.

Two years earlier, in June, 1792, the 22 year old lieutenant had been sent on a special mission to the Chickasaw Bluffs in Tennessee. Clark commanded a unit of 24 men who travelled down the Mississippi and managed to pass successfully by the Spanish fort at Natchez. They brought trade goods, corn, and arms to the Chickasaw, and returned with a Chickasaw chief and 8 warriors who would act as scouts and spies for the legion. Clark received high praise for this mission: General Wayne wrote to Secretary Henry Knox—

> This young Gentleman has executed his orders . . . with a promptitude & address which does him honor & which merits my highest approbation."[113]

Despite this high praise, Clark continued to remain a member of Wilkinson's faction in the army.

After almost two years of preparation, some 2,000 members of the legion were joined by 1,500 mounted Kentucky volunteers and the army set off to fight the Indians in late July, 1794. This would mark the beginning of the end of the Indian Wars in the Ohio Valley, until they were resumed during the War of 1812.

Clark began keeping a journal when they left Fort Greene Ville on July 28th, apparently intending to document the errors of General Wayne's command, perhaps in anticipation of a later congressional investigation or military court martial. Wilkinson most likely had asked Clark to keep a journal.

On August 3rd, Clark recorded what appears to have been

attempt on General Wayne's life. Wayne called Wilkinson "that vile assassin" in a letter to Secretary of War Knox a few months later. Whether Wayne was referring to this incident or something else is not known, but it seems the most likely reason he would have accused Wilkinson of being an assassin.[114]

The incident happened on the afternoon of August 3rd. It was very hot and General Wayne and his aide de camp, Captain Henry DeButts, were napping in a marquee tent. Outside, the troops were chopping down big trees to use in constructing the new fort near the St Mary's River, to be named Fort Adams. William Clark's journals records what happened next:

> —This day fortune in one of her unaccountable pranks, (I say unaccountable for [no] one will pretend to know why She Should at this juncture, at this critical moment, but an illtimed, prank perturb the minds of Hundreds & create for a moment in theire Breast such unthought of Sensations & open So large a field for Speculation) had nearly deprived the Legion of its *Leader*, had nearly deprived 'Certain individuals of theire *A. W.* [Anthony Wayne] & Particular persons of their consequence—the downfall of Some would have been equale to the tumble our Chief occasion'd by the fall of a large Beech Tree, which fell about his Marquee [tent], had fortune directed its course a few feet more to the right or left—but (Fav'd Fort'n) His Excellency'cy was more Sceared then hurt. All ended well, though there was considerable concorse—[115]

An anonymous journal writer of the campaign wrote:

> About 3 O'Clk this afternoon a tree fell upon the tent of the C in C and that of his A D C. Capt. De Butts, both being in their beds—the escape of each was miraculous, the Bed of the aid being crushed to pieces and the C in C escaped, by about six inches being mashed to Death by the Body of the tree—he was dragged out of the tent, having received a severe stroke on his left leg & ankle—his pulse was gone.—but by the application of a few volatile drops he was soon restored to good spirits and comfort.[116]

Wayne wrote to his son Isaac Wayne that he was:

> . . .almost deprived of life. . . by the falling of a tree . . . & was only preserved by an old trunk near where I was struck down—which supported the body of the tree from crushing me to attoms.[117]

Beech trees can reach a height of a hundred feet or more. Wayne's life was saved because a tree trunk near the tent broke the impact of the falling tree. The falling tree wasn't chopped down. It had been set on fire by a fire kindled near its base—which made it almost impossible to prove it was deliberate.

After that, Wayne ordered all of his off duty officers to report to duty and help get the fort finished, because they were leaving the next day. The next morning General Wayne continued on with his army, despite being in severe pain. He already suffered from gout in his leg. From that time on, he had to be helped onto his horse.[118]

Wayne surprised his own officers, the Indians, and the British, by taking an unexpected route of attack. He had intentionally misled them. The attack was expected on Kekionga, a group of Miami Indian villages at the junction of the Maumee, St Joseph's and St Mary's Rivers. Instead he headed north along the Auglaize River to its junction with the Maumee, where they stopped for eight days and built Fort Defiance on the site of a deserted Shawnee village of log cabins and huts. Fort Defiance was in between the Miami Indian villages and Fort Miamis on the Maumee River.[119]

An army deserter, Robert Newman, warned the village that the army was coming and its residents fled the day before the troops arrived. It is thought that Newman was acting as an agent of Wayne's in alerting the village.[120] The general was out to fight a war to end the Indian Wars, not to harm those who were left behind in the villages. A lieutenant wrote about arriving at the Auglaize village:

> This place far exceeds in beauty any in the western country. . . Here are vegetables of every kind in abundance, and we have

> marched four or five miles in cornfields down the Oglaize, and there is not less than one thousand acres of corn round the town.

Nearby villages were the homes of the Shawnee leader Blue Jacket and Miami Chief Little Turtle, who had been leaders of the confederation that defeated St. Clair's army in 1791.[121]

Lieutenant Clark wrote that when the army was within 12 or 14 miles of the Indian settlements, their spies returned with the news that the settlements were being evacuated:

> everybody was flushed with the Idea of Supprising them in the moment of providing for theire Wives & Children the Scheeme was perposed & certain Suckcess insued if attempted—General Wilkinson Suggested the plan to the Comd^r in Chief but it was not his plan, nor perhaps his wish to Embrace So probable a means for Ending the War by compelling them to peace—this was not the first occasion or oppertunity, which itself to our observent Genl. for Some grand stroke of Enterprise, but the Comd^r in Chief rejected all & every one of his plans—[122]

A Shawnee captured by spies told Wayne:

> the Indians talked on stepping to one side and letting the army go on to the British or that if the British did not now come forward to assist them they would join the Americans.

Wayne sent the Shawnee captive back to the confederation on August 13th with a message urging that the hostile tribes meet with him and establish the preliminaries of a lasting peace with the United States which would—

> ...eventually and soon restore to you ... your late grounds and possessions, and to preserve you and your distressed and hapless women and children from danger and famine during the present fall and ensuing winter. [123]

The battle was about to begin. All the sick and the lame of the regular troops were left behind at Fort Defiance. About 70 Chickasaw and Choctaw warrior scouts refused to advance and stayed at the fort.[124] The army left the fort on August 15th, and the next day a messenger brought the Shawnee's answer— they wanted a 10 days halt to the army's advance in order to consult with other nations. It was agreed they were stalling for time, and the army continued to advance.

The British had sent additional troops to Fort Miamis under a new commander. It was uncertain what role the British would play in the battle and whether the fort would be attacked by the Americans. Thousands of women and children were in refuge camps north of the fort.

On August 18th, Wayne built one last temporary fort to hold excess gear. The soldiers carried only a blanket and two days of rations into battle—traveling light was one of his key strategies. Fort Deposit was similar to what Wayne's army built every day while on the march: grounds marked off for tents for the four sub-legions; the whole enclosed with a breastwork of logs; and smaller outposts beyond the breastworks for protection of the sentries. If necessary, it would serve as a retreat.[125]

Lieutenant William Henry Harrison, one of Wayne's aides, wrote that Wayne's strategy and tactics:

> . . . united with the apparently opposite qualities of compactness and flexibility, and a facility of expansion under any circumstances, and in any situation, which rendered utterly abortive the peculiar tact of the Indians in assailing the flanks of their adversaries. [126]

The Battle of Fallen Timbers took place on August 20th. The battlefield was covered with trees struck down by a tornado. It was just five miles south of Fort Miamis, giving the British every opportunity to participate in the battle—although it was not really a battle but a skirmish which lasted for 60-85 minutes.[127] The total casualties were 27 killed and 87 wounded of the estimated 1,000 American troops who participated. An estimated 1,300 warriors were in the area, but about 400 Indians and 60 Canadian militia were active in the battle.[128] The Indian casualties are unknown; however, 30–40 Indians were found dead who had not been carried away by their comrades.[129] The Indian tribes who participated were Wyandotte, Mingo, Delaware, Shawnee, Potawatomi, Ottawa, Chippewa, Miami and Mohawk.[130] A Frenchman who fought on the Indian side explained their defeat by saying that the soldiers—

> would give them no time to reload their pieces but kept them constantly on the run.[131]

The gates of the British fort were kept tightly shut, not allowing the Indians or Canadian militia to enter the fort as they retreated. The Indians never forgot this and it did more to ensure peace between the Indians and Americans than anything else. The British refusal to help—after all their promises—broke the confederacy. Wayne and the fort commander wrote belligerent letters back and forth, and Wayne thought about laying siege to the fort, but he didn't have enough provisions or artillery for it, and the army went back to Fort Defiance.[132]

After the victory, Wilkinson renewed his attacks on Wayne. Wilkinson commanded the right wing of the troops during the battle. William Clark served under Colonel John Hamtramck, who commanded the left wing. Wilkinson wrote to his friend Judge Harry Innes in November with this opinion of Wayne's campaign:

> The whole operation presents to us a tissue of improvidence, disarray, precipitancy, Error & Ignorance, of thoughtless termerity, unseasonable Cautions, and shameful omissions, which I may safely pronounce, was never before presented to the view of mankind; yet under the favor of fortune, and the paucity & injudicious Conduct of the enemy, we have prospered beyond calculation, and the wreath is prepared for the brow of a Blockhead.[133]

Before the battle, Wilkinson had asked for a military court of inquiry covering his own conduct and made five charges against

Wayne in letters he sent to Secretary of War Knox in June and July. In December he received a reply from Knox who had forwarded Wilkinson's charges to Wayne. Knox asked Wilkinson to be more specific. Wayne wrote to a friend that if he was guilty of Wilkinson's charges, he "ought to be hanged." In January, 1795 Wilkinson backed down publicly, yet continued to write to Knox. Wayne could not afford to get involved in a fight with a subordinate while negotiating peace treaties with the tribes, and so he ignored Wilkinson's charges until the treaties were accomplished.[134]

Wilkinson continued to agitate—the army would soon be reduced and either Major General Wayne's position or Brigadier General Wilkinson's position would be abolished. He had more to worry about than his job, however—Wayne was charging Wilkinson with being a British agent. In December, 1794 Robert Newman had returned to Fort Greene Ville, and said he had direct proof that Wilkinson was intriguing with Canadians for a union of Kentucky with Canada.

Historians disagree as to whether Newman was working for Wilkinson or Wayne. According to Robert Newman's brother, General Wayne had told Newman to:

> become associated with such persons as might be suspected of being in favor of the Indians or British.

Newman was in fact working for Wayne, when he appeared to be working for Wilkinson. The story was that Wilkinson gave Newman a letter to carry to Alexander McKee, the trader at Fort Miamis, and Wayne read the letter. But Wayne must have substituted another letter before it was given to McKee, because both McKee and the Lieutenant Governor of Upper Canada, John Simcoe, suspected Newman was part of a "sinister plot." Simcoe wrote that the letter contradicted information they had already received from Wilkinson—proving they were in correspondence with Wilkinson. After Newman returned to Fort Greene Ville and was interviewed by Wayne, Newman came into a possession of a large amount of money in brand new army bills, which he used to pay a debt he owed.[135]

Anthony Wayne (1745–1796), pastel drawing from life by James Sharples, Sr., 1796.(Wikimedia Commons and NPS Museum Collections, Independence National Historical Park, IN 11922)

James Wilkinson (1757–1825) oil painting by Charles Willson Peale, 1797. (Wikimedia Commons and Independence National Historical Park, IN 14166)

General Wayne, of course, believed that Wilkinson was genuinely working with the British. He had seen correspondence in which Wilkinson asked an army contractor to delay providing supplies to the army in 1793. The question is—was Wilkinson acting as a traitor, or as a double agent? The federal government must have believed he was acting as a double agent because they didn't pursue the matter.

Wayne wrote privately to Secretary of War Henry Knox on January 29th, 1795 warning him of Wilkinson's treachery. The letter which is torn in several parts (an early form of censorship) said:

> I am much obliged by your [torn] it has been consistant [torn] my own honor, & the true of interest of my Country to accomodate with that vile assassin *Wilkinson*, I most certainly would have made an attempt in compliance with the President's wishes But it is impossible—for I have a strong ground to believe, that this man is a principal agent, set up by the British & *Demoncrats* of Kentucky to dismember the Union.
>
> N______n [Robert Newman] says that to the best of his knowledge & belief the letter to Colo McKee was in W______n's [Wilkinson's] hand writing & that it was put into his (N______ns) hand [torn] Indeed I have cause to beli [*sic*, believe] [torn] & tampering both by the British & Spaniards. [136]

Some "*Demon-crats*" in Kentucky—led by George Rogers Clark—had been actually engaging in a conspiracy to start a new country for France. Clark and his brother-in-law, Dr. James O'Fallon, wrote a letter to Edmond Genêt, the newly appointed French minister to America, volunteering to raise a force of 1,500 men to seize the Spanish posts on the Mississippi and New Orleans for France. The letter was waiting for Genêt when he arrived in Charleston, South Carolina on April 8, 1793.[137]

The King and Queen of France, Louis XVI and Marie Antoinette, were executed by guillotine in January, just before

André Michaux and Edmund-Charles Genêt
(Wikimedia Commons)

Genêt left the country. After their executions, the new French Republic declared war against England in February and against Spain in March. In August, it abolished slavery in its richest colony, Saint Domingue (Haiti).

When Edmund Genêt arrived in Charleston, he commissioned four American ships to serve as French privateers, authorizing them to attack and capture British ships, and began making arrangements for 1,500 Georgia volunteers to seize Spanish East Florida.[138] Citizen Genêt was greeted by large and enthusiastic crowds as he made his way from Charleston to Philadelphia, but on April 22nd, President Washington issued a Proclamation of Neutrality forbidding American citizens from aiding any of the nations at war with each other.

The French Republic planned to finance invasions of the Spanish Floridas, Canada and Louisiana by receiving an advance on money it was owed from loans France had made to America during the American Revolution. President Washington refused to release any money, saying the future of the French government was uncertain. It proved to be a wise move. In June the officials who had appointed Genêt were executed and the Reign of Terror began in which thousands were executed by the new Jacobin government.

André Michaux—a distinguished French botanist and explorer—was living in Charleston where he had established gardens to study and collect American plants for the royal botanical gardens. A year earlier, in 1792, Michaux had applied to lead an expedition to the Pacific Ocean which was being organized by Thomas Jefferson and the American Philosophical Society. Meriwether Lewis also "warmly solicited" Jefferson for the chance to go to the Pacific with Michaux, but because he was 18 years old, he was turned down.[139]

Michaux was chosen, and Genêt also appointed Michaux a secret political agent for the French Republic. Michaux left Philadelphia in July, 1793, accompanied by two French artillery officers. Botanizing along the way, Michaux met with George Rogers Clark in Louisville on September 17th and gave him his commission as "Commander in Chief of the Independent and Revolutionary Army of the Mississippi."[140]

Clark planned to live in France after he led the French Revolutionary Army. He was given blank commissions to recruit officers, and by January, 1794 it was said that over 1,000 men had joined the army. Clark published advertisements in western newspapers seeking volunteers for "the reduction of Spanish posts on the Mississippi." He offered land grants of 1,000 acres at enrollment; 2,000 acres after one year of service; and 3,000 after two years. If they wanted, instead of land grants, enlisted men could receive one dollar a day pay. Officers would receive French army pay and share in "all lawful plunder." Members of the army would need to expatriate themselves and become French citizens.

In March, 1794 French embassy officials in the United States ordered a halt to the filibuster plans and Genêt was recalled by his government. Facing certain execution, Genêt was granted political asylum and later married the daughter of Governor Clinton of New York. President Washington followed up with a proclamation on March 24th, warning American citizens not to participate in the French filibuster.[141]

At the same time, Secretary of War Knox ordered General Wayne to "immediately" build a new fort on the site of the old French fort, Fort Massiac. Fort Massac (the American version of the name) was located on the north bank of the Ohio, eight

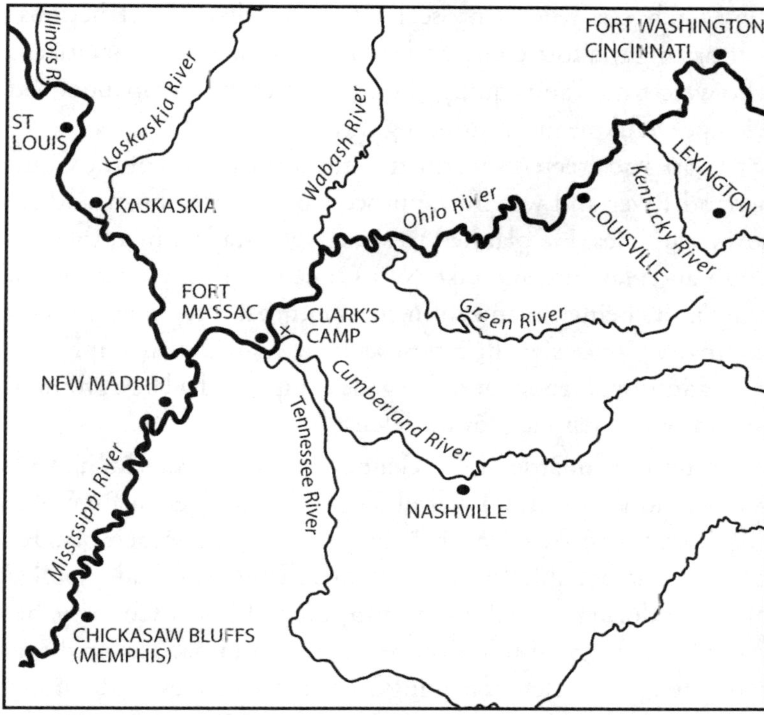

Ohio and Mississippi Rivers, Fort Masaac.

The reconstructed Fort Massac is located in Metropolis, Illinois.

miles downstream from its junction with the Tennessee River. On March 31st Knox sent "secret and confidential instructions" authorizing the fort commander to fire upon American citizens who were armed and equipped for war, after the commander had tried peaceful means of stopping them.[142]

Clark had erected a fortified camp at the mouth of the Cumberland River and was in the process of building boats and securing supplies. He planned to seize New Madrid first, then St. Louis and Natchez, and take New Orleans with the assistance of army units being organized in the southeast. The Spanish took the matter seriously: their post at New Madrid was reinforced with additional troops and five galley ships, and a large group of Indians was assembled for its defense.[143]

After the collapse of the Genêt project, Clark, who had advanced money for it, was owed $4,680 by the French Republic. Clark was also owed money by the Virginia Assembly. He was personally liable for debts to merchants who had supplied his army during the 1778–79 invasion of Illinois Country, because he had signed notes on behalf of Virginia. The creditors came after Clark. Before making his offer to the French Republic, Clark applied to the Virginia legislature in November, 1792 for $20,500 in compensation. The assembly rejected his claim, saying he should have presented it earlier. It was an insulting reply, as his original expense receipts had been given to them in 1781 and his reconstructed receipts in 1783. The lack of gratitude embittered him and he wrote:

> I have given to the United States half the territory they possess and for them to suffer me to remain in poverty in consequence of it, will not redound much to their honor later . . . I must look out further for my future bread.[144]

General Wilkinson, however, profited from this affair. It seems likely that Wilkinson may have used William Clark—who was a great admirer of Wilkinson at this time—as a source of information about his brother's filibuster. George

Rogers and James O'Fallon, their brother in law, first approached the French government in the winter of 1792–93. The 24 year old William may have revealed their plans for a French invasion army either to Wilkinson or to a fellow officer who informed the general.

It would explain why Wilkinson was colluding with army contractors in 1793 to delay the supplies and transportation needed for Wayne's army to advance. General Wayne was eager to begin a war against the Indians in 1793, but unaccountable delays made it impossible.

George Rogers Clark didn't receive his commission as a major general in the French army until September, 1793. The plans for a French army invasion became public knowledge in October. If Wilkinson had prior knowledge of it, he would have wanted to keep Wayne's army in southern Ohio to prevent the invasion. Wilkinson was serving both the United States' and the Spanish government's interests, and making General Wayne look bad—accomplishing three things at once in his usual style. It is also likely that Alexander Hamilton, the Secretary of the Treasury, was interfering and arranging for problems with the contractors to keep the army in southern Ohio near the French Revolutionary Army.[145]

General Wilkinson billed the Spanish government for $12,000 on June 20, 1794. He said he had papers to prove he had spent $8,640 in breaking up Clark's plot to invade Louisiana; and he deserved a pay increase to $4,000 annually for his secret services. He wrote he was employing an informer to learn about Clark's plans, and $200,000 should be sent to him from Philadelphia to use as bribe money to win Kentuckians over to Spain. The general's monthly army pay was $104, which didn't begin to cover his expenses.

The new Governor of Louisiana, Baron de Carondelet, agreed to send $12,000 by Wilkinson's two couriers, Henry Owens and Joseph Collins. Owens would set off first with $6,000 and Collins would follow. A Spanish vessel carried Owen and the 6,000 silver dollars from New Orleans to New Madrid where the dollars were transferred into three small barrels. At the mouth of the Ohio, Owen and the barrels were transferred to a pirogue manned by a Spanish crew of six oarsmen and a pilot. Soon afterwards, Owen was murdered and the crew fled with the money.

Three crew members were later arrested and taken to Lexington, Kentucky, where Judge Harry Innes—Wilkinson's lawyer and his partner in the Spanish Conspiracy—interviewed them and sent them to General Wilkinson at Fort Washington, who attempted to send them back to the Spanish authorities at New Madrid. However, while en route to New Madrid, the Fort Massac commander arrested them and sent them back to Kentucky. When they were returned to Judge Innes to stand trial, he freed them for lack of evidence and the money was never recovered.[146]

Collins set out from New Orleans by sea in late August, but he spent much of the money on travel expenses and investing in land around Fort Pitt before he arrived in Ohio in April, 1795; Wilkinson reported to the Spanish that only $2,500 was left of the $6,334 that Collins had started out with. Wilkinson, —whatever his involvement was in destroying Clark's French army conspiracy—had little to show for it.[147]

General Wayne negotiated the Treaty of Green Ville ending the Indian Wars in the Northwest, signed by 80 chiefs and agents on August 3rd, 1795. The grand council was attended by 1,130 Indians, representing the Wyandot, Shawnee, Delaware, Ottawa, Miami, Eel River Miami, Chippewa, Potawatomi, Wea, Kickapoo, Piankashaw and Kaskaskia tribes. As part of the treaty provisions, the tribes agreed to vacate the southern 2/3's of Ohio, leaving it open for American settlement. The number of Americans on the Ohio frontier grew rapidly after that, from an estimated population of 5,000 in 1795 to 230,000 in 1810.[148]

Ensign Meriwether Lewis joined the 4th Sub-Legion in May, 1795. The records are fragmentary for both Lewis and William Clark during this time period, but it seems likely that Lewis and Clark first became friends during the summer of 1795. Their time together was brief, however, as Clark was gone most of the time during his final months in the legion.

Lewis served in a peacetime army, coming into the legion after the fighting had ended. Four years separated the two friends in age, but William Clark had acquired a wealth of experience in the last four years—fighting Indians in three separate campaigns, leading men, and traveling widely throughout the region. During this time, he also had the unenviable distinction of having a famous older brother who was organizing a French army to invade Louisiana.

Within a few months, the 21 year old Lewis got into a quarrel which might have ended his military career. He and another ensign called at a house where a Lieutenant Devin was attending a private party. Lewis asked to speak with Devin and they went out on the porch, where a heated discussion took place. Lieutenant Elliott, who was hosting the party, spoke to Lewis about his behavior. In later court testimony Elliott said Lewis was intoxicated when he arrived, but it must not have been very serious, as he offered the young men a brandy to calm them down.

Lewis felt his honor had been insulted and challenged Lieutenant Elliott to a duel. Dueling, or "affairs of honor," were a common practice among gentlemen of the upper classes and considered a mark of prestige.[149] They were, however, illegal under military regulations, and Elliott responded with a demand for a court martial. The trial ended with Lewis being acquitted "with honor." Anthony Wayne wrote:

> The Commander in Chief confirms the foregoing sentence of the General Court Martial, and fondly hopes that this is the first, that it may also be last instance in the Legion, of convening a Court Martial for a trial of this nature.[150]

After the trial ended on November 7th, 1795, Lewis was reassigned to the Chosen Rifle Company, an elite group of 75 riflemen commanded by William Clark, who had just arrived back at the fort.

Clark had been sent on a mission by General Wayne to deliver a letter of protest to Spanish Governor General Manuel Gayoso de Lemos at New Madrid. The letter objected to a Spanish fort being built on American territory at the Chickasaw Bluffs, some hundred miles south of New Madrid. Clark, who had visited Chickasaw Bluffs on a special assignment in 1792, delivered the letter to Governor Gayoso under a flag of truce. Accompanied by a sergeant, a corporal and 15 privates, he was away on his mission from September 15th–November 4th.

According to information Clark received from other travelers, the fort at Chickasaw Bluffs was 2/3's completed and had about 500 men, considerable cannon, and three galleys (warships) and one galliot (a smaller ship) attached to it. After it was completed, another fort, close to the junction of the Ohio and Mississippi Rivers, was going to be built.

William Clark delivered Wayne's letter to Governor Gayoso, containing an extract of Knox's secret instructions to the Fort Massac commander, which authorized the use of force to prevent American citizens from invading Louisiana—that is, George Rogers Clark's French Revolutionary Army. He most likely was not aware of its contents. Wayne wrote to Gayoso:

> . . . the United States are disposed to preserve peace and harmony with all the powers of Europe and with Spain especially. As a convincing proof of this fact, I take the liberty of inclosing a printed proclamation of the President of the United States of America of March 24, 1794, together with an extract from the orders given by me to the commandant of fort Massac at the same time.

Gayoso's superior, the Spanish Governor of Louisiana and West Florida, Baron de Carondelet, reported to his superior in Spain that Wayne's extract of Knox's secret instructions to him said:

> You know that a number of lawless people living on the shores of the Ohio, affronting the national authority, have conceived the bold design of invading the Spanish territory. The atrocity of this attempt and its effects are manifest in the enclosed Proclamation of the President of the United States.
>
> If they should persist in the same design or desire again to attempt it, and if any party present itself in the vicinity of the garrision, if you find yourself well informed that they are

> armed and prepared for war with the criminal intent reported in the Presidents Proclamation, you shall send them some person in whose veracity you can trust, and if he is a justice of the peace he will be the most appropriate messenger for making them comprehend their unlawful procedure and for preventing them from passing the fort at their risk; but, if, in spite of the pacific efforts made to persuade them to abandon their criminal design, they still persist in trying to descend the Ohio, you must make use of all military means in your power to prevent them. For which these instructions will serve you as sufficient justification, provided you have beforehand employed all the pacific steps first prescribed.

Secretary of War Henry Knox ordered Fort Massac to be built on the site of the old French Fort Massaic in order to repel any advance by Clark's army.[151]

Fortunately, the Treaty of San Lorenzo (Pinckney's Treaty) signed on October 27, 1795 in Madrid, ended the conflict. The treaty opened navigation on the Mississippi to Americans, and there would be no deposit fees collected in New Orleans to transfer goods between ships. The treaty also provided for a joint boundary line commission to survey the boundary line separating Spanish West Florida and East Florida from United States territory at 31 degrees north latitude.

After he made his report to General Wayne, the 25 year old Clark requested a leave of absence to take care of business in Kentucky and Virginia. His furlough extended from December through May, 1796. Upon his return in May, he wrote to his sister Fanny in Louisville—

> . . . my reception from the general at my arrival was more favorable then I had any reason to expect, from my inattention to his orders to return at a State time—but as he is a reasonable as well as a Galant man, and had some Idea of my <u>Persute</u>, he treated my inattention as all other good fathers would on the Same acasion (Galentrey is the Pride of a Soldur—and attachment followers in corse.)[152]

Clark's "Persute" must have been a romantic one, as most of his letter is concerned with the young ladies of Louisville. The general who was so "Galant" was Wilkinson, because Wayne was in Philadelphia. Obviously—despite a decade of deceit—the Clark family was still unaware of Wilkinson's treacheries.

William Clark resigned from the Legion of the United States on July 1, 1796 and returned to Louisville to begin helping George Rogers take care of his financial affairs. The time that Lewis and Clark spent together in the legion was only a "fiew months" as Clark reported to the editor of the Lewis and Clark Journals in 1811.[153] Their close friendship must have developed during Clark's visits to Washington while Lewis was serving as President Jefferson's private secretary in 1801–03, but again there is a lack of records.

The quarrel between Generals Wayne and Wilkinson continued. General Wayne went back east first. When he arrived at Philadelphia, the seat of the federal government, on February 6, 1796, he was greeted with a triumphal arch over his parade route, fireworks displays, and banquets.[154] The general remained in Philadelphia for several months. His wife had died in 1793 and he was courting Mary Vining, an old friend whom he planned to marry.

In May, Wayne received a letter from the new Secretary of War, James McHenry, warning him that President Washington had learned there were three foreign agents going west from Philadelphia: a French general, Victor Collot, his map making adjutant, Joseph Warin, and an agent of the Spanish named Thomas Power, who was suspected of being involved with Wilkinson. Wayne was urged to keep a close watch on all three. He followed them west, returning to Fort Greene Ville in July.

En route, on June 27th,Wayne reported from Pittsburgh to Secretary McHenry he had learned that Collot had arrived at Fort Washington to meet General Wilkinson, with a "bushel of introductory letters" and Collot had purchased horses in order to overtake Wilkinson, who was on his way to Fort Detroit. He also learned Thomas Power was traveling incognito with Judge Sebastian (Wilkinson's associate and a Spanish pensioner) on his way to Kentucky. General Wayne wrote:

> I shall follow close in the rear of those incendiaries & doubt not of making the necessary discoveries in due season & without alarm.[154]

After he reached Fort Washington, on July 8th, Wayne reported to McHenry that Wilkinson had met with Power for several days at Fort Greene Ville, and an informer said that Wilkinson had promised Power a reward of $1,000 if he would either bring Robert Newman to Wilkinson, or provide Wilkinson with Newman's answers to a set of questions—but only if Newman answered them satisfactorily. Wayne wrote that the informer told him Power had—

> . . . shewed [him] certain instructions from *Wil_____son* [Wilkinson] to put a suit of interrogatories to a Certain *Newman* & if he answered them to satisfaction, he wou'd reward him *powers* with One thousand Dollars either for them or the person of Newman. Powers had a number of letters for W_____son and gave out that his object in visiting Greene Ville was to obtain materials for writing a history of this country &c.[155]

Wayne also learned General Collot had changed his mind about following Wilkinson north and instead had sent by "safe express" a "great Number of letters" to the "great & Popular Genl" and intended to wait a week or ten days for his reply at the Falls of the Ohio.[156]

In Philadelphia, Secretary McHenry wrote a private letter to General Wayne on July 9th, advising him:

> Conciliate the good will and confidence of your officers of every rank; even of those who have shewn themselves your personal enemies. Gen. Wilkinson has entered upon a specification of his charges against you both old & new, and will press for a decision inquiry or court martial. I shall, unless I should

> be of opinion on reflexion that it is improper, send them to you in their condensed form, that you may prepare to meet them should it become necessary.[157]

Wilkinson's offer to pay $1,000 to Thomas Power for access to Robert Newman shows that he desperately needed to know what Newman had revealed about Wilkinson's dealings with British officials in Canada. The Governor of Canada, John Graves Simcoe, wrote that Newman "had been sent for some sinister purpose," and Alexander McKee, who owned the trading post near Fort Miami, said "Newman was not the character he represents."[158]

Newman was acting as Wayne's secret agent. What seems evident—from Wilkinson's distress—is that Newman had given Wayne a letter written by Wilkinson to British officials, and that Wayne had substituted another letter containing a mix of false information for Newman to deliver to the British. The British Governor Simcoe said the letter contradicted information supplied by Wilkinson in previous letters.

If Wayne had possession of Wilkinson's original letter, Wilkinson's career in the army would be destroyed. It seems likely Wayne did have this letter, and it would be presented as evidence during the impending military court of inquiry.

Wilkinson was able to offer Power $1,000 for information because he was about to receive money from the Spanish government. Power had come to the fort to arrange for delivery of $9,640 to Wilkinson from Governor Carondelet. It was restitution for the money he had not received in 1795, due to the murder of Owen and the dereliction of Collins. $640 of the amount was owed to Power for the delivery.

The 9,640 silver dollars were packed in barrels of coffee, sugar and tobacco at New Madrid. An informant at New Madrid warned Wayne that money was being sent by boat to Wilkinson, but a search by the commander at Fort Massac failed to reveal the contraband. Wayne fumed that he should have realized that sending tobacco to Kentucky was absurd because "it is one the principal stapel's of that State."[159] The barrels of silver dollars were put in the care of Philip Nolan, the general's young protegé, who delivered them to a store in Frankfort where the barrels were unpacked in front of Power and Nolan.[160] (This all became a matter of sworn testimony and documentary evidence during congressional investigations in 1808.)

When General Wayne returned to his command in the Ohio Valley, he had survived a political attack in Philadelphia. Wilkinson—knowing the Secretary of War supported Wayne—had lobbied Republican members of Congress to replace Wayne with himself, by establishing that a brigadier general would command the peacetime army (his own rank). The House of Representatives voted to reduce the authorized strength of the army by 60% to 2,000 men and 1,000 artillerists and engineers, and to eliminate the rank of major general. The Senate agreed to the reduction of the army, but refused to abolish Wayne's position. The actual size of the army remained unchanged, and Wayne was guaranteed his position until the next session of congress met in early 1797.[161]

There was strong sentiment in favor of state militias and abolishing, or severely limiting, the national army. The federal government had been in existence for only seven years, and it was just twenty years since the colonies had declared independence from Great Britain. A standing army in peacetime was perceived as a threat to liberty. Ironically, however, the Military Act of 1796 actually marks the beginning of the peacetime regular army.[162]

Wilkinson renewed his attacks on Wayne by pressing for a military court of inquiry regarding Wayne's leadership and tactics in winning the Battle of Fallen Timbers. Since only the commanding general—Wayne himself—had the power to convene a court of inquiry, President Washington and his cabinet procrastinated. Wilkinson came to Philadelphia in October, 1796 and threatened to bring the matter before Congress if the administration didn't act. The attorney general finally ruled that Wayne could be tried if President Washington authorized the court of inquiry.[163]

General Wayne, who was on a tour of the forts acquired from the British, had established headquarters at Fort Detroit. By the 20th of September, he was able to announce that the United States had taken possession of all the British forts in the Northwest: Michilimackinac, Detroit, Miamis, Niagara, and Oswego. He had hoped to visit Michilimackinac, but the approaching season of winds and ice on the river made travel impossible until spring.[164]

On October 28, 1796 General Wayne wrote to Secretary of War McHenry charging that his impeachment trial was criminally motivated, and that Wilkinson was colluding with Thomas Power and others:

> Thus Sir the three week visit to Greene Ville by P. . . .r in June last, was not idly spent, but rather improved in maturing the plan for a seperation of the Western Country from the Atlantic States. The frequent and long interviews between W. . . and P. . . .r & J. Br-wn, have not for the object the peace & happiness of the Union, in that which took place at Fort Washington the latter end of August, it was (among other *necessary* preparatory business) determined to remove me from the head of the Army by Impreachment, for piracy in order to make way for W . . . . . . son & thereby facilitate the accomplishment of their favorite Object, a seperation from the Union.[165]

General Wayne gave Ensign Meriwether Lewis a leave of absence until March 4th, 1797. On November 7th, Lewis left Fort Detroit to bring dispatches to Philadelphia containing the service records of the late Legion of the United States and Wayne's reorganization of the infantry and dragoons in the new United States Army.[166] Historian Grace Lewis Miller speculates Lewis may have carried a second set of Wayne's papers regarding Wilkinson, as it was common practice to send duplicates by a separate route.[167] It was a mark of regard for Lewis's abilities to be given such an important assignment at 22 years of age.

A four month leave of absence gave Lewis the opportunity to return home to Albemarle County for the holidays. It had been a little over two years since Lewis left home with the Virginia militia to camp at McFarlane's farm during the Whiskey Rebellion. He had joined the Legion in May, 1795, been present at the Treaty of Greene Ville with the northwest tribes, gone through a court martial over his dueling challenge, become friends with William Clark, and now was entrusted by the commanding general to carry the army's most important records of service.

On November 12th, General Wayne notified McHenry the dispatches were on their way, and he would be traveling by the sloop *Detroit* to Fort Presque Isle on Lake Erie, where he would write to him again. Wayne was planning to establish his permanent headquarters at Pittsburgh, as it was equally distant from Fort Washington at Cincinnati and Fort Detroit, and convenient to the government at Philadelphia. It was the last letter that Wayne wrote to the War Department.

The death of Anthony Wayne occurred at Fort Presque Isle on December 15th, 1796. He was 51 years old. Presque Isle which means "almost an island" in French, was the piece of land with a narrow neck which formed a harbor for the new town of Erie, Pennsylvania. The fort, located on the bluff overlooking the harbor, had originally been a French fort and then a British fort. It was now under the command of Captain Russell Bissell.

Wayne was suffering from gout in his leg, his old complaint aggravated by the injuries sustained by the falling tree, when he arrived at the fort on November 18th. On November 29th, Captain Bissell wrote to Major Isaac Craig, at Pittsburgh:

> . . . the Gen has been exceedingly ill ever since his arrival but is now recovering. I expect in a few days he will be on his way to Pitt.

A week later, the general was suffering from violent stomach pains and high fever—and it was reported that the gout had settled in his stomach. His adjutant, Henry Debutts, wrote on December 14th:

> The Gout fixed itself in the Generals stomack about a week since & continues with unabated violence—how long he continues to suffer such torture is hard to say—but it appears to me that nature must soon sink under such acute affliction.[168]

Wayne was attended by Dr. George Balfour, a surgeon's mate, at the fort. Balfour, in announcing the death of the general, described the general's illness on December 14th:

> . . . you have heard of General Wayne's arrival here from Detroit of his having the gout very badly. 11 days ago he was attacked with gout bilious in the stomach & bowels . . .

In his brief letter, he ends by requesting a quarter of a cask of wine be sent to him.[169]

Before a year passed, Balfour was recommended by General Wilkinson to become one of the first surgeons in the United States Navy, serving on board the frigate *Constellation* under Captain Thomas Truxton. Truxton wrote that General Wilkinson had recommended Dr. Balfour as "an officer of professional skill, honor, and strong sensibility."[170]

General Wayne asked for his physician, Dr. J. C. Wallace, to come to Presque Isle from Fort Fayette at Pittsburgh and a messenger was also sent to Dr. Benjamin Rush in Philadelphia. Dr. Wallace arrived a few hours after Wayne died.

Gout is a painful form of arthritis that attacks the joint of the big toe, or perhaps the joints of the knees, wrists or fingers. There is no such thing as "stomach gout." However, arsenic was the poison of choice for many centuries—called the "King of Poisons" and the "Poison of Kings." The symptoms of acute arsenic poisoning are headaches, violent stomach pain, convulsions and diarrhea. It is odorless and tasteless, and ingesting a small amount over a few days will kill a person. There was no test to detect its presence in the body until reliable tests were developed in the 1830's. After that, arsenic as a method of poisoning declined in use. By the middle of the 1800's it became a standard practice to use arsenic to embalm a body.

General Wayne, at his request, was buried at the foot of the flag pole at Presque Isle in his uniform with his boots on. In 1808, his daughter, being in ill health herself, asked to have her father's remains reburied at the family cemetery in Radnor, Pennsylvania near Philadelphia. In the spring of 1809 his son Isaac went to Presque Isle with a horse and small carriage to bring his father's remains back home. Dr. Wallace was now practicing in Erie, and Wayne's son asked Dr. Wallace to take care of the exhumation. He did not want to witness it himself.

Thirteen years had passed since the general died, but upon opening the grave, to everyone's astonishment, Wayne's body was perfectly preserved, except for his bad leg and foot, which was partially decomposed. Dr. Wallace proceeded to boil the flesh from his bones so that the bones could be packed in a small box and taken to his new burial ground. The remains of his flesh were reburied at the fort in the same grave.[171]

A lock of the general's hair was saved as a souvenir, and is still part of the collection of the Erie County Historical Society. Was it arsenic poisoning? "Arsenic mummification" was believed to occur during the time when arsenic was used as a poison. According to embalming practices, the small amount needed to kill a person, would not be nearly enough for embalming. But arsenic was in common use as treatment for syphilis, as a face powder, in bullet manufacture, in paint manufacture, and as a recreational drug at that time.[172]

Perhaps a forensic examination of the general's hair will reveal the presence of arsenic. The circumstantial evidence certainly supports the idea of another assassination attempt by General Wilkinson—and a successful one. There was a motive, a strong motive, and there seems to have been an agent, George Balfour, a surgeon's mate at an obscure army post, who was rewarded with the plum assignment of becoming a naval surgeon of the United States Navy. There was also an opportunity for General Wilkinson to have arranged for the assassination at Fort Presque Isle when he traveled to Philadelphia in October.

Assassinations, if cleverly done, are meant to seem like accidents, illnesses, or random killings by deranged individuals or petty criminals. The very word, "assassination" means a killing of a prominent person, usually by a secret attack. Humphrey Marshall suspected Wilkinson of arranging the assassinations of the two army officers Hardin and Trueman, and he said that he himself had been the object of an attempted assassination by someone who was a "catspaw" of powerful individuals. General Wayne called General Wilkinson "that vile assassin" in his letter to the Secretary of War. Was Wayne referring to the tree falling on his tent?

General Wilkinson was a man under extreme pressure. He couldn't win his court of inquiry case against Wayne which was based on the charge that General Wayne had mishandled the winning of the Battle of Fallen Timbers. Wayne would probably have produced the letter carried by Robert Newman—the one Wilkinson would pay $1,000 to learn about. However, with the death of his superior, James Wilkinson would become the Commanding General of the United States Army, and that is what happened next.

Anthony Wayne statue at Valley Forge National Historic Park looking towards his home in Paoli, Pennsylvania, 5 miles from Valley Forge. The family home was built by his grandfather in 1774. Historic Waynesborough is open to visitors. His grave is at Old St. David's Churchyard in nearby Radnor, PA. www.historicwaynesborough.org (Wikimedia Commons)

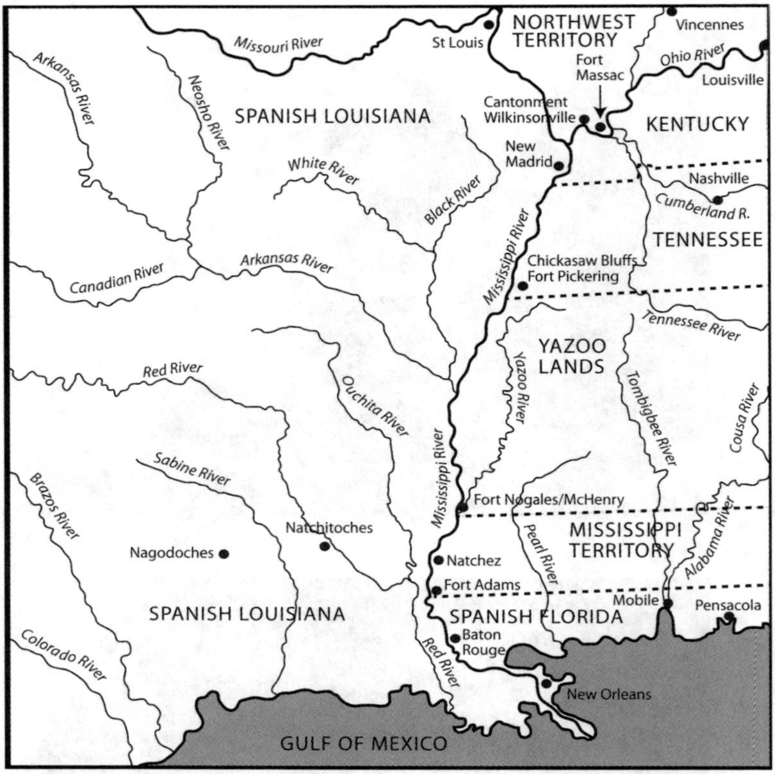

Mississippi Valley in the 1790's

- Northwest Territory was in existence from 1787 until 1803. In preparation for Ohio becoming a state in 1803, Indiana Territory was established in 1800 in the western part of the territory.
- Kentucky became a state in 1792; Tenessee, in 1796.
- Yazoo Lands were the subject of land speculation.
- Mississippi Territory was established in 1798; its northern boundary was extended to the Tennessee border in 1804.
- The Treaty of San Lorenzo in 1795 established the 31st parallel north as the boundary line between Mississippi Territory and Spanish West Florida.
- Spanish West Florida was seized by the United States in two stages: the Mississippi River to the Pearl River in 1810; and the Pearl River to the Perdido River in 1813. The Perdido River is located at Pensacola, in West Florida. Spain ceded East Florida to the United States in 1819.
- Spain acquired Louisiana from France in 1762. France took it back in 1800 and sold it to the United States in 1803.

Chapter 7

Mississippi Valley Conspiracies: 1797-98

The "Sweet Spot" on the globe was New Orleans and the land surrounding it—Spanish Louisiana and West Florida and the Chickasaw and Choctaw Indian territories that would become Mississippi Territory in 1798. Both Britain and France coveted the lower Mississippi Valley as the ideal location for establishing a new country. Spanish officials wanted to extend their empire, or establish a new independent country. All three foreign powers actively sought the collaboration of American citizens. The country that controlled New Orleans could use it as a base to establish a new empire on the Gulf of Mexico that would take control of the Mexican silver mines. The wealth of the silver mines could finance any number of revolutions in the Spanish colonial empire in Central and South America. It was the Age of Revolutions, and the world was in an uproar.

General Wilkinson arrived in Philadelphia in October, 1796 with his wife Ann and 10 year old Joseph. Their two older sons, John and James, had been attending school in Philadelphia since 1792. En route, they learned the devastating news that 15 year old John, their oldest son, had died of yellow fever while they were on their way to see him.

Philadelphia was still serving as the nation's capital city, and the Wilkinsons remained there for several months. President Washington had published his famous Farewell Address in September, almost two months before the first contested election for the American presidency was held in November, 1796. The address warned of the dangers associated with the newly established political parties. Alexander Hamilton had founded the

Federalist Party in 1791, and Thomas Jefferson and James Madison had founded the Republican Party in opposition to it in 1791–1793.

The two parties represented a deep split in the American people, loosely characterized as being either pro-British or pro-French. The Federalist Party was committed to a strong centralized government, a national bank, manufacturing, and commercial ties with Great Britain. Its supporters were generally found in New England and the business community. The Republican Party was the party of farmers and frontiersmen, who opposed a strong central government; and favored states rights, an agricultural economy, slavery, and the French Revolution.

The development of parties had not been anticipated by the writers of the constitution. In the election of 1796—to the consternation of both parties—Federalist John Adams was elected president and Republican Thomas Jefferson was elected vice president. It was not until after the election of 1800 that a constitutional amendment allowed voters to choose a presidential and vice presidential candidates running together on the same ticket.

Wilkinson was given a seat of honor at the inauguration of President Adams on March 4th, 1797.[173] Because of General Wayne's death, the court of inquiry was not held, and congress abolished the rank of major-general which had been held by Wayne. Wilkinson had to be content with remaining a one-star brigadier general, although he was now the Commanding General of the United States Army.

The European wars of revolutionary France continued with an ever changing series of alliances between nations. After Spain and France became allies in August, 1796, King Carlos IV of Spain declared war against Great Britain. How did the revolutionary wars in Europe affect the United States? The Quasi-War between the United States and France took place between 1797 and 1801. It was a undeclared naval war, fought mainly in the Caribbean. The French were upset by the United States' renewal of commercial ties with Great Britain under the Jay Treaty; and by the United States' refusal to continue paying on debt it owed to France for its support during the American Revolution. The United States contended the debt had been owed to the French Crown, not to its post-revolutionary government.

French privateers seized 316 American ships engaging in trade with British colonies in the Caribbean in 1795. The Spanish also captured American ships. (Privateers were ships authorized by governments to capture enemy ships during wartime and split the proceeds with the sponsoring government.)[174]

Regarding the territories of Spanish Louisiana and the two Floridas, the Spanish minister Manuel Godoy summed it up neatly: "You can't lock up an open field."[175] France was considered the most likely candidate to acquire Louisiana and the Floridas—either by negotiation with its ally Spain, or by force—which affected the decisions made by all four governments and domestic conspirators. Various plots were either to take possession of the lands before France did so, or to aid France in acquiring them. A leading historian of the time period, Arthur Whitaker, colorfully wrote that "the atmosphere of uncertainty brought forth a tropical profusion of plots."[176]

Because Britain and Spain were at war, there was also a threat of a British invasion force coming down from Canada through the Mississippi Valley to seize Spanish territories. General Wilkinson, after settling his family in Pittsburgh, began a tour of forts in the Northwest reaching Detroit in June, 1797. In September, the Spanish agent Thomas Power met with him to deliver a confidential message from Baron de Carondelet, the Spanish governor of Louisiana and West Florida, who was based in New Orleans.

The Baron offered $100,000 in funds for Wilkinson to persuade the residents of Kentucky and Tennessee to secede from the union and form a new Spanish province east of the Mississippi. The province would extend from the mouth of the Yazoo River to the Ohio River. Wilkinson himself would be granted land in Illinois Country and receive a $4,000 annual pension.

Wilkinson replied that the opening of the navigation on the Mississippi had caused Americans to lose interest in such schemes. He told Power to inform Carondelet he had destroyed his cipher code and would no longer be acting as their agent. The General—who had received approximately $26,000 from Spanish authorities in the years 1790–1796—would stay off the Spanish payroll for the next seven years. However, he promised Carondelet the United States Army would oppose any attempt by Britain to seize control of the Floridas and Louisiana, as these were also his instructions from the Secretary of War.[177]

Governor Carondelet was attempting to block the plans of Zachariah Cox, an American speculator in the Yazoo lands east of the Mississippi, and/or the British plans to seize the Spanish territories. The lands on the east side of the Mississippi River were the homelands of the Chickasaw and Choctaw Indians (which later became Mississippi Territory) and Spanish West Florida with its access to the Gulf of Mexico. The area could become the center of a new British colony using New Orleans as its capital city; or it could become a rival to Mississippi River business interests by linking together the rivers which flowed into Mobile Bay. The Indian lands had not yet been acquired by the United States and were a magnet for real estate speculation and conspiracies for years to come.

In 1796, Zachariah Cox established a company to create a commercial water route from the Ohio River to the Gulf of Mexico at Mobile Bay. Cox's plan started with establishing two settlements, one located at the mouth of the Cumberland River —on the Ohio, and the other at Muscle Shoals near the Great Bend of the Tennessee River. A short 35 mile carriage route from Muscle Shoals would connect the Tennessee River to the Tombigbee River which flowed into Mobile Bay. Cox's route would shorten the travel distance from the Ohio River to the Gulf of Mexico, and would be in direct competition with New Orleans business interests.[178]

A partner of Cox, John Smith T. (the "T." stood for Tennessee), established Smithland, Kentucky in 1796 at the mouth of the Cumberland River. Cox led a group of settlers to Muscle Shoals in 1797. These activities incurred the wrath of General Wilkinson who had Cox arrested at Natchez in August, 1798.

The day before Wilkinson arrived at Natchez, Cox escaped from jail and fled to New Orleans.[179] Secretary of War McHenry believed that Manuel Gayoso, the newly appointed Spanish governor of Louisiana and West Florida, "probably was in connivance" with Cox [180] as Gayoso refused to return him to American territory.[181] Cox's escape may well have been another instance of escaping an assassination plot by Wilkinson. However, a few years later, General Wilkinson and Smith T. became friends and associates in the St. Louis area.

General Wilkinson, in July, 1797, issued a proclamation at Detroit declaring Martial Law:

> To guard the National Interests against the Machinations of its enemies, secret or covert, Foreign or Domestic—to baffle the Arts of Seduction, which have led to numerous desertions from the Public Service and to restrain licentiousness and the infamous habits of drunkenness encouraged among the Troops by the disorderly conduct of the vendors of Ardent Spirits. The Commander-in-Chief thinks it is a duty appurtenant to the high trust reposed in him, a duty of indispensable obligation to declare Martial Law within the line of the Guards and limits of the Fortifications of the place . . . [182]

Martial Law was needed because of a dramatic news story which began to unfold in July—the first impeachment trial of the federal government. The month before, in June, President Adams had come into possession of an incriminating letter revealing that the Senator William Blount from Tennessee was engaged in a conspiracy with a foreign power, Great Britain, to invade Louisiana and the two Floridas.

Blount, the first Governor of Territory South of the Ohio, was elected a U. S. Senator from the new state of Tennessee in 1796. Tennessee had been granted statehood that year, despite Federalist opposition to admitting a state that would vote Republican in the upcoming national election. Aaron Burr, then a Federalist senator from New York, helped secure the vote for

Tennessee Senator William Blount, by W. B. Cooper (Wikimedia Commons)

statehood. In 1796, both Blount and Burr switched their party affiliations from Federalist to Republican, and Blount supported Burr as the Republican vice presidential candidate, with Jefferson as the Republican candidate for president.[183] Due to the election procedure at that time, Thomas Jefferson—who received the second highest number of votes—became vice president, while the Federalist candidate John Adams became president.

Senator Blount—a major speculator in western lands— was in financial difficulty in 1797 and became involved in a conspiracy to seize Spanish territories for the British. British diplomat Robert Liston reported to Lord Grenville, the British Foreign Secretary, on January 25, 1797 that an Indian agent named John Chisholm had met with him in Philadelphia to tell him there were 1,000–1,500 British Loyalists and Indians in the Southwest Territory who were ready to invade the two Floridas and take possession of them for England. Liston sent Chisolm to England. In May, after Chisolm had left for London, Liston reported to Grenville there were suspicions that the British government actually intended to attack the Spanish possessions on the Mississippi. Many loyalists had settled in the Natchez area and the two Floridas during and after the Revolution. [184]

British officials, worried by rumors that France would soon acquire Louisiana from Spain, were preparing to launch major expeditions to seize the Spanish territories. General Collot, the French spy who traveled through the Mississippi Valley making maps in 1796, reported he had heard in Natchez in October,

1796 that a mixed group of whites and Indians would come from the north and capture St. Louis. From there, one group would march to Sante Fe and the other would capture New Orleans.[185]

George Rogers Clark wrote on March 2, 1797 that he had been approached by British agents from Canada who offered him a colonelcy to lead an expedition of 2,000 men to seize Sante Fe. General Elijah Clarke in Georgia—who had also participated in the Genêt plot—received an offer to lead an expedition to seize the Floridas. Both men, ardent Republicans, turned the British agents down.[186]

After the Blount Conspiracy was exposed, John Chisholm reported the outline of the conspiracy to Congress on November 29, 1797. He said men would have come from the New York and Pennsylvania area, including a party led by Joseph Brant, the Mohawk leader. The northern men—

> . . . were to attack New Madrid, leave a Garrison in it, and proceed to the Head of the Red River and take possession of the Silver Mines. Mitchell and Craig with their Associates were to descend the Ohio in the character of Traders. The people of Tennessee, Whitley's men from Kentucky, with those of the Natchez and the Choctaws were to attack New Orleans; no precise arrangement was made concerning the Command of this Party; but I supposed it would be headed by Blount.

The conspiracy got derailed by a man who received an indiscreet letter written by Blount. The senator was in Tennessee when he learned he had to return to Philadelphia for a special session of Congress. Changing his plan to visit an associate named James Carey, Blount wrote to him on April 21, 1797:

> Among other things that I wished to have seen you about, was the business Captain Chesholm mentioned to the British Minister last Winter in Philadelphia. . . .
>
> I believe, but am not quite sure, that the plan then talked of will be attempted this fall; and if it is attempted, it will be in a much larger way than then talked of; and if the Indians act their part, I have no doubt but it will succeed. A man of con-

> sequence has gone to England about the business, and if he makes arrangement as he expects, I shall myself have a hand in the business, and probably shall be at the head of the business on the part of the British. . . . When you have read this letter over three times, then burn it. . . .

Carey, who had recently become an employee at the Tellico Blockhouse, a government trading post near Knoxville, asked another clerk what he should do. His friend told him to remember his oath to the government, and the letter was turned over to the manager of the trading post, James Byer, who was an enemy of Blount's. The news spread quickly in Tennessee. A copy of the letter was sent to George Washington and the original letter was brought to Philadelphia.[187]

President Adams, after receiving the letter in June, conferred with administration officials. It was decided it was a high misdemeanor crime and Blount could be impeached. Impeachment is a process by which civil officers of the United States government can be removed from office. It would be the first time the impeachment process was used in the ten years since the constitution was established in 1787.

The letter was read before closed sessions of the House of Representatives and Senate on July 3rd, 1797. By July 5th it was public knowledge. On July 8th Blount was expelled from the senate, despite the protest of his lawyers. Blount fled to Tennessee, forfeiting his bail, and Congress adjourned until November. A congressional investigation committee met throughout the summer. In the end, Blount escaped impeachment by a vote of 14 to 11 in the Senate, on the grounds that senators were not subject to impeachment.[188] Blount was elected to the Tennessee State Senate in 1798, and served as its speaker until his death in 1800.

The question of the relationship between the United States and France underlay all of these issues. It was the beginning of the undeclared naval war with France in the Caribbean. A crisis involving French refugees had been brewing for sometime. By the mid 1790's, 45,000 French "aliens" had arrived in

America fleeing the Reign of Terror in France and the slave revolts in Saint Domingue (Haiti). The nation's capitol, Philadelphia, which had a population of 44,096 in 1790, gained a French refugee population of 5,000 by 1793.[189] The 2,000 refugees from Saint Domingue brought the deadly yellow fever with them. Yellow fever is a disease caused by the bite of an infected mosquito—something that was not known at the time. Members of Congress left town along with 20,000 other residents and a quarantine was established.[190] Ultimately, 4,044 people died of yellow fever during August–November, 1793 in Philadelphia.[191]

Most of the French refugees were temporary residents in America fleeing the horrors of civil war in their homelands. In 1793, the 2,000 refugees from Saint Domingue in Philadelphia received $14,000 in private charity, and aid from the federal government in 1794.[192] Politics played a big part in their life, with opposing factions continuing their quarrels.

One of the most important refugees was Prince Talleyrand, the French aristocrat and diplomat who was a leader of the Enlightenment Era, and an excommunicated Catholic priest. Talleyrand had spent two years in England before coming to the United States in 1794. Talleyrand and his friend Moreau de St. Méry were both elected to the American Philosophical Society during their stay in Philadelphia. They spent many hours at Moreau's bookstore in Philadelphia discussing Louisiana and planning to become its governors when it was once again acquired by France.[193]

Talleyrand greatly admired Alexander Hamilton, and said this, comparing him to Napoleon and the British statesman, Charles James Fox:

> . . . he had known nearly all the marked men of his time, *but that he had never known one on the whole equal to him*.[194]

However, he didn't understand why Hamilton, who had been Secretary of the Treasury under Washington, still needed to work as a lawyer to support his eight children. He wrote, after seeing him working late at his law office:

> I have seen a man who made the fortune of a nation laboring all night to support his family. [195]

Talleyrand had no such scruples about money. While he was in the United States he sold secret plans of the British government which he had acquired. The plans—sold to the Spanish envoy for $8,000—were for a joint attack on Spanish America which Britain had proposed to France in 1790. Sir Archibald Campbell would lead the attack against Spanish territories in the Caribbean and Gulf and at New Orleans.[196]

When Talleyrand returned to France in September, 1796, he became Foreign Minister. President Adams sent three diplomats to France in July, 1797 to resolve issues regarding the French seizure of American merchant ships doing business with the British. Talleyrand demanded a bribe of 50,000 pounds and a loan before he would formally enter negotiations. Informal talks were held, but by March, 1798 all hope of progress was gone.

Republican opponents of President Adams demanded that Adams release the diplomatic dispatches. The release of their letters reporting the demand for bribery led to calls for war. The "X, Y, Z Affair"—so called because the names of three French negotiators were taken out and X, Y, and Z substituted—was the match that lit the fire for anti-French sentiment in the United States. The words of the Federalist congressman from South Carolina summed up popular sentiment—"Millions for defense, but not one cent for tribute!"

Six "super-frigates" were being built in shipyards. They were of innovative design, fast, and constructed with wood able to withstand cannon fire. Three of the ships were launched in 1797 to fight in the war with France known as the Quasi-War.

The 1790's were a complex time for the early American Republic. The Quasi-War with France was one of many crises. The new nation considered itself a collection of states—"these United States," rather than "the United States"—and the development of political parties added to the tension. Would the states hold together, or would they split apart?

Prince Talleyrand
(1754–1838)

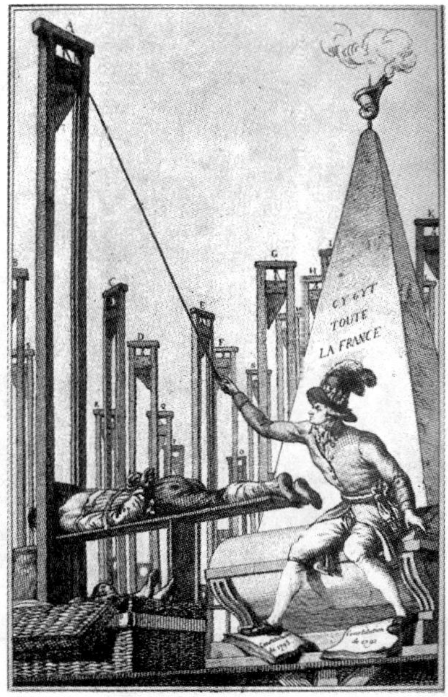

Execution of King Louis XVI on January 17, 1793.

Over 16,000 people were executed by the guillotine during the Reign of Terror in France. The Terror lasted from September,1793–July, 1794. An additional 25,000 people were killed by other means. Robespierre is shown executing the executioner in this cartoon. Robespierre, a Jacobin leader, was executed himself by the guillotine in July, 1794.[196] (Wikimedia Commons)

The super frigate *Constitution,* "Old Ironsides," sailed again for her 200th birthday in 1997. The ship at the Charlestown Navy Yard in Massachusetts is open to visitors year round. (Wikimedia Commons)

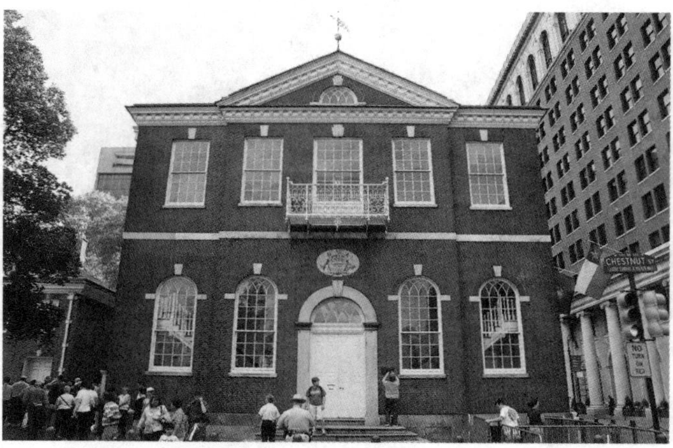

Congress met at Congress Hall from December, 1790 until May, 1800. Philadelphia was the temporary capitol of the U.S. while the new seat of federal government was being built in Washington, D. C. (Wikimedia Commons)

Despite rivalries and partisanship, both Republicans and Federalists agreed that France could not be allowed to reestablish its former empire in Louisiana. France had owned most of the continent of North America before losing it to England in the Seven Years War (1756-1763). Spain kept the land west of the Mississippi River undeveloped in order to guard its silver mines in the Sierra Madre Mountains against intrusion.

France wanted to take back Louisiana in order to grow agricultural products for Saint Domingue (Haiti), its wealthiest colony. Saint Domingue, occupying one third of the island of Hispaniola, was the world's leading producer of sugar and coffee. Land in Saint Domingue was too precious to be used to feed its own people.

Saint Domingue underwent a slave revolution in 1791, becoming the second American colony to fight for its independence. Before the revolution, the French colony's population was 452,000 slaves, 40,000 whites and 28,000 free blacks and mulattoes—African slaves making up nearly seven times the number of its free population. The fight for independence would take 13 years, ending with the establishment of the black republic of Haiti in 1804, the second successful colonial revolution in the Americas, and the only slave revolt to establish a new nation.

Before the revolution, some 500 U. S. ships were trading in Saint Domingue. They were supplying almost all the sugar and molasses imported into the United States. The island was the second largest market for American exports, surpassed only by Great Britain. When British troops occupied Saint Domingue in 1793, American ships brought supplies to the British occupiers. By 1797, there were over 600 American ships engaged in trade with Saint Domingue. Federalists, the party in power, favored either Britain controlling Saint Domingue or black independence. Britain lost at least 7,500 troops—or perhaps three times as many—to yellow fever while occupying Saint Domingue, before it finally retreated from the island in 1798.[197]

President Adams reopened trade with Saint Domingue in 1799, supplying arms and ammunitions to the black rebels. The American navy maintained a base on the island. Jefferson and the Republicans, however, were opposed to aiding the rebels, fearing the spread of slave revolts in the southern states.[198]

The surrender of the Spanish forts on the Mississippi and the surveying of a boundary line between Spanish and American territory was established by the Treaty of San Lorenzo (Pinckney's Treaty) signed in 1795, which guaranteed U. S. navigation rights on the Mississippi and the free transfer of goods at New Orleans. but the actual relinquishment of Spanish forts on the Mississippi took longer than expected.

Andrew Ellicott (1754-1820) was the leading surveyor in the United States. He taught surveying methods to Meriwether Lewis before Lewis led the expedition to the Pacific Coast. (www.explorepahistory.com)

Andrew Ellicott arrived in Natchez in February, 1797. The distinguished scientist was the U. S. Commissioner for the joint nation survey to establish the boundary line at 31 degrees north latitude between Spanish West Florida and the new Mississippi Territory. Ellicott had surveyed the western border of Pennsylvania, which established the prime meridian for Northwest Territory, and the boundary lines for the nation's new capitol at the District of Columbia.

It was an inauspicious time for the surrender of forts—the Blount Conspiracy was still underway and the threat of British troops coming down from Canada to seize Spanish forts on the

Mississippi was real. Ellicott and his party encamped on the top of a hill near the fort at Natchez and raised the flag of the United States, despite Spanish threats to cut it down. Ellicott noted in his journal that he had earlier seen a confidential letter written by Natchez Governor Gayoso in June, 1796 that said:

> . . . the treaty was not intended to be carried into effect, and that delay on their part would reduce it to dead letter. And . . . that the country either was, or would be ceded to the republic of France. [199]

The surrender of the Spanish forts was delayed for more than a year as Spanish officials anticipated either a British invasion or the cession of Louisiana and the Floridas to France. Things quieted down after the Blount Conspiracy was exposed, and Ellicott finally began the boundary line survey in April, 1798. Ellicott remained the senior American official in the area until General Wilkinson and Winthrop Sergeant, the first Governor of Mississippi Territory, arrived in September, 1798.

On November 14, 1798, Ellicott wrote in cipher code to Secretary of State Timothy Pickering that he had discovered General Wilkinson and several others were in the pay of Spain, and he accused them of plotting with Spanish officials—Governor Gayoso and Baron de Carondelet—to start a revolution. He reported their plan was to detach Kentucky and Tennessee, and after that, establish a new empire. Ellicott wrote:

> The first object of these plotters is to detach the States of Kentucky and Tenesee from the union and place them under the protection of Spain. If that could have been effected this season the treaty would never have been carried into effect; and to ascertain the probability of such an event Mr. Power was sent in the beginning of last June into the States before mentioned.
>
> The design of detaching the western country from the union is but a small part of the general plan which is very extensive and embraces objects of immense magnitude; nevertheless to ensure success, this point must be first carried; which being

> effected and by the system of promotion adopted by the court of Madrid, Govr Gayoso will be at Quito and the Baron de Carondelet at Mexico about the same time; so as soon as this arrangement takes place or sooner if the necessary officers can be corrupted a general insurrection will be attempted, and cannot fail of success if the first part succeeds. Genl Wilkinson is to proceed from Kentucky with a body of troops through the country by way of the Illinois into New Mexico which will be a central position—the route has already been explored.
>
> Nine tenths of the officers of the Louisiana [American] regiment are at this time corrupted and the officers of the Mexican regiment which is now in this country are but little better. The apparent zeal of the Spanish officers on the Mississippi for the dignity of the Crown, is only intended to cover their designs till the great plan which is the establishment of a new empire is brought to maturity. Their principles are highly revolutionary.[200]

Two kinds of conspiracies—foreign and domestic—existed in the Mississippi Valley. The governments of Great Britain and France had competing plans to invade the Spanish Floridas and Louisiana and establish a new country which would include Kentucky and Tennessee. Both governments needed the support of western leaders to succeed. They each had previously owned the territory as part of their North American empires and both governments still owned islands in the Caribbean.

Spain was in a different position. It owned Louisiana and the Floridas, but its colonial officials were disloyal and plotting to establish a new empire with the assistance of Americans. Baron de Carondelet, the Governor of Louisiana and West Florida, proposed creating a Spanish province east of the Mississippi which would include Kentucky and Tennessee. This was the plan his agent Thomas Power brought to General Wilkinson in Detroit in 1797.

Ellicott said that in 1798, nine-tenths of American military personnel were ready to invade Spanish territory and that Spanish officials were supporting invasion plans. He described their principles as "highly revolutionary" and that they were making plans of "immense magnitude." The sticking point in this conspiracy was the first condition, that the states of Kentucky and Tennessee secede from the union and become part of a Spanish province, before New Mexico would be invaded by American troops. After the invasion, citizens of New Spain would establish local governments independent of Spain; and once the conspirators were in control of the silver mines, the revolution would be exported to other countries in the Spanish colonial empire.

General Wilkinson visited Andrew Ellicott at his surveying camp for three days in October, 1798. Ellicott complained to him about his assistant surveyor, Thomas Freeman. With Wilkinson's consent, Ellicott expelled Freeman from his surveying party, calling him "an idle, lying, troublesome, discontented, mischief-making man." Freeman was often drunk and disobedient. He enjoyed reciting lines from Shakespeare, wielding a sword, and playing Falstaff, Shakespeare's comedic fool.

Wilkinson and Ellicott maintained a cordial relationship as the two high ranking U. S. officials in the territory, but Ellicott naively wrote to Wilkinson on December 16th warning him that he had seen a letter written by Thomas Power—

> in which your name is mentioned in a manner which astonishes me. I dare not commit any part of it to paper. . . . I am confident it is false.

Even though he assured the general he trusted him, Ellicott had already sent the information to Secretary of State Pickering in Washington.[201]

Within six months, Wikinson hired Thomas Freeman as a civilian contractor to build Fort Adams. The fort, located six miles above the 31st degree boundary line of Spanish Florida and Mississippi Territory, became the southernmost point of entry into United States territory. It was built on a 100 foot bluff overlooking the Mississippi River called Loftus Heights.[202]

Fort Pickering was established by the United States in 1799, 200 miles north of Fort Adams, on the site of the Spanish Fort Fernando de las Barrancas at the Chickasaw Bluffs.[203] The

Chickasaw Bluffs were four strategic bluffs on the east side of the Mississippi River which marked the western boundary of the Chickasaw Indian nation and had been the location of French and Spanish forts for centuries. Memphis, Tennessee would be founded here in 1819, and Meriwether Lewis would spend some of the last days of his life at Fort Pickering in September, 1809.

The two other Spanish forts acquired by the United States were Fort Natchez at Natchez, Mississippi and Fort Nogales at the mouth of the Yazoo River. Fort Nogales was renamed Fort McHenry for the Secretary of War, and Natchez became the territorial capitol of Mississippi.

William Clark had gone to the Chickasaw Bluffs in 1793 to recruit Chickasaw scouts to serve in Wayne's army during the Ohio Valley Indian Wars. He visited it again in March, 1798, while bringing a flatboat of the Clark family's tobacco down to New Orleans. During his travels, he sketched Fort Ferdinand and the Governor's Spanish war vessel. Clark noted on his drawing of the boat:

> The Galley that Govr Gayosoa de Lemoz stays abord of 28 men, 24 oars, one 6 pounder & 18 swivels . . . In 1798.

The boat appears to have been the model for the Lewis and Clark expedition boat, which is sometimes mistakenly called a keelboat, but which they called a barge.

Gayoso became the Governor General of Louisiana and West Florida in 1797, replacing the Baron de Carondelet. He granted permission for Clark to travel from New Orleans to Philadelphia by ocean ship in July, 1798. It was eight months before Clark returned to Louisville in December, 1798, having traveled 4,400 miles on his journey. [204]

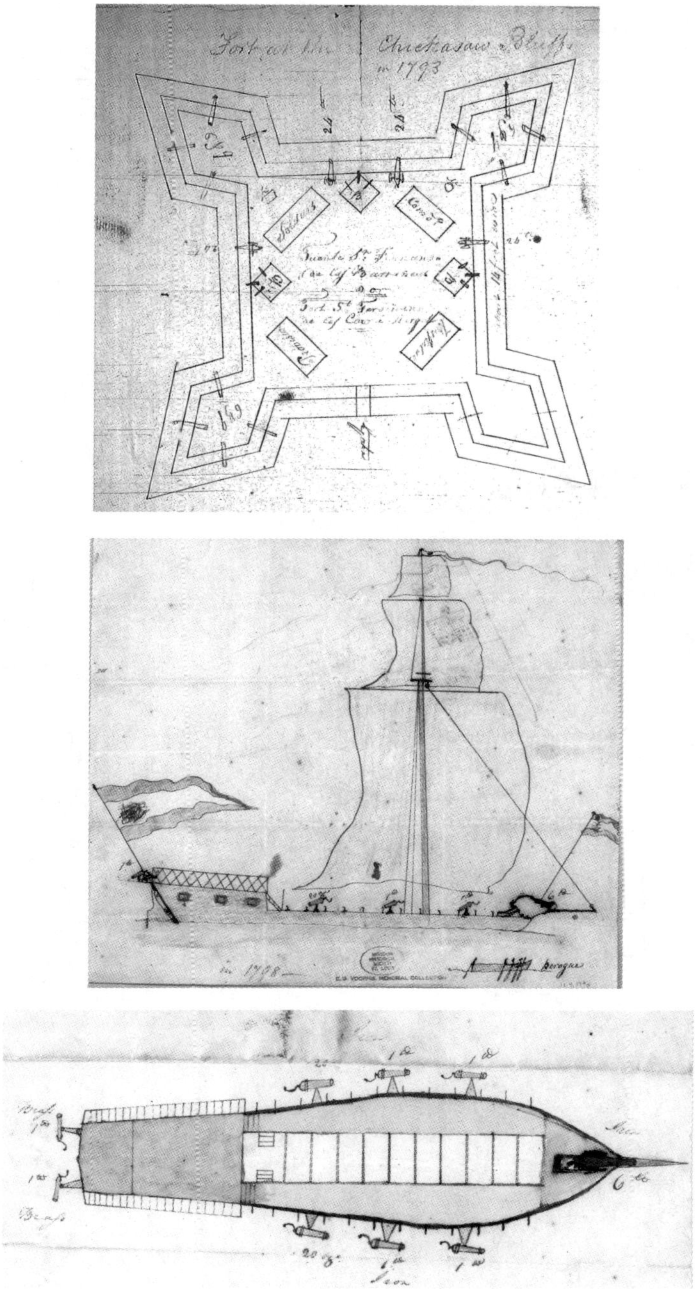

William Clark's 1798 sketches of Fort Ferdinand and Governor Gayoso's war vessel. (Missouri History Museum, St. Louis)[204]

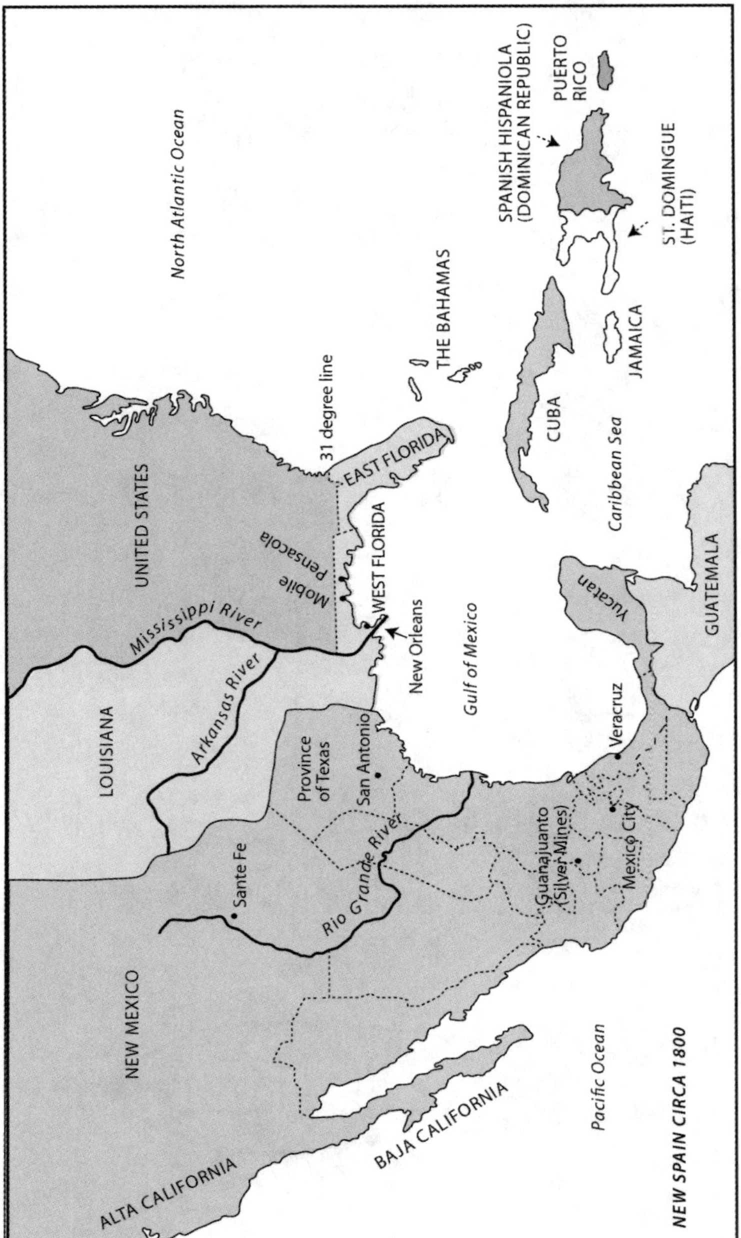

New Spain, circa 1800

Chapter 8

Conflicts and Secrets: 1798–1801

Meriwether Lewis, after delivering the army records, arrived home in Albemarle County in December, 1796. He was elected to membership in the local Masonic Lodge in January. Lewis rose rapidly in the secret association, becoming a Master Mason and then a Past Master Mason by April, 1797 at the age of 22. He was active in the Door to Virtue Lodge while he lived there. On October 31, 1799 Lewis became a Royal Arch Mason in the newly established Widow's Son Lodge of Milton, Virginia near Jefferson's Monticello.[206]

During 1797 he was on leave of absence from the army, traveling to Kentucky to secure land titles inherited by the children of his mother's second marriage to John Marks, and to look after his own land holdings. He bought 2,600 acres in Kentucky at 20 cents an acre; and then went to Georgia to assist in bringing slaves belonging to his Locust Hill estate back to Virginia.[207]

On January 1, 1798 Lewis became the Commanding Officer in charge of recruiting at Charlottesville. He hired a musician to play the fife and paid for circulars. Over the next fifteen months, 39 men received enlistment bounties for joining the army—undoubtedly, they were all good Republicans.[208]

When Jefferson returned to Monticello in July of 1798, it seems likely Lewis would have been helping him with Republican party activities. Jefferson—the Vice President of the United States—remained at home at Monticello until mid December, when he went to Philadelphia for the winter session of Congress.[209] He was deeply disturbed by the Alien and Sedition Acts which had been passed by Congress that summer.

The Alien Acts applied mostly to French and Irish emigrants. It allowed for aliens to be deported by the president whenever they were suspected of being dangerous to the peace and safety of the United States; and in the event of a declared war, they were to be apprehended. In addition, the time required for living in the United States before applying for citizenship was changed from 5 years to 14 years.

The Sedition Act was aimed at Republicans. Persons could be charged with a crime of high misdemeanor for conspiring together to oppose the government by writing, publishing, printing, or uttering anything slanderous, malicious or false against the government. Both the Alien and Sedition Acts would expire in March, 1801 at the end of Adam's presidency. The suppression of dissent was aimed at preventing a Republican victory in the upcoming presidential election of 1800. It turned out that the backlash against the Sedition Act probably helped Jefferson gain the presidency rather than hurt him.

Over the next few months, Jefferson and James Madison secretly drafted the Kentucky and Virginia Resolutions, arguing that because the constitution was created by a "compact of states," the states had the right to declare federal laws unconstitutional and not legally binding. It was aimed at nullifying the Alien and Sedition laws, Jefferson wrote the Kentucky Resolution adopted by the state of Kentucky in 1798, and Madison wrote the Virginia Resolution adopted by Virginia in 1799. Though their endorsement was requested, no other states in the union supported these resolutions. Jefferson and Madison remained anonymous, knowing they were on dangerous constitutional grounds. The arguments for state's rights were later adopted by the southern states during the Civil War, citing Jefferson's precedent.

Jefferson didn't trust the postal service. Jefferson wrote to a friend on February 13, 1799:

> I avoid writing to my friends because the fidelity of the post office is very much doubted.[210]

Historian Adrienne Koch writes about the period when Jefferson was drafting the Kentucky Resolution in 1798—

> In the critical months directly after Congress passed the suppressive legislation [the Alien & Seditions Acts] that dismayed the Republicans, there was not a single entry to or from Madison in Jefferson's ledger between July 21 and October 26. Yet when correspondence was resumed, it is clear that Madison was fully aware of what Jefferson had done.[211]

Lewis was living in Albemarle County. He probably served as the courier between Jefferson at Monticello and Madison at Montpelier in Orange County. It was a distance of 28 miles.[212] Lewis was an ardent Republican, and General Wayne had used him to carry the official records of the army two years earlier. He appears the logical choice to have acted as their go-between, and to have performed other confidential Republican party services.

The 24 year old Lewis was promoted to Lieutenant in the 1st Regiment of the Infantry on March 3, 1799. In May, he moved with his detachment of recruits to Camp Allegheny in the Shenandoah Valley near Staunton, forty miles west of Charlottesville.[211] At Staunton, Lewis became adjutant, or executive assistant, to Captain Alexander Gibson, commander of Camp Allegheny. Lewis served as adjutant from July 1, 1799 until July 31, 1800.[213]

Staunton was a commercial center and a Federalist stronghold. A newspaper reported that in early April, members of the Greenbrier County Court reacted to the Virginia Resolutions by tearing them up and trampling them underfoot.[214] Perhaps Lewis and his detachment were stationed there to keep an eye on things. Matters were at a boiling point in Virginia.

The Governor of North Carolina, William R. Davie, wrote to his friend James Iredell, a Justice of the Supreme Court, on June 17th, 1799, about the political climate in Virginia:

> Virginia is the only State of which I despair. . . . My opinion, collected from some gentlemen who have lately been travel-

> ling in that State, and others who were at the Petersburg races, presents a melancholy picture of that country. These gentlemen returned with the firm conviction, that the leaders there were determined upon the overthrow of the General Government; and if no other measure would effect it, then they would risk upon the chance of war. I understand that some of them talked of "seceding from the Union," while others boldly asserted the policy and practicality of "severing from the Union;" alleging that Pennsylvania would join them; that Maryland would be compelled to change her politics with the situation; that the submission and assistance of North Carolina was counted on as a matter of course; and that the two Southern States would follow.[215]

In January, 1798 the Virginia legislature had authorized the creation of an armory to manufacture arms at Richmond, and two arsenals to store arms at Richmond and Fredericksburg. The bill also authorized the purchase of 20,000 guns. Taxes were almost doubled to pay for all of this. Was secession intended? The military build up was planned for several years before the legislation was enacted, but there is no doubt that Virginia—the largest and oldest of the original colonies—felt entitled to determine its own course of action.[216]

The two Clark brothers, George Rogers and William, were traveling in 1798. During the summer, George was in Philadelphia lobbying Congress against expanding the Quasi-War with France. George still considered himself a French citizen and was waiting to be reimbursed for his expenses by both the French and American governments. In June, 1798 he wrote to his friend Samuel Fulton in Paris that he should—

> . . . assure the Directory of my unalterable attachment to the cause of France and her Republican constitution and that I await with great impatience further instructions.[217]

When Clark was in Philadelphia he was threatened with arrest under the Alien and Sedition Acts. He wrote to Fulton that he was asked to renounce his French citizenship, and—

> . . . invited to give my resignation to the counsel general of France or to retire from the United States. . . . I refused to do the former . . . but I believe that I shall be obliged to subscribe to the second proposition."[218]

When he returned to Louisville, there was an order for his arrest. His friends disarmed the officers serving the arrest warrant and George fled to the safety of St. Louis, which was then under Spanish rule and an ally of France. While he was living in St. Louis, he studied Spanish military defenses and declared that in the event of a war between France and the United States, Spanish forces would support France and not resist a French invasion of Louisiana. In September, 1798 he wrote to French officials, urging haste, because—

> . . . in nine months, at most, it will be too late, this country will be conquered by America or England.

By June of 1799, he was back in Louisville. So it must have blown over—George Rogers Clark was, after all, one of the most famous men in the American West.[219]

It is unknown how deeply his younger brother William was involved in George's plan to establish a new French nation in the Mississippi Valley. George was an alcoholic by this time. William may have been humoring his brother's frustrated ambitions, or he may have truly believed in the idea of establishing a French nation on the gulf coast. No evidence has turned up revealing William's intentions, other than the drawings he made of the Spanish fort and war vessel on the Mississippi River at the Chickasaw Bluffs in Tennessee.[220]

In 1797, William had been dealing with lawsuits and meeting with creditors in an attempt to resolve his brother's financial affairs. He wrote:

> I have rode for Bro. George in the course of this year upwards of 3000 miles . . . continually on the pad, attempting to save him.[221]

In March, 1798, William Clark began a long journey down the Mississippi to sell the family's tobacco crop. It was on this voyage he made the sketches at Chickasaw Bluffs. After selling the tobacco in New Orleans, he went by ship to Baltimore, arriving in late September. He visited Washington, still in the process of being built, and then went on to visit family in Virginia. Altogether he traveled about 4,400 miles on his journey.

When he arrived home in Louisville on Christmas Eve, 1798, he found that his mother had died on the very same day. Ann Rogers Clark was 68 years old. His father John Clark died soon afterwards in July, 1799. William was the principal beneficiary of their estate. He inherited 3,318 acres of Kentucky land, 23 slaves, the family's Mulberry Hill home, and a grist mill. Nothing was left to George, who was increasingly impaired by alcohol, and whose assets could be seized by creditors.

Among the slaves were York, who would accompany William to the Pacific Coast, and York's father and mother, old York and his wife Rose. York had been William Clark's companion since they were boys. Whether York accompanied Clark as his servant during his army days is unknown.[222]

John Adams faced some very difficult decisions regarding the military during his years as president in 1796–1800. The country associated a federal army with tyranny, but in 1796, Congress had established the beginnings of a small peacetime army under the command of General Wilkinson, in recognition of the need to guard the western territories.

It was still very much a union of "these United States" rather than "the United States." The country's lack of military defense was seen as positive evidence of neutrality in the European wars. In addition, American commerce was profitting by its neutral status, supplying food stuffs to the warring nations, even though its merchant ships were being captured.

President's house in Philadelphia, where John and Abigail Adams lived from 1796–1800. They moved to the White House in Washington in November, 1800. (Wikimedia Commons)

Alexander Hamilton believed in military preparedness to secure peace. He had organized the 13,000 volunteers and state militias that put down the Whiskey Rebellion in 1794. After the congressional elections in the fall of 1794, he resigned from the office of Secretary of Treasury to re-enter private practice of law in New York City. He was not a rich man and needed to rebuild his own finances.

Hamilton didn't like John Adams and supported Thomas Pinckney, the other Federalist candidate, in the 1796 elections. It was a narrow victory for Adams, who received 71 votes to Jefferson's 68 votes in the electoral college. The "French influence" was the biggest campaign issue, as Pierre Adet, the French minister to the United States, made an *Appeal to the American People* urging them to vote for Jefferson. It most likely had the opposite effect to what he had intended.

Adams retained Washington's cabinet members, ensuring a smooth succession, but the cabinet members kept in close touch with Hamilton, particularly his former deputy Oliver Wolcott who replaced him as Secretary of Treasury. When Adams learned of Hamilton's disloyalty in supporting Pinckney instead of himself on the Federalist ticket, he was furious.

France—which was winning the war against Britain—was capturing American ships, even in American waters, in an attempt to starve Britain into submission. Merchant ships were forbidden by law to arm themselves for defensive purposes, and were helpless against attacks by small ships armed with a few guns. Over 300 American merchant ships had been sunk or captured by the French by June of 1797.[223]

The Quasi-War with France in 1797-1801 was never officially declared and was fought on water, but it was feared that France would invade America. The country was now convinced that military preparedness was necessary. Hamilton urged that a regular army of 50,000 and a provisional army of 50,000 be called up in anticipation of an invasion by the French army. Congress responded by establishing an "additional army" of 10,000 men in May, 1798. Congress also established a provisional army of volunteer corps of 50,000 men to be called up in case of invasion or imminent danger of one. The Department of the Navy was established, and ships were rushed to completion. [224]

In the years 1798-1799 there were five separate armies authorized by Congress. The first was the original "western army" of four regiments under the command of Wilkinson, known as the old army. A "new army" was created of twelve additional regiments and six mounted dragoons in July, 1798. The third was the "provisional army" of 50,000 volunteers to be called up in a crisis. The fourth was the 10,000 men "additional army." The fifth, the "eventual army," was created in March, 1799 giving the President the authority to appoint officers for 28 additional regiments in case of emergency. Only the old army and the new army existed in reality. The others were on paper.[225]

Hamilton's friends in the cabinet—Secretary of State Timothy Pickering, Secretary of War James McHenry, and Secretary of Treasury Oliver Wolcott—were consulting him in a continual stream of correspondence between Philadelphia and New York during this time.[226]

The nation was in hysteria. Military historian Richard A. Kohn in *The Eagle and the Sword* writes:

> Yet conspiracies *were* afoot in the 1790's, and for sheer number and range of seditious plots this decade was perhaps unique in American history. . . . Party and faction were nearly synonymous, and faction was understood to operate normally in the conspiratorial mode. . . . The generation of the Federalists had witnessed too much fundamental political change. . . . To doubt in 1798 the viablity of their own institutions and the indepence of the nation was only realistic.[227]

It was against this background of factions and plots that Adams was forced to make his decisions. To Hamilton's surprise, Adams was firm in his support of military preparedness. But the president remained as hostile as ever to Hamilton himself. In 1797 Adams had written to his wife that Hamilton was—

> A proud, spirited, conceited, aspiring Mortal . . . As Great a Hypocrite as any in the U. S. . . . I shall take no notice of his Puppy head but retain the same opinion . . and maintain the same conduct towards him I always did, that is to keep him at a distance.[228]

Adams asked George Washington to lead the new army despite his cabinet members wanting Hamilton to be in charge. In retirement, Washington agreed to assume command, but everyone knew he would serve as a figurehead. Hamilton, his most trusted aide during the Revolutionary War, was elevated from the rank of colonel to inspector general of the new army, with the rank of major general. Two other major generals were also appointed. Adams balked at allowing Hamilton to be placed in charge, but Washington insisted and Hamilton became his second in command in July, 1798. After Washington's death in December, 1799, Hamilton became the ranking general of the U. S. Army.

John Adams and Alexander Hamilton (Wikimedia Commons)

Hamilton began talking about taking New Orleans in early 1798. If an open war with France came about, he wanted—with the help of the British fleet—to take East and West Florida and Louisiana from Spain and perhaps, venture into South America. He was corresponding with Francisco de Miranda, a Venezuelan, who had intrigued for years with both the French and English governments to liberate Venezuela and other Spanish colonies in South America. Miranda was now proposing a British and American alliance.[229]

In January, 1799 Hamilton was writing to Congressman Harrison Gray Otis, chairman of the defense committee:

> If universal empire is still to be the pursuit of France, what can tend to defeat the purpose better than to detach South America from Spain, which is the only channel through which the riches of *Mexico* and *Peru* are conveyed to France? The executive ought to be put in a situation to embrace favorable conjectures for effecting that separation.[230]

A month earlier he had written to Otis saying that:

> . . . with a view to the possibility of internal disorders alone, the force authorized is not too considerable.[231]

Hamilton was referring to the possibilities of an armed insurrection in Virginia, and slave revolts in the U. S. inspired by the revolution taking place in Saint Domingue under the leadership of Toussaint Louverture.

President Adams must have been aware of Hamilton's intentions, because on February 18th, 1799 he took the bravest step of his career according to his biographer David McCullough—he sent a message to congress nominating William Vans Murray to be appointed as Minister Plenipotentiary of the United States to the French Republic, with instructions to arrange a peace with France. He consulted no one—neither his wife, nor his cabinet—in making this decision. His son, John Quincy Adams, who was serving as a diplomat in Europe, had alerted him to the fact that the French government was ready to negotiate. Adams then retired to his home in Braintree, Massachusetts on the outskirts of Boston, where he spent the next seven months.[232]

In October, 1799 the president went to Trenton, New Jersey where the government had gone to escape another yellow fever epidemic in Philadelphia. The peace mission, expanded to three men, had not yet left for Paris. Hamilton was in Trenton, and he argued strenuously with Adams, urging him to suspend the mission and enter into an alliance with Great Britain in the war against France. Adams reported:

> I heard him with perfect good humor, but never in my life did I hear a man talk more like a fool. [233]

The commission departed for France in November, 1799. The Treaty of Mortefontaine was signed in September, 1800 ending the former Treaty of Alliance and recognizing the neutral status of the United States. News of the outcome didn't reach the U. S. until after the presidential election, when Adams lost his bid for a second term, running against Thomas Jefferson and Aaron Burr. Adams received 65 votes; and both Jefferson and Burr received 73 votes. News of a peace treaty might well have changed the outcome.

John Adams at the end of his life said he wanted only one inscription on his gravestone:

> "Here lies John Adams, who took upon himself the responsibility of the peace with France in the year 1800."[234]

John Adams asked for the resignation of Secretary of War James McHenry and fired Secretary of State Timothy Pickering in May, 1800, accusing them of disloyalty to him. The new army was disbanded by Congress on June 15, 1800, and Hamilton ended his army career on the same date.

It would appear that Adams had vanquished the new army and its commander Alexander Hamilton, but—without his knowledge—there were events taking place behind the scenes in the remaining months of his administration.

There is a hidden history regarding General Wilkinson and the establishment of a secret military base called Cantonment Wilkinsonville. Wilkinson's three volume *Memoirs of My Own Times*, published in 1816, has hundreds of pages about his military career, but significantly, he wrote not one word about Wilkinsonville.

Hamilton had asked the general to come to New York City in the fall of 1799 to brief him on western military affairs. In his *Memoirs*, Wilkinson wrote that during their initial meeting on August 1st, he asked Hamilton whether he would object to his meeting with "an old friend who I have not seen in several years." The friend was Aaron Burr. Hamilton had no objections, and Wilkinson met with Burr on the same day.[235]

After Wilkinson submitted a lengthy report to Hamilton in September, the two generals spent the last part of 1799 making plans for a new military base on the western frontier. Hamilton preferred the Falls of the Ohio at Louisville for the location, while Wilkinson argued for a location closer to the Mississippi River—the location that became Cantonment Wilkinsonville.

Wilkinson returned to Natchez in February, 1800 to bring his family back east. His sons, James and Joseph, were enrolling in the College of New Jersey. Aaron Burr's father had founded the college, which would later become known as Princeton University. By June, the general and his wife were in residence in Georgetown. Since Adams had asked for McHenry's resignation as Secretary of War in May, the general was effectively in charge of the War Department in the brand new Federal City of Washington for the next six months.[236]

Without the knowledge of President Adams, the logistics of moving troops to the site of the new military base were carried out by Wilkinson. The troops were en route to the base, starting in October. Over 700 troops arrived there on the first day of the new year, January 1, 1801.

Samuel Dexter, a congressman from Massachusetts, served as temporary Secretary of War from June, 1800 through January, and then served as temporary Secretary of Treasury from January through May of 1801. He was officially in charge during the time that Wilkinson was actually running the War Department.

A mysterious fire broke out in the War Department Office on November 8, 1800, destroying the building and all of its records. A second fire broke out in the Treasury Office on January 20, 1801, destroying some of its records. It was generally believed the fires were the work of arsonists. Republicans charged that the Federalists wanted to destroy proof they had sent arms to Toussaint, the leader of the slave revolution in Saint Domingue, and to destroy evidence of their "corruption and lies."[237]

The reader is probably ahead of me here. As I have said, Wilkinson was a master of the paper trail. There is no proof that he had anything to do with the fire in the war department, but he certainly would benefit from the destruction, or removal, of incriminating records. The two fires had the advantage of being able to report that records were no longer available.

In any event, the fire obscured the deployment of troops to Cantonment Wilkinsonville. A cantonment is a military camp without the walls of a fort. The existence of the camp was not known until historian Norman Caldwell published a journal article in 1949. Caldwell wrote:

> The present writer is at a loss to explain how such an important military establishment could be omitted from the annual reports of the Secretary of War and other official publications as printed in the *American State Papers* series. It may be suggested that the secrecy of the project was the reason why no official public reference to the venture was made.

Caldwell found that, three weeks after the fire, Wilkinson asked for official approval of the cantonment on December 1, 1800. Secretary of War Dexter granted approval the next day. [238]

The general and Ann celebrated Christmas at Fort Fayette in Pittsburgh. For the next two months, Wilkinson busied himself with renovating the fort, adding housing for new recruits and fixing up his own quarters. Thus he was conveniently absent from Washington during the tumultuous three months of December, January and February, while the election of the next president remained undecided. In March, 1801 he returned alone to Washington to greet the newly-elected President of the United States, Thomas Jefferson.[239]

Fort Fayette in Pittsbugh was 942 river miles from Cantonment Wilkinsonville, or four to five weeks travel distance on the Ohio when speed was not an issue.[240] Colonel David Strong arrived at the cantonment with about 700 troops on January 1, 1801. In the following months, they constructed a town of 200–300 huts laid out in regular streets. By the summer of 1801, there were about 1,500 soldiers in residence, most belonging to the Second Regiment of Infantry. This was about half of the U. S. Army which then numbered about 3,600 soldiers. Wilkinsonville was in use from 1801-1802, but after that, the troops were moved to other locations on the Ohio where the climate was healthier and they were employed in building military roads and forts south of the Ohio River.

The abandoned town of Wilkinsonville was taken over by Cherokee Indians and other squatters. In 2003–2005, Southern Illinois University conducted archeology digs at the site.

In researching military records, archeologist Mark Wagner discovered at least five Lewis Clark Expedition members were stationed here in 1801–1802: Patrick Gass, Hugh McNeal, John Robinson, Richard Windsor and Thomas Howard.[241]

On March 7, 1801 Meriwether Lewis wrote to Ferdinand Claiborne, his "old acquaintance and messmate" who was stationed at Wilkinsonville, about his belongings. Lewis was in Pittsburgh and had just come from Detroit. He told Claiborne he could keep Lewis's frying pan and Dutch oven, etc, as he had found a letter from Jefferson waiting for him at Pittsburgh asking him to accept the job of private secretary to the newly elected President of the United States. If Jefferson hadn't called Captain Lewis to Washington, he most likely would have followed his frying pan to Wilkinsonville.[242]

So, what was the purpose of Wilkinsonville? It started out as a plan for a military base for 3,000 troops who would take Louisiana and the Floridas from Spain in the event of an attack by France on the U. S. This was the proposal that Hamilton and Wilkinson were working on in the fall of 1799. General Washington and President Adams refused to support any offensive actions against Spain without a declaration of war, and Hamilton's and Wilkinson's proposals to station troops at Natchez and to appoint army officers from the region were denied.

Alexander De Conde, author of *The Quasi War,* says "the effort faded into failure" to liberate the Floridas and Louisiana—but De Conde didn't know about the existence of Wilkinsonville.[243] The mission ultimately did end in failure, but there is a "rest of the story" involving Wilkinsonville, Philip Nolan, and the contested election between Jefferson and Burr in the winter of 1800-1801.

How are they related?—Wilkinsonville was a secret military base; Philip Nolan was the leader of an illegal expedition into Spanish Texas with a secret mission; and the election of 1800 unaccountably ended in a tie between Thomas Jefferson and Aaron Burr, running for president and vice president on the Republican party ticket. All three events happened at the same time. The case will be made for a planned invasion of Spanish territory in 1801 or 1802 that would have happened if Aaron Burr had become the President of the United States.

Philip Nolan was Wilkinson's trusted assistant—the young man who could carry a bag of 2,000 Spanish silver dollars with one hand in 1789. He lived in the Wilkinson household in Kentucky as a teenager. He was born in Belfast, Ireland in 1771. How he got to Kentucky is unknown, but he was a "young man of talents, good breeding and moderation," according to Governor Carondelet. Nolan later told Spanish officials, that he was related to the former Irish governor of Spanish Texas, Hugh O'Conor, or "Hugo Oconór."

Nolan was a skilled map maker—Andrew Ellicott had instructed him and given him instruments. He made maps of Spanish Texas which he shared with Wilkinson. He led armed parties to capture, or acquire by trade, thousands of wild mustang horses on expeditions into Spanish Texas. In December, 1799, he married into a wealthy family in Natchez, which had connections throughout the Spanish Floridas and Mississippi.[244]

In January, 1800 Jefferson wrote to William Dunbar, a scientific correspondent in Natchez, about Nolan, who was planning to travel east with his mustangs:

> . . . in that case as my residence [Monticello] is on his best route, I may have the pleasure of seeing him personally and perhaps purchasing one of his fine horses for the saddle, which I am told are so remarkable for the singularity and beauty of their colors & forms.[245]

Wilkinson wrote a letter of introduction for Nolan, dated May 22nd at Fort Adams on the Mississippi, and dated by Jefferson as received on November 3rd—but Nolan was in Texas by the month of November. The letter of introduction said:

> In the Bearer of this letter—Mr. P. Nolan, you will behold the Mexican traveller, a specimen of whose discoveries, I had

> the Honor to submit to you in the Winter 1797, Mr. N____s subsequent excursion have been more extensive, & his observations more accurate, He feels pride in offering Himself to your investigation, and I am persuaded you will find pleasure, in his details of a Country, the Soil, clime, population, improvements & productions of what are so little known to us.

Jefferson was perfectly capable of postdating the receipt of a letter to obscure a meeting with a "Mexican traveller." A meeting must have taken place because he had the letter in his possession. Some historians have wanted to deny—based on the date of receipt—that a meeting occurred.[246] However, it seems most likely the reason Nolan went east was to meet with supporters of Aaron Burr who were planning an invasion of Texas.

By late August, Nolan was back in the Natchez area recruiting men to join his fifth expedition into Spanish Texas.[247] On October 21, 1800 at Nacogdoches, Nolan wrote to his good friend Jesse Cook asking him to join him:

> I am taking a large quantity of goods. Everyone thinks that I go to catch wild horses, but you know that I have long been tired of wild horses. I expect to return by January. . . .Take my advice my friend. Don't lose one moment in setting out. . . . I have good men and never will be taken. Come in, my friend, if you value your own safety. . . Burn this letter as soon as you receive it.

The letter was intercepted by the Spanish and never reached Cook. It is thought Nolan used a courier whom he knew would give the letter to Spanish authorities and its information was deliberately misleading. The Spanish commandant was receiving reports of Nolan and thirty men all armed with "pistols, saber and carbine," and that Nolan had a "saber, four pistols and two carbines." Nolan and his men left Natchez on October 31st or November 1st. [248]

One member of Nolan's expedition, Mordecai Richards, deserted the expedition in December along with two companions.

Richards had worked for Nolan in Natchez managing his horse herds. His son and a nephew remained with Nolan, but Richards was sufficiently alarmed by what Nolan had revealed to him that he was determined to inform the Spanish authorities.

After Richards deserted the expedition, Nolan and his men spent a week trying to find him, claiming that Richards and his companions were "lost," but they had no success. When Richards reached the Spanish authorities, they reported that he told them Nolan had more on his mind than catching horses—

> . . . for he [Nolan] had maps of all the rivers, marked all the distances and explored everything more attentively than a man would do who had no other ends in mind than catching horses—that he had told him many times that there were mines in this country, and moreover that he had assured his men that if the Spaniards killed a single one of them he would take a commission as an English general (which he could have) and would take possession of all this country—that Philip Nolan had said upon leaving Natchez that he would return within three months but later explained that his business would detain him for more than a year.[249]

In another testimony, Richards declared Nolan told him:

> My plan is to follow a northwestern course to pass through the country of the Caddos, and at a certain distance to erect a fort, as a protection from any attack; from there we can make our expeditions to examine the mines and other precious things which I know are to be found in those places; from time to time we may catch some horses. After we have reconnoitered the country and have secured the number of horses which we desire, we shall start our march to Islas Negras [Illinois] and continue to Kentucky without finding any obstacles in our way. After we arrive there we shall find friends and others who await our arrival, which I hope will be next August or Sep-

> tember. I hope then to receive a commission from the British minister at Philadelphia, with whose authorization there are many whom I know will come to conquer those lands or at least to gather the riches which are there. For this reason, I am taking the horses in order that they all may come mounted. At the end of two years we shall return to our homes rich and without any evil befalling us because we have the support of the British government from which I already have indirect assurance. [250]

On December 22nd, orders were sent to Natchitoches to arrest them. On March 4th, 1801, 70 regular Spanish army troops and 50 militia set out to locate their camp. On March 21st they found Nolan and his men, "ragged, bearded and hungry," where they had stayed for nearly five months, near today's Blum, Texas south of Fort Worth, Texas. In a day long siege, Nolan was killed by a cannonball—or perhaps a "chance shot to the forehead." Both accounts were reported. No one else was killed.[251]

Spanish troops across the frontier were patrolling daily for signs of Nolan and his men. Historian Noel Loomis writes that:

> . . . along that river [the Rio Grande] was one of the greatest mobilizations of Spanish soldiers the northeastern frontier had seen. . . The practical Spaniards smelled a filibuster in the making and had visions of British and Americans pouring into Texas so they mobilized against the possibility.[251]

Fifteen Americans, two slaves and seven Spaniards were arrested at Nolan's camp and imprisoned by the Spanish. A few weeks later, three Americans and one slave escaped from the prison at Nacogdoches. One of them, Robert Ashley, would once again escape capture in February, 1807, when he was escorting Aaron Burr on his flight to escape arrest in Mississippi Territory. This is one piece of evidence linking Philip Nolan to Aaron Burr.

The strongest evidence is a report by a Federalist lawyer in Kentucky, Joseph Hamilton Daviess, who was so alarmed by the filibustering activities of Burr in 1806 that he wrote to

President Jefferson several times warning him about Burr's plans to invade Mexico. Since Jefferson was playing his own game in 1806—giving Burr enough time to provide Jefferson with indisputable evidence of conspiracy and treason—Daviess was ignored. Angered by Jefferson's apparent lack of interest, Daviess published a book in 1807 containing their correspondence. He had written to Jefferson about meeting Wilkinson in St. Louis in 1806. He said he would take an oath on the truth of this report and informed Jefferson that Wilkinson—

> . . . took out a map of the country of New Mexico, which I think was in manuscript; and after some conversation about it, tapping it with his finger, told me in a low and very significant tone and manner, "that had Burr been president, we would have had all this country before now."[252]

We do know Burr and Wilkinson were in communication in 1800. Burr wrote to Charles Biddle (who was a cousin of Wilkinson's wife Ann) on November 18th, saying "Just before I left Albany, I received two letters from Wilkinson." The editor of the *Political Correspondence and Public Papers of Aaron Burr* says no letters have survived from Wilkinson to Burr from this early date.[253] Wilkinson, in his *Memoirs of My Own Times*, states that he began corresponding in cipher with Burr in 1800. He was defending his use of a cipher during the Burr conspiracy of 1805-07 as being a longstanding practice.[254]

Was Wilkinson involved in a conspiracy with Aaron Burr to invade Mexico, in the event that Burr was elected president? It seems a real possibility—both Blount's conspiracy in 1795-97 and Hamilton's plan in 1798-99 were to seize the Floridas and Louisiana with British help. Wilkinson's protégé Philip Nolan expected to receive a commission from the British minister at Philadelphia in 1800. Americans were involved in a series of ongoing conspiracies with Britain to seize Spanish territories before France could retake her former territories.

Now the cotton lands of Texas and the silver mines of Mexico were added to the Floridas and Louisiana. Nolan planned to map the location of the mines—which produced almost 2/3's of the world's silver in 1800[255]—and there was immense wealth to be had in cotton exports. With the invention of the cotton gin in the 1790's, land suitable for growing short-staple cotton suddenly became very valuable. Exports of cotton to British textile mills grew sixty times between 1790 and 1800.[256]

Natchez, the capitol of Mississippi Territory in 1798, was the center of wealth in the Deep South, with cotton plantations worked by thousands of African-American slaves. The area had been settled by British loyalists from the original thirteen colonies in the 1770's. Plantation owners were eager to expand across the Mississippi into the cotton lands of Louisiana and Texas. Whether the land across the river became a British province or an American territory wasn't important—compared to taking possession of the land before the French could.

Nolan married Fannie Lintot, the daughter of Bernard Lintot, a Natchez plantation owner, in December, 1799. Lintot, a native of England, had moved from Connecticut to West Florida in the 1770's, before moving to Natchez in 1792.[257] He was well connected throughout the region. Nolan was related to Hugh O'Conor, the Irish former governor of Spanish Texas, and there were many Irish bureaucrats, military officers and priests living in Spanish territory as Spanish citizens.

There were multiple ways conspiracies could be organized and outcomes imagined to liberate the Spanish colonies. Alliances and schemes were constantly shifting. It is likely some of the same people involved in the invasion plans of William Blount and Alexander Hamilton were now involved in a new conspiracy based on the expectation of Aaron Burr becoming president. Federalists would have been part of all the plots.

Any conspiracy would have involved federalist ship owners and merchants in the northeast who were engaged in trade with Britain. Their trade in the Caribbean and South America reached a new height in 1800 due to the neutral status of U. S. ships during the war between France and Britain. The author of *The Quasi-War*, Alexander De Conde, writes:

Many a New England fortune was founded on the trading of this period. Taking advantage of war conditions, American merchant ships carried goods for practically anyone all over the world, ran blockades, and eluded frigates and privateers. Despite laws against it, New England merchants even carried on a lucrative trade in smuggled goods to French colonies.[258]

Thomas Jefferson and Aaron Burr (Wikimedia Commons)

The election of 1800 ended in a dramatic tie between Thomas Jefferson and Aaron Burr in the electoral college. Burr refused to concede the election to Jefferson, even though he had run as the vice presidential candidate on the Republican ticket. The contest between Jefferson and Burr moved to the United States House of Representatives, where the rules were different. Each state's delegation had only one vote, no matter how big or small their delegation was. The voting in the House took 36 ballots to break the deadlock before Jefferson was elected the third President of the United States on February 17th.

The tie vote was known in mid December. During the two months before the congressional balloting began on February 11th, there must have been discussions about using the secret military base to invade the Spanish territories. Aaron Burr, a former Federalist who turned Republican, had supporters in both parties. Some of them would have known about the 700 soldiers who began traveling to Wilkinsonville in October, 1800.

The Federalists in Congress held the key to the contest between the two Republican candidates. There were sixteen states in the union and a majority of nine states was needed to elect the president. Jefferson had eight votes, Burr had six, and two state delegations were deadlocked. One state would have to come over to Jefferson. Eventually a deal was brokered and the final vote was 10 for Jefferson, 4 for Burr, and 2 states abstaining.

> JEFFERSON: Georgia, Kentucky, Maryland, New Jersey, New York, North Carolina, Pennsylvania, Tennessee, Vermont, and Virginia.
> BURR: Connecticut, Massachusetts, New Hampshire, and Rhode Island.
> ABSTAINING: South Carolina and Delaware.

The election of 1800 has been the subject of numerous books by historians and political scientists. Because the secret military base was concealed in the historic record, the possibility of an invasion of Spanish territories has never before been linked to Burr's attempt to become president, except for Wilkinson's comment to Joseph Hamilton Daviess in 1806:

> . . . had Burr been president, we would have had all this country before now. [259]

Alexander Hamilton was lobbying hard against Burr—urging his fellow Federalists to support Jefferson as the lesser of two evils. He wrote to Oliver Wolcott, Jr. on December 16th, when the tie vote was known:

> As to *Burr* there is nothing in his favour. His private character is not defended by his most partial friends. He is bankrupt beyond redemption except by the plunder of his country. . . .

> If he can, he will certainly disturb our institutions to secure to himself *permanent power* and with it *wealth.*

He said about Jefferson:

> There is no doubt but that upon every virtuous and prudent calculation Jefferson is to be preferred. He is by far not so dangerous a man and he has pretensions to character.[260]

A power struggle was going on behind the scenes in the Federalist Party—could Burr be trusted, and had the time come to seize the Spanish territories with the help of Britain? Britain was already cooperating with the United States in its Quasi-War with France. The British naval blockade of the Spanish navy at the Port of Cadiz from 1797-1802 allowed American ship owners and merchants to profit greatly in their trade with Latin America. It was natural to think of taking it a step further.

Usurpation and civil war were talked about while the election remained undecided. Usurpation means taking power by force. Historian Michael Bellesiles writes:

> Rumors of assassination, arson, military preparations, revolution and civil war swirled around the new capitol of Washington, D. C. Most fearful were the southern Republicans. They were certain Federalists had secretly sent arms to Toussaint Louverture for his slave rebellion in Saint Domingue. Many southern Republicans were convinced that two mysterious fires, one at the War Department in November, the other at the Treasury in January, were part of an effort to cover up this support for slave rebellions, perhaps even in the United States.[261]

Slave rebellions were a real concern. Gabriel, a literate slave who worked as a blacksmith, organized a large rebellion in 1800 in the Richmond, Virginia area. Participants would be armed with swords made from scythes. The plan was to burn the warehouse district in Richmond as a diversion; take Governor James Monroe hostage; and seize the militia arms being stored in the state capitol building. Gabriel estimated he had "5 to 600 associates" and expected 1,000 men would participate.[262]

The date was set for August 30, 1800. On that day, there was a sudden and very violent thunderstorm which flooded the city and washed away bridges. One of the slaves grew frightened and informed his owner of the conspiracy. Gabriel and 25 of his associates were convicted and hanged within two months of the failed revolt.

Gabriel had instructed his followers not to harm Methodists, Quakers and Frenchmen, as these groups were anti-slavery. Urban Republicans were members of Democratic-Republican clubs and worked alongside black laborers hired out by their owners. Newspapers were read in the taverns to those who were illiterate and the political situation was widely discussed. The captured conspirators were nearly unanimous in stating that two Frenchmen were involved in planning the rebellion, but records of their testimony and other evidence were never released by Governor Monroe.[263]

Republicans accused the Federalists of organizing the rebellion and Federalists accused the Republicans. If the presence of Frenchmen had been proven, it would have had a devastating effect on the Republican party. Undoubtedly, the rhetoric of freedom and equality led the slaves to revolt. In 2002, the city of Richmond honored Gabriel and his followers; and in 2007, the governor gave them an informal pardon.[264]

Fear of slave revolts was real. However, it is not the explanation for the fear of civil war or usurpation that took place during the weeks the election remained undecided. Republicans were afraid that the deadlock in the House was intentional. If the deadlock wasn't broken, then the president would be chosen by the president pro tempore of the Senate, the most senior senator of the majority party. Since Federalists controlled the Senate, Aaron Burr would become president.

Republicans warned that this usurpation of power would lead to civil war. Governor Thomas McKean of Pennsylvania had

22,000 men ready to oppose usurpation and Governor Monroe of Virginia was prepared to call out his militia and had secured the state arsenal against any seizure of arms by the Federalists. A Federalist newspaper warned that 70,000 militia from Massachusetts and militias from New Hampshire and Connecticut would support the Federalists.[265]

The nation was definitely preparing for a showdown. It is significant that Governor McKean's daughter Sally had married the Spanish ambassador to the United States, Marqués de Casa Yrujo, in 1798. At the height of the crisis, Jefferson reported to McKean that "Virginia was bristling up" in response to the threat of usurpation. After the election, he wrote to McKean on March 9th, that one of the main reasons Federalists backed down in the contest to elect Burr as president was because of —

> "the certainty that legislative usurpation would be resisted by force of arms."[266]

The fact that Governor McKean was the father-in-law of the Spanish ambassador indicates Republican leaders were following a policy of non-intervention in Spanish affairs. The proof is that after Jefferson became president, the troops stationed at Wilkinsonville weren't used to invade Texas. Instead Jefferson doubled the number of troops in 1801 (to approximately 1500) to defend against any potential French invasion in the Mississippi Valley and/or slave insurrections.

Aaron Burr disappointed his followers when he refused to take a public role in seeking the presidency. He was in upstate New York, serving as a state senator in the legislature during the election crisis. On February 2nd at the state capitol in Albany, his 17 year old daughter Theodosia married Joseph Alston of South Carolina. South Carolina had unexpectedly voted Republican in the electoral college votes, producing the deadlock between Burr and Jefferson. It was charged that Burr had bartered his daughter for votes and money, as he later borrowed a great deal of money from his new son-in-law, a wealthy rice plantation owner, who owned more than 200 slaves.[267]

During the deadlock in the House of Representatives, after the 35th ballot on February 16th, letters arrived from Aaron Burr in which he "explicitly resigned his pretensions" to the presidency, according to Speaker of the House Theodore Sedgewick. These letters have since been lost. Sedgewick, a Federalist, wrote, "The Gigg is, therefore, up." James Bayard wrote:

> Burr has acted a miserable, paltry part. The election was in his power.[268]

But it seems more likely it was a wise decision to avoid civil war. On the 36th ballot, the deadlock was broken and Jefferson was elected president. The nation had learned it could survive a very heated political contest and not break apart.

Meriwether Lewis became paymaster for the First Regiment and spent the months of September, 1800 through February, 1801 traveling from headquarters in Pittsburgh to Cincinnati, Fort Wayne, and Fort Detroit.

He was promoted to captain on December 5th. Both biographer Richard Dillon and researcher Grace Lewis Miller state that Lewis had a company under his command, although he was never able to be with them. According to Miller's notes, the company continued in operation through 1806.[269]

He had just returned to Pittsburgh in early March, when he received the letter from Jefferson asking him to become his private secretary. Captain Lewis remained on active duty while serving as Jefferson's aide and secretary. His army pay was suspended, and Jefferson paid Lewis's salary during this time. He would now become the trusted aide and confidante of the new president during the first years of his administration.

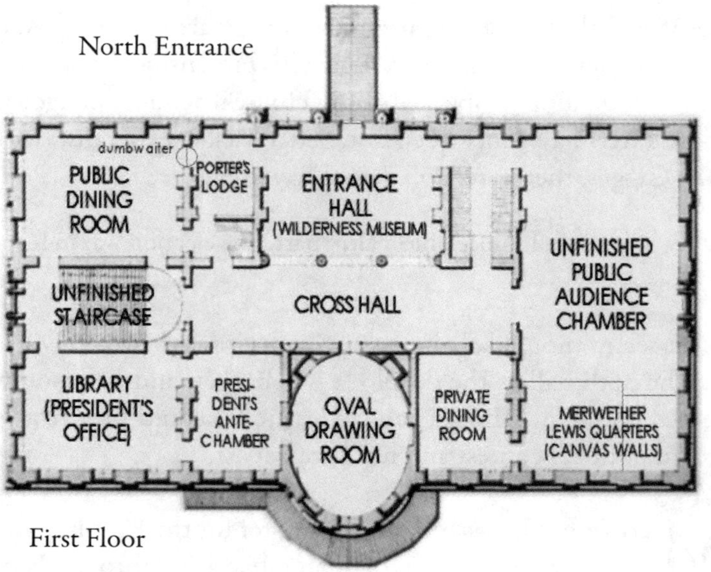

First Floor

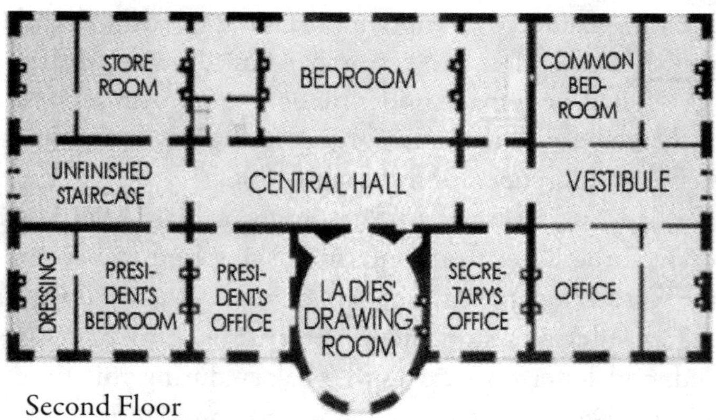

Second Floor

Thomas Jefferson's White House, based on 1803 drawings by architect Henry Latrobe and an unknown artist. Originals in Library of Congress. "Residence Overview: Jefferson's Enhancements: 1801-1809" at www.whitehousemuseum.org.

Chapter 9

The President's House: 1801-1803

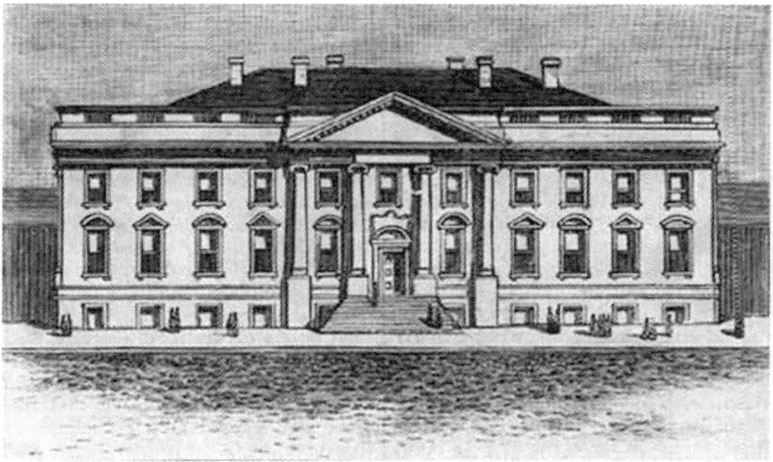

During the two and a half years Meriwether Lewis lived with Jefferson in the President's House—as the White House was first called—the building was unfinished. The huge East Room measured about 80 by 37 feet with a 20 foot ceiling. Its walls and ceiling were still unplastered and the ceiling leaked. Next to the private dining room, two rooms with canvas walls were erected for Lewis on the south corner which served as his office and living quarters. The East Room also housed Jefferson's collection of mounted animals and Indian artifacts, which were later moved to the entrance hall.

A temporary wooden walkway and staircase led to the entrance on the north side of the building (shown above). The Adams's had used the Oval Room as an entrance during their brief stay, but Jefferson shut off the entryway and followed the original architectural plans for a north entrance. It was the largest house in the United States until the end of the Civil War.[270]

Margaret Bayard Smith, a frequent visitor and friend, whose husband Samuel Harrison Smith was the publisher of *The National Intelligencer*, wrote the following description of the room where Jefferson's cabinet officers met, which was located on the southwest corner with a view of the Potomac River:

> It was a spacious room. In the centre was a long table, with drawers on each side, in which were deposited not only articles appropriate to the place, but a set of carpenter's tools in one and small garden implements in another from the use of which he derived much amusement. Around the walls were maps, globes, charts, books &c. In the window recesses were stands for the flowers and plants which was his delight to attend and among his roses and geraniums was suspended the cage of his favorite mocking-bird, which he cherished with particular fondess, not only for its melodious powers, but for its uncommon intelligence and affectionate disposition, of which qualities he gave surprising instances. It was the constant companion of his solitary and studious hours. Whenever he was alone he opened the cage and let the bird fly about the room. After flitting for a while from one object to another, it would alight on his table and regale him with its sweetest notes, or perch on his shoulder and take its food from his lips. Often when he retired to his chamber it would hop up the stairs after him and while he took his siesta, would sit on his couch and pour forth his melodious strains. How he loved this bird! How he loved his flowers! He could not live without something to love, and in the absence of his darling grandchildren, his bird and his flowers became the object of his tender care.[271]

Meriwether Lewis's salary was paid by Jefferson, $500 and then $600, annually.[272] He worked as an aide de camp, and courier for the president, and liked to hunt for game in the early morning hours to serve at the dinner table. He did not serve as a secretary for Jefferson, despite his title. Instead, he was there as a protégé for Jefferson, being trained in the sciences to lead the expedition to the Pacific.[273]

Jefferson and Lewis lived alone in the White House with a small staff of servants. For a few months in the spring of 1801, James and Dolly Madison lived with them before moving into their own house. Jefferson wrote to his daughter:

> Mrs. Madison left us two days ago to commence housekeeping, so that Capt. Lewis and myself are like two mice in a church.[274]

Jefferson had a unique style of political influence and promoting harmony. He hosted dinner parties up to five times a week in his private dining room, which had a large, oval-shaped, dining table which could seat 15 guests when it was expanded. When he hosted small dinner parties of not more than four people, he used "dumbwaiters"—serving shelves on casters which held all the dishes required for the entire meal, and told no tales of private conversation.[275]

He followed a policy of *pêle-mêle* in seating arrangements. There was no hierarchy or social pecking order. British Foreign Minister Anthony Merry and his wife Elizabeth were shocked by the lack of protocol, and after their first experience, refused to attend any more dinners. Jefferson wrote:

> They [the British minister and his wife] claim at private dinners (for of public dinners we have none) to be first conducted to dinner and placed at the head of the table above all other persons, citizens or foreigners, in or out of office. We say to them no; the principle of society with us, as well as of our political constitution, is the equal rights of all; and if there be an occasion where this equality ought to prevail preeminently, it is in social circles collected for conviviality. Nobody shall be above you, nor you above anybody; pêle-mêle is our law.[276]

Lewis would have sometimes served as a substitute host at the dinner table, according to William Seale, author of *The President's House.* Later records indicate Jefferson's family, his secretary, and house guests were automatically included in the dinners.[277] Mrs. Smith wrote about his dinner parties:

> While Congress was in session, his invitations were limited to members of this body, to official characters and to strangers of distinction. But during its recess, the respectable citizens of Alexandria, Georgetown and Washington were generally and frequently invited to his table.
>
> The whole of Mr. Jefferson's domestic establishment at the President's House exhibited good taste and good judgement. He employed none but the best and most respectable persons in his service. His maitre-d'hôtel had served in some of the first families abroad, and understood his business to perfection. The excellence and superior skills of his French cook was acknowledged by all who frequented his table, for never before had such dinners been given in the President's House, nor such a variety of the finest and most costly wines. In his entertainment, republican simplicity was united with Epicurean delicacy; while the absence of splendour ornament and profusion was more than compensated by the neatness, order and elegant sufficiency that pervaded the whole establishment.[278]

No dinner lists have been found for 1801–1803, but records for 1804–1808 show Jefferson carefully noted who was invited and how often they attended. Women were rarely invited. The dinners were Jefferson's means of exerting political influence and guests were invited according to party affiliation. In rotation, only Republican or only Federalist congressmen attended the dinners, generally eight at a time.[279]

This was an amazing opportunity for Lewis to participate in the inner workings of government. If he had lived beyond the age of 35, he could well have become one of the Virginia dynasty of presidents.

One of Lewis's duties was to help "Republicanize" the army. Jefferson's invitation to join him said Captain Lewis was being recruited—

> ... to aid in the private concerns of the household, but also to contribute to the mass of information which is interesting to the administration to acquire, your knoledge of the Western country, of the army and of all it's interests & relations has rendered it desirable for public as well as private purposes that you should be engaged in that office.[280]

As soon as Jefferson became president, efforts began in earnest to establish an academy for training officers—something Hamilton and Adams hadn't been able to accomplish during Adams's administration. Within the month, Secretary of War Henry Dearborn appointed Jonathan Williams to become the head of the new West Point Military Academy on the Hudson River. The school would recruit and train a corps of young Republican officers.[281]

In July, 1801 the War Department supplied a list of 269 commissioned officers to the President. Using a set of secret signs—circles, crosses, symbols, and dot combinations— Meriwether Lewis evaluated officers in the following manner:

(1) "1st Class, so esteemed from a superiority of genius and military proficiency."

(2) "Second Class, respectable as officers, but not altogether entitled to the first grade."

(3) "Unworthy of the commissions they bear."

(4) "Republican."

(5) "Opposed to the administration, otherwise respectable officers."

(6) "Opposed to the administration more decisively."

(7) "Opposed most violently to the administration and still activte in its vilification."

(8) "Professionally the soldier without any political creed."

(9) "Political apathy."

(10) "Officers whose political opinions are not positively ascertained."

(11) "Unknown to us."[282]

Lewis made no evaluation of General Wilkinson, whose name was the only one left blank in the evaluation column. Of the six officers assigned to Wilkinson's General Staff, five received the ranking of "opposed most violently to the administration and still active in its vilification," and the sixth was "opposed to the administration more decisively." Three were discharged and one was reassigned. Their positions were filled by civilians, because there were not enough high ranking Republicans to replace them.

The reduction of the army as a cost-cutting measure allowed for the removal of officers who were either incompetent and/or opposed to the administration. Enlisted men were retained. Captain Lewis knew all but 25 of the 269 officers listed. He was unable to identify the politics of 139 officers (10 & 11), but could offer an opinion on the remaining 130.[283] By creating a new grade, that of ensign, below the rank of 2nd lieutenant, twenty new Republican officers were added. Altogether, 88 officers were dismissed and the reorganized corps of about 162 officers now contained at least 29 Republicans.[284]

Where was General Wilkinson while all this was happening? In March, 1801, he had applied to become Governor of Mississippi Territory at Natchez, but his request was denied.The general was absent from Washington during the reorganization of the army. He and his friend, Major Jonathan Williams, spent the spring of 1801 on an inspection tour of the northeastern frontier. At Fort Niagara, they made plans to build a military road between Lakes Erie and Ontario, and then went on to army headquarters in Pittsburgh.

The next year, after Congress passed the Military Peace Establishment Act of March 16,1802, Williams would become

Jonathan Williams
(Wikimedia Commons)

the first Superintendent and Chief Engineer of the United States Military Academy at West Point. He had received an army commission in November, 1800 upon the recommendation of General Wilkinson, although he had no prior military experience. He was an expert on artillery and fortifications, a member of the American Philosophical Society, and the grandnephew of its founder, Benjamin Franklin.

The two friends would then go onto to the semi-secret military base on the Ohio. Major Williams traveled to Cantonment Wilkinsonville in July in advance of Wilkinson. On July 2nd, the day of his departure from Pittsburgh, Wilkinson instructed Williams, who was traveling with the general's musicians—

> Silence your music in passing & while at all villages . . . . the hills resound and the vallies ring—but give no occasion for those who listen with invidious pleasure to fasten upon us the foul imputations of Ostentatious Pomp & parade at the public expense—a word to the wise—hasten or I shall overtake you before you reach the falls.[285]

Wilkinson and Major Williams arrived together at the cantonment on July 29, 1801. Wilkinsonville had reached its maximum strength of about 1,500 men, or almost half of the troops of the U. S. Army during that summer. 21 units of the army were stationed there, and a large amount of artillery and ammunition was sent to the cantonment before the Major's arrival.[286] It was obviously planned as a training exercise for the army in anticipation of a French invasion of Louisiana or a slave revolt—both real possibilities in 1801–1803. The general remained at the cantonment until August 4th, when he left to conduct treaty negotiations with the Indians. Major Williams stayed behind, in charge of the artillery units.[287]

Lieutenant Colonel David Strong was commander of the cantonment. He was a Revolutionary War veteran, who had served at the Battle of Lexington and wintered at Valley Forge in 1777 and 1778. He served as Commander of the Second Sub-Legion at the Battle of Fallen Timbers. He was commander of Fort Wayne in 1796-98, and of Fort Lernoult (Detroit), before assuming command of Wilkinsonville. He had undoubtedly known General Wilkinson since the early days of the Revolutionary War, and they must have been enemies, because he was highly regarded by General Wayne.

Colonel Strong, died suddenly after a fall from his horse on August 19, 1801. The question is—was he another victim of an assassination by Wilkinson?[288] What are the indications, and what would be the motive?

Colonel Strong arrived with the first troops on January 1, 1801. The camp was struck by a tornado on March 13th, killing one man and injuring several. By late April, surrounded by wetlands, as many as three or four men were dying daily from fever and dysentery.[299] The camp lacked medicine and hospital supplies. In June, newspaper stories caused the new Secretary of War to order the cantonment abandoned.The cantonment had been built near the Grand Chain of Rocks, a six mile stretch of rocks in the middle of the Ohio River. Low lying river land was associated with malarial fevers, but it was not suspected the bites of infected mosquitoes caused the fevers during the "sickly season."

When the general heard the news of Colonel Strong's death from Major Williams, he reported to Secretary of War Henry Dearborn on September 8th, that the colonel had died from:

> an acute nervous affection which rendered an almost immediate insensibility . . . [different from that] ascribed from the climate or to the causes which have affected [the other soldiers at the cantonment].[290]

The Western Spy newspaper in Cincinnati announced the death of Colonel Strong in its October 10th issue:

> Died at Wilkinsonville on Wednesday, the 19th of August, after an illness of only 40 hours, as universally regretted as known, DAVID STRONG, lieutenant colonel commander of the 2nd United States' regiment of infantry.[291]

The Strong Family Papers cite a letter from the Inspector General of the Army, dated October 9, 1801:

> I have just heard of the death of Col. Strong. He fell from his his horse on the anniversary of the day when he bravely fought in that memorable battle under the gallant Wayne in the Maumee Valley in August, 1794. [292]

So it would seem, based on the three accounts, that the colonel was sick for 40 hours; that it was not malarial fevers or dysentery, but a "nervous affection;" and that he fell from his horse and died. As an assassination theory, an associate of Wilkinson could have administered arsenic to Colonel Strong. Arsenic can cause death within a matter of hours or a few days. Wilkinson was always careful to keep one step removed from any wrong doing, and he always had an alibi.

What would have been his motive in arranging for an assassination? Strong must have known of a Burr-Wilkinson conspiracy to create a military base for an invasion of Spanish territory. Such information would have ended the general's army career, as he already was widely accused of being a Spanish agent. This is what all of his possible victims of assassination have in common—they were his rivals in the army, and/or they could have jeopardized his career by revealing damaging information.

There is a very odd letter written by General Wilkinson in reply to a letter written by Major Williams announcing the death of Colonel Strong at the cantonment. On September 6th, General Wilkinson replied to Williams from the Indian Commissioners's camp at Southwest Point, Tennessee:

> I the last evening at 9 oclock received your affecting detail of the Death of our friend & Brother Soldier Col Strong, which produced a twinge of anguish I have seldom experienced, but a moments reflecion dissapated my concern for the Dead and excited all my cares & anxieties for the living. The good widow has all my sympathies & I have given orders for whatever she may need. . . .
>
> Having submitted to my discretion, your Eulogium over our gracious, noble minded, brave Friends remains; I obey the obligations of Candor & affection when I observe to you that, in my judgement, it has been too hastily conceived to be submitted to the press—it has been written under the strongest impulses of the affections & does Honor to your Heart, but the colourings are too strong for the cool examiner and an ardent imagination has overleaped the bounds of congruity & obscured essential distinctions—I shall offer no apology for this liberty, being assured I shall find one in your own Bosom.[293]

Either Major Williams had already sent the eulogy to *The Western Spy* newspaper in Cincinnati, or he disregarded the general's request to not publish it in the press. It appeared in the newspaper on November 7th. The eulogy began:

> That robust body of which health had appeared to have firm possession; that piercing eye, expressive of military ardor, and that attentive ear, hitherto open to every just complaint, now appears before you, a lump of lifeless clay, and you will soon see it mixed with its mother earth. . . . Take, my fellow soldiers,

> your last look. see where he lies—the sad but faithful picture of human frailty! Alas! His body moves not. . . . But two days ago you saw him mounted on yonder steed; from that (now vacant) seat he gave his orders; and when he spoke you heard the dignity of command combined with the mildness of affection. He loved you all—himself a veteran, he could not fail to be the soldiers friend. His heart was by nature kind, the blood that runned from it through all its various channels, was the very milk of humanity.[294]

The eulogy continued to extoll the virtues of the colonel. General Wilkinson may not have wanted to have on record in a newspaper article, "That robust body of which health appeared to have firm possession; that piercing eye . . . ," rather than jealousy over whether Williams was praising Colonel Strong too highly.

The other odd thing about Wilkinson's letter is his use of a cipher code over the word "Dead." Wilkinson used a cipher in corresponding with Aaron Burr and Spanish officials.These markings—a series of three symbols—appear to be a cipher, but what it means is unknown. It could have something to do with the scratched out word above the cipher, where the words "moments reflecion" are found. For whatever reasons, it definitely appears to be a cipher. The discolorations are due to writing showing through from the back side of the paper.

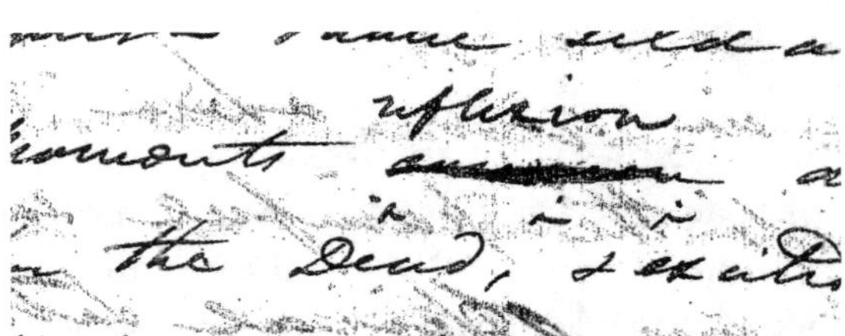

I leave it to the reader to decide whether the death of Colonel Strong should be added to the list of possible assassinations by General Wilkinson. Admittedly, it is speculative. The colonel was 57 years old when he died. What qualifies it, in my opinion, is Wilkinson's attempt to minimize his death by telling Williams not to publish his eulogy; the odd cipher marks in his letter; the contradictory reports of how Strong died; and that he might have revealed a conspiracy to invade Spanish territory under a Burr presidency. The colonel's log books for Wilkinsonville are missing.

Major Williams served as commander of the cantonment until he returned to the east coast in late October. He supervised the cleaning up of the camp and the care of the men who were too sick to be moved. Most of the troops were moved to new camps, but the artillerists and engineers remained at Wilkinsonville in addition to a company of the 2nd Infantry. On October 13th, General Wilkinson returned to conduct military exercises with the assembled troops.

Major Williams had selected a new site, six miles west of the cantonment, where two companies of artillerists were stationed after his departure. Troops were scattered along the Ohio from the Cumberland and Tennessee Rivers to the new camp, 13 miles east of the Mississippi and Ohio confluence. The cantonment was still being used as a "customs house," or port of entry, for the United States in April, 1802 when a small contingent of troops were stationed there. When the Lewis and Clark Expedition headed west in the fall of 1803, at least five members who had been stationed at Wilkinsonville were recruited for the expedition at Kaskaskia, Illinois.[295]

With the western side of the Mississippi under the control of Spain, the army could not rely on moving troops and supplies on the Mississippi River in the event of a war with France. As soon as possible, treaties needed to be signed with Indian tribes, military roads built through Indian territories, and forts established.

General Wilkinson and the two other Indian Commissioners, Benjamin Hawkins and Andrew Pickens—who were both highly respected Indian agents—first met with the chiefs of the

Cherokee on September 4, 1801 at Southwest Point.The chiefs had gone to Washington in June, where they had met with Secretary of War Dearborn. They refused to sell any of their land or allow roads to be built.

The commissioners next met with the high chiefs and principal men of the Chickasaw Nation at Chickasaw Bluffs on October 21-24, 1801. The commissioners asked permission to improve the path from the white settlements at Natchez to settlements on the Cumberland River. This is the road that became known as the Natchez Trace.

After a day's discussion, Major Colbert, a mixed blood chief of the Chickasaw, replied they were very glad to hear they would not be required to cede any land. They agreed to the building of a wagon road through the nation, but refused to allow building houses for travelers. They wanted time to consider how they could benefit from it, but in the meantime they would always supply provisions to travelers. It was agreed that the ferries over the rivers would be the property of the Chickasaw. They were given $702 in goods in return for their permission to build the road. On October 27th, Wilkinson reported to Secretary of War Dearborn he had ordered eight companies of the 2nd Infantry to begin work on the road.

The commissioners next met with the Choctaw Nation at Fort Adams, whose chiefs agreed to a road being built between Fort Adams near Natchez to Walnut Hills on the Yazoo River (present day Vicksburg). In addition, the Choctaw gave up possession of almost two million acres of land in return for $2,000. The land acquired from the nation was bounded by the 31st parallel on the south and the Yazoo River on the north, establishing the United States as the owner of that most desirable piece of property. The Treaty of Fort Adams was signed on December 21, 1801.

The commissioners then met with the Creeks who agreed to give up valuable agricultural land in exchange for a $3,000 annual payment and $25,000 in trade goods. The Treaty of Fort Wilkinson was signed on June 16, 1802. Fort Wilkinson was a trading house located on the Oconee River on Georgia's Indian boundary line.[296]

Wilkinson separately negotiated another treaty with the Choctaw, signed on October 7, 1802 at Fort Confederation (Fort Tombecbe), located about 270 miles north of Mobile on the Tombigbee River. A new boundary line survey, road construction, and the establishment of a trading house were supervised by the general in 1803.[297]

The tribes were the "Five Civilized Tribes"—the Cherokee, Chickasaw, Choctaw, Creek and Seminole nations—who were descendants of the Mississippian culture whose towns had been inhabited by thousands of people before the arrival of Europeans in 1500. They had adopted many ways of Euro-American culture and were generally peaceful.

General Wilkinson, in summing up his experiences as an Indian Commissioner in 1802-03, wrote that he had:

> . . . traversed a trackless wilderness four times, from the borders of Louisiana to the frontiers of Georgia, through the Choctaw and Creek nations; having traveled on public service in the years 1802–3, through forests, and by inland navigation, more than 16,000 miles.[298]

He reported he spent six months in which he had not slept under a roof and suffered from severe malarial fevers and other illnesses during his travels.[299]

The years of 1802-03 clearly indicate that General Wilkinson was willing to sacrifice personal comfort for the safety of the nation. He was not "in exile" from the seat of power, he *was* the power in the interior of the continent. In discussing the conspiracies he was involved in, it is challenging to follow his devious activities. But his years of treaty making with the Indians and road building through the early Southwest prove that he was indeed a patriot—even if, an unhappy one.

President Jefferson sent Meriwether Lewis to Capitol Hill with two copies of his State of the Union address on December 8, 1801. It broke precedent, as both George Washington and

John Adams had delivered their messages orally before the joint session of Congress. On the same day, the text was published in Smith's newspaper, *The National Intelligencer*.[300]

On New Year's Day 1802, there was an open house at the President's House. To the dismay of the ladies of Washington, Jefferson had abolished the weekly receptions of his predecessors. He chose instead to hold open houses on New Year's Day and the Fourth of July, where he shook hands with all who wanted to attend, ending the custom of formal bowing.[301]

At this open house Jefferson was given a 1,235 pound Mammoth Cheese from the citizens of Cheshire, Massachusetts. The cheese bore the inscription: *Rebellion to Tyrants is Obedience to God.* It was the idea of a Baptist minister who wanted to emphasize the message of religious liberty in Calvinist New England and Baptist support for the new president. Over 900 Republican cows provided the milk to make the cheese. No Federalist cows were allowed.

The Mammoth Cheese was about four feet in diameter and 15 inches tall. It was kept in the East Room, where Meriwether Lewis's canvas room was located. Finally at the open house on July 4th, 1803, in celebration of the Louisiana Purchase, the cheese was broken open and served to guests. Federalists said it was wretched, while Republicans raved about it.[302]

The president was generally accessible to visitors in the morning. He rose before dawn. For a couple of hours in the afternoon, he went for a solitary ride around Washington or out in the countryside where he visited farmers. Dinner invitations were for 3:30, when the daily session of Congress ended. The guests were gone by 8:00, and after that Jefferson could devote himself to his personal studies and correspondence.[303]

As his aide and secretary, Meriwether Lewis issued the dinner invitations, using printed cards that only needed the guest's name and date of the dinner. The invitations were used after the chore of writing them became too burdensome.[304] There would have been an endless series of tasks for the 27 year old, exposing him to the entire range of Jefferson's multifaceted life.

Meriwether Lewis by St. Memin (1802), Library of Congress. Compare this St. Memin portrait with the frontispiece, which would have been made in 1807, after the expedition.

On April 19th, 1802, a young lawyer from Philadelphia, Mahlon Dickerson, was a dinner guest at the President's House. He wrote to his brother about meeting the president:

> He is accused of being very slovely in his dress, & to be sure is not very particular in that respect, but however he may neglect his person he takes good care of his table. No man in American keeps a better.[305]

Dickerson was four years older than Lewis. The two young men struck up a friendship, and the next month after Congress ended its session, Lewis traveled to Philadelphia while Jefferson was taking a two week vacation at Monticello.

Philosophical Society Hall in 1800.
(Wikimedia Commons)

The two friends spent several evenings visiting with Pennsylvania Governor Thomas McKean. One evening they went with a group of friends to see a magician's performance. They also visited Dr. George Logan, a founder of the American Philosophical Society, and attended a party at Alexander Dallas's home.

They had their portraits made. On May 18th, Dickerson wrote that he "sat to St. Memin for my profile." It's likely Lewis sat for his profile on the same day.[306] The artist, Charles Balthazar Julien Févret de Saint-Mémin, was a refugee from the French Revolution. During his exile in Philadelphia, he created portraits using a physiognotrace, a gadget invented by a French artist in 1783-4. A physiognotrace used a metal arm to trace the edge of a person's profile while he was seated behind the instrument; the arm produced an image on paper using an identical movement. It took a skilled artist to effectively utilize it. The physiognotrace was similiar to the polygraph instrument that Jefferson would begin using in 1804 to make mechanical copies of his letters. Within the year, Charles Willson Peale would begin selling physiognotrace silhouettes at his new museum in the State House (Independence Hall).[307]

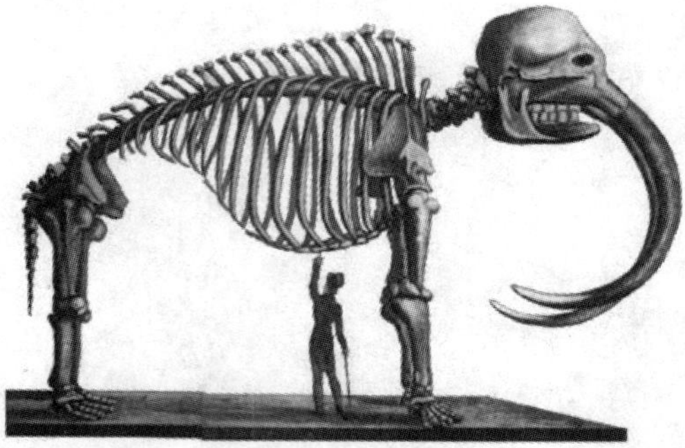

Peale's Mammoth by de Montulé (www.common-place.org)

Meriwether Lewis must have gone almost immediately to see the mastodon skeleton at the American Philosophical Society's museum. In 1801, the society had sponsored Charles Willson Peale's excavation of the prehistoric bones of the American *incognitum* in the Hudson River Valley. The skeleton, described as the "carniverous elephant of the north," had gone on display on December 24, 1801. A national mammoth craze began while the excavation was taking place—including the making of the Mammoth Cheese. Admission was 50 cents to view the skeleton in the Mammoth Room at the Philosophical Hall. Lewis met with Peale during his visit and they talked about a skull Jefferson was sending to him.[308]

The assembled skeleton stood 11 feet tall at the shoulder, and was 15 feet from the chin to its rump. The tusks were 11 feet long. The tusks had been mounted upside down in the belief that the creature was a meat eater, and they remained that way until at least 1816 when a French traveler, Edouard de Montulé, sketched the mastodon.[309]

Jefferson was president of the American Philosophical Society from 1797 to 1814. The giant mastodon became part of the national identity of the United States. It was America's answer to European theories of "American degeneracy." A French naturalist, the author of a 36 volume encyclopedia, the Comte de Buffon, asserted that the animals and people of the New World were inferior to the Old World—that they were smaller and less virile than their counterparts in Euope and Asia. Buffon's encyclopedia was the first attempt to deal with a large number of evolutionary problems and his theory of "American degeneracy" was widely accepted in Europe.[310]

Jefferson had challenged Buffon's theory in his book *Notes on the State of Virginia,* published in 1787. He made a chart, "A Comparative View of the Quadrupeds of Europe and of America," and included the mammoth in his chart, writing:

> It may be asked, why I insert the Mammoth as if it still existed? I ask in return, why I should omit it, as if it did not exist? Such is the oeconomy of nature, that no instance can be produced or her having permitted any one race of her animals to become extinct; of her having formed any link in her great work so weak as to be broken. To add to this, the traditionary testimony of the Indians, that this animal still exists in the northern and western parts of America . . . Those parts still remain their aboriginal state, unexplored and undisturbed by us, or by others for us. He may as well exist there now, as he did formerly where we find his bones.[311]

Did Meriwether Lewis think that he might be sent on a mission to discover whether mammoths still roamed in West? Perhaps he did. Within the year he would be back in Philadelphia making preparations to lead an army expedition called "The Corps of Volunteers for North Western Discovery."

Jefferson had been alerted by a letter from a member of the American Philosophical Society to the American publication of Alexander MacKenzie's *Voyages from Montreal, on the River St. Laurence, through the Continent of North America, to the Frozen and Pacific Oceans; In the Years 1788 and 1793*. The book had been published in England in 1801.[312] Mackenzie's Tour, as it was called, was an account of two expeditions across Canada to reach the Pacific Ocean.

"Alex MacKenzie from Canada by land, 22 July 1793,"a Canadian National Historic Site near Bella Coosa, British Columbia. The MacKenzie Rock inscription was engraved by surveyors from its original paint. (Wikimedia Commons)

His first expedition in 1788 had reached the Arctic Ocean in 1789, and his second reached the Pacific Ocean in 1793. MacKenzie, a Scottish fur trader and partner in the North West Company, traveled in a 25 foot long birchbark canoe with nine others: his cousin and second-in-command, Alexander MacKay; two Indian hunters/interpreters; six French voyageurs, and a dog.

Jefferson ordered MacKenzie's book in June, along with a a 1795 map by the English cartographer, Aaron Arrowsmith, "Exhibiting All the New Discoveries in the Interior Parts of North America."[313] Jefferson spent the months of August and September at Monticello in Albemarle County, avoiding the "sickly season" on the Tidewater and Lewis stayed nearby at the home of Dr. William Bache, located on the road to his mother's home at Locust Hill.

They must have started talking about Lewis leading an expedition across the continent that summer. Jefferson had been attempting for twenty years to organize an expedition to the Pacific Coast—beginning with George Rogers Clark in 1783, then John Ledyard in 1785, and André Michaux in 1793. The 18 year old Lewis had applied to Jefferson to join Michaux's expedition, but Jefferson turned him down.

Winter Quarters of the *Columbia Redidivia* on the Pacific Coast in early 1792. (Wikimedia Commons)

All of these plans had the same objective: to discover the location of a navigable water route from the west side of the Rocky Mountains to the Pacific Ocean.

On May 11, 1792 an American sea captain from Rhode Island, Robert Gray, discovered the mouth of the "Great River of the West" and named it the Columbia River for his ship the *Columbia Redidivia.* For years, ships had unsuccessfully searched for the river mouth, concealed by rain and mist and a barrier of sand and gravel six miles long and three miles wide known as the Columbia Bar. The entire region is called the "Graveyard of the Pacific" for the thousands of ships which have been wrecked along its coast.

Alexander MacKenzie wrote that George Vancouver had discovered that the Columbia River was the only navigable river along the entire coastline, and that it was the key to a globe-circling trade in furs and fish. He wrote in possessing the Columbia—

> . . . the entire command of the fur trade of North America might be obtained from latitude 48 North to the pole, except that portion of it which the Russians have in the Pacific. To

this may be added the fishing in both seas, and the markets of the four quarters of the globe.[314]

MacKenzie had crossed the continent, but failed to find the Columbia. Gray had found the mouth of the river and named it, but had not traveled more than 15 miles up it. Under international law, discovering and claiming the mouth of a river established a nation's right to claim ownership of all the land drained by the river and its tributaries. Gray's discovery established the United States claim to the entire watershed of the Columbia River.

The Doctrine of Discovery dated back to 1455, when the Pope authorized European nations to claim lands inhabited by non-Christians who were not already the subjects of Christian monarchs. The right of first discovery allowed indigenous people to *occupy* their land, but they could only *sell* the land to the nation claiming first discovery. They could not sell their land to individuals or to other nations. The nation also had sovereign and commercial rights over the indigenous people.

The second stage of the Doctrine of Discovery required a nation to establish possession of the land by proving they had been there—burying coins in a bottle, lead plates in the ground, marking rocks and trees, planting flags and crosses, keeping journals and records, making charts and maps. The best way to confirm possession was to establish a settlement.[315]

Britain, Spain, Russia and the United States all had merchant ships on the Northwest Coast trading for sea otter skins to bring to China. Britain and Spain had almost gone to war in 1789 over their competing claims.The Nootka Sound Conventions of 1790-94 established San Francisco Bay as the northern boundary of Spanish control and that the Northwest Coast would be open to both Britain and Spain. It became increasingly clear that the right of first discovery had to be backed up by actual occupancy of the land.

After the Nootka Sound Conventions, Spain realized the interior had to be mapped in order to guard against intrusion into their silver mines region, and a chain of forts needed to be established to reinforce their claims to the Upper Missouri. It was feared that Britain and the United States might find an easy access route to the Spanish silver mines in New Mexico when they explored the Upper Missouri.[316]

The "Commercial Company for Discovery of the Nations of the Upper Missouri" was formed in 1794 under the direction of a wealthy St. Louis merchant, Jacques Clamorgan. The first expedition of ten men in a pirogue, led by Jean Baptiste Trudeau, ended in failure, having been robbed of their cargo by Indians. Another expedition sent out to aid Trudeau was also robbed. The two expeditions cost the company a total of $104,000. To encourage another effort, King Carlos IV of Spain offered a reward of $3,000 to the first Spanish subject to reach the Pacific by travel on the Missouri, and to pay the salaries of up to 100 men to guard the forts to be established on the Upper Missouri.[317]

The Missouri Company sent out a third expedition led by James Mackay in 1795. Mackay was a Scotch trader who had become a Spanish citizen. He was accompanied by John Evans, a young Welshman, whose passage to America had been paid by a radical Welsh society seeking to determine whether the Mandan Indians were the "Welsh Indians" of legend.

In 1792, Zenon Trudeau, Lieutenant-Governor of Upper Louisiana, wrote to his superior, Baron De Carondelet, Governor of Louisiana and West Florida, that a Frenchman, Jacques D'Eglise, had gone up the Missouri to trade in 1790:

> . . . and although the commandant has forbidden all trading with the nations we know, he has dared to make his way in his hunting more than eight hundred leagues up the Missouri. There he found eight villages of a nation about which there was some knowledge under the name of Mandan, but to which no one had ever gone in this direction and by this river. All were provided with English arms. A Frenchman [Menard] . . . told de la Iglesia that the Mandans trade directly with the

> English. This nation received the said man, dela Iglesia, well and welcomed him. They [Mandans] are white like Europeans, much more civilized than any other Indians, and always live in tribes fortified against the numerous nations of Ciux [Sioux] with whom they are perpetually at war. It seems that these Mandans also have communications with the Spaniards, or with nations that know them, because they have saddles and bridles in the Mexican style for their horses, as well as other articles which this same de la Iglesia saw.[318]

John Evans came from Wales in October, 1792 to discover whether the Mandan Indians were the Welsh Indians of legend. He was 22 years old. While spending a few months in Baltimore and Philadelphia, he initially planned to travel through British Canada to the Mandan villages, which were located in present day North Dakota on the Missouri River.[319]

Changing his mind, he made a fateful decision to go to St. Louis in Spanish territory and travel on the Missouri River to reach the Mandan. While enroute to St. Louis, he stopped at Cincinnati and met General Wilkinson who advised him to go to New Madrid, where George Morgan had established an American colony in Spanish territory, rather than attempt to enter St. Louis.[320]

Evans arrived in New Madrid in the spring of 1793. At New Madrid, he took an oath of allegiance to the Spanish government. Stricken with malarial fevers, he was cared for by a Welsh family in New Madrid, and then by a fellow countryman in Kaskasia, Illinois, John Rice Jones, who introduced him to William Arundel, a court clerk and merchant in Cahokia, Illinois. Evans stayed with Arundel during his months in Illinois territory.

But around Christmas, 1794 he crossed over to St. Louis, having learned about James Mackay's proposed expedition to the Upper Missouri. Evans realized he was taking a great risk, and wrote:

> Now or never as I thought, it was my time to make applica-

> tion, so I went over the Mississippi. I thought within myself that it was rather a ridiculous busyness as it was a Critical time on the Spanish side on account of the report of Clark's armie [George Rogers Clark's French army in Kentucky] and I not able to speak one word with anybody, they speaking French. However, I went and was taken for a Spy, Imprisoned, loaded with iron and put in the Stoks besides, in the dead of winter. Here I suffered very much for several days till my friends from the American side came and proved to the Contrary and I was released.[321]

Evans biographer, Gwynn A. Williams, a Welsh historian, writes:

> It was members of Wilkinson's Conspiracy who got John Evans out of trouble in his early days in Spanish St. Louis . . .[322]

After James Mackay arrived in St. Louis in the summer of 1795 he hired Evans as his second in command and field operator. It appears that John Evans was a Freemason, or became a Freemason during his stay with William Arundel in Cahokia. Before his departure, Arundel wrote to Evans on August 28th, 1795:

> Your intended journey will separate us for a long time, but altho' at a Distance we may find ways and means of communicating and giving intelligence one to the other. Which I am in hope will always be in your memory. My Level & your Square must strike the necessary Ballance and Best Rules in the World be the guide among friends.[323]

John Evans had started out on a mission sponsored by a Welsh radical society to discover whether the white-skinned Mandans were the "Welsh Indians" of legend. It had been a subject of general interest for centuries, but it was also of enormous geopolitical significance for Britain, Spain and the United States. When Evans abandoned his original goal and joined the Missouri Company's expedition to the Pacific, his silence on the question of Welsh Indians was effectively bought by the Spanish government.

The Welsh Indians were thought to be descendants of Prince Madoc of Wales and his men who visited Mobile Bay in 1170. After returning to Wales, they made a second voyage from which they never returned.[324] If a Welsh prince and his men had discovered the New World in 1170, and through the centuries their descendants had become "White Indians," then the Spanish right of first discovery, based on Columbus's voyage of 1492, was no longer valid.

In the 1580's, during the reign of Queen Elizabeth I, the queen's advisor John Dee had argued for the priority of British claims to the New World based on the story of Prince Madoc. Now in 1795, the poet laureate of England Robert Southey was writing an epic poem, *Madoc*.[325] Whatever the truth of the matter, it was in the best interests of both Spain and the United States to deny the existence of Welsh Indians in order to discredit any British claim to the right of first discovery. For Welsh radicals, however, the idea of Welsh Indians had inspired an "America fever" with the goal of planting a Welsh colony in America and preaching the gospel to Indians.[326]

Evans's mission had changed since joining the Missouri Company. After they wintered on the Missouri River, he was going to lead a detachment across the Rocky Mountains to find a navigable water route to the Pacific Ocean. The Mackay-Evans expedition consisted of 32 men in four pirogues carrying merchandise worth $50,000. One pirogue was loaded with merchandise for the Arikara. Another had merchandise for "buying their way" past the Sioux. A third was intended for the Mandans and the fourth was for the Indians of the Far West. The expedition was supposed to last for six years.

At the mouth of the Platte River, Mackay left a few men with merchandise and built a small trading post to obtain the goodwill of the Otoe Indians. Near the Omaha village of Chief Blackbird he built a stockaded trading post, Fort Charles, named for the Spanish king. While at Fort Charles, Mackay introduced the seeds of the watermelon, musk melon and other vegetables to the

Omaha, which "succeeded very well." In 1804, the Otoe chiefs would bring a gift of "Water Millions" to the council Lewis and Clark held with the Otoe at Council Bluffs. The watermelons most likely originated from Mackay's initial gift of seeds to the Omaha.[327]

The problem in ascending the river was that the tribes wanted to keep all the merchandise for themselves, and prevent tribes living further up the river from receiving any merchandise. British traders—illegally operating in Spanish territory and wanting to keep the trade for themselves—encouraged the tribes to rob and plunder the traders from St. Louis.

Evans wrote about his first attempt to travel to the Pacific Coast, when his party was discovered by the Sioux in the vicinity of the White River in South Dakota—

> In February 1796 started on my voyage to the West but at a distance of 300 miles from the Mahas [Omahas] was discovered by the Enemy the Sioux and obliged to retreat and return to the Mahas.[328]

Mackay wrote to Evans on February 19, 1796 that if he found a suitable river route to the west that was lower down the Missouri than the Mandan villages, he should take that route.[329] But when Evans set off for the Pacific for the second time in June, he was instructed to visit the Mandan villages in North Dakota and take possession of them for the Spanish government.[330]

Mackay gave John Evans instructions for his journey to the Pacific that were remarkably similar to the instructions that Jefferson would give to Lewis in 1803. He was to keep journals noting the latitude, longitude, weather, winds, minerals, vegetables, plants, animals and provide detailed information about Indian tribes. He was to take possession of all the land he traveled in the names of King Charles IV and the Missouri Company by carving both of their names and the day, month, and year in large letters on stones or trees near river junctions.[331] Mackay commanded him to—

> Reveal your Plans & projects to no one whatever, not even to Mr. Truteau except what is necessary to forward your expedition & your information . . . you will at all events try to see the white people on the coast if it should answer no other purpose than that of corresponding by letters it will be of great service as it will open a way for a further discovery.[332]

Evans set off on his voyage to the Pacific on June 8th, 1796, having received "the necessary Instructions, as well as men, Provisions, and Merchandizes" from Mackay. When his detachment arrived at the Arikara Village on August 8th, they were forced to surrender most of their merchandise before they were allowed to continue.

When they arrived at the Mandan Village on September 23rd, the British traders were not at the village. Evans gave the Mandan chiefs flags and medals and hoisted a Spanish flag over the stockaded North West Company trading post, renaming it "Fort Mackay." The trading post belonged to René Jessaume, a French trader employed by the British North West Company.

On October 8th some North West traders returned to the village, but within days Evans had forced them to leave, sending them back to their headquarters with a document provided by Mackay. It was a declaration—

> "forbidding all strangers whatever to enter any part of his Catholic Majesty's Dominions in this Quarter under any pretext whatever."

On March 13, 1797, Evans received an answer. Jesseaume arrived with several engagés and merchandise to distribute as gifts to the Mandans in order to counteract the Spanish takeover. Jessaume had an Indian wife and children at the village, and it was his trading post that Evans was occupying. Evans reported that Jessaume urged the Indians:

> ... to enter into my house under the Mask of Friendship, then to kill me and my men and pillage my property.

Friendly chiefs intervened and guarded Evans and his men. But Evans reported that some days later:

> Jussom came to my house with a number of his Men, and seizing the moment that my Back was turned to him, tried to discharge a Pistol at my head loaded with Deer Shot but my Interpreter having perceived his design hindered the Execution—the Indians immediately dragged him out of my house and would have killed him, had I not prevented them—this man having refused me Satisfaction for all the Insults he had given me.

Evans summed up what he had learned about the British traders among the Mandan, writing that they intended—

> . . . to spare no trouble or Expence to maintain a Fort at the Mandaine Village. Not that they see the last appearance of a Benefit with the Mandanes, but carry their views further, they wished to open a trade by the Missouri with the Nations who inhabit the Rocky Mountains, a Trade, that at this Moment is Supposed to the best of the Continent of America.[333]

Mackay and Evans returned to St. Louis in the summer of 1797 due to lack of merchandise. Despite their planning, they had been forced to give up almost all of their merchandise to the Indian tribes they encountered before Evans reached the Mandan villages. When Mackay was at Fort Charles he may have gone up north to meet with Evans, because historian Abraham Nasatir writes:

> Mackay himself says that he explored up to the Mandan; however this may be, he was back in St. Louis in May, 1797.[334]

When he returned to St Louis in May, Mackay was said to have "definite information" that Evans was on his way to the Rocky

Mountains. However, Evans returned to St. Louis by July 15th, reporting that his travel down river had taken 68 days.[335]

A lasting benefit of Mackay's and Evans's expedition was the map of 1800 river miles of the Missouri from St. Louis to the White River that Mackay produced from John Evans's charts and journal entries, and from Mackay's own travels. Mackay was a skilled cartographer and his map circulated widely. Lewis and Clark carried a copy of the map and relied on it during their travels. It was the best map of the region until William Clark's map replaced it in 1810.[336]

James Mackay was rewarded by the Spanish government, being appointed commandant of St. Charles and St. André, a district west of St. Charles. He also received several large land grants totalling about 46,000 acres (72 square miles).[337] In Mackay's report to the Spanish government he claimed exclusive credit for Evans's discoveries, writing:

> I also found means to drive the English out of His Majesty's Territories & took possession of a Fort which they built at the mandan nation . . . I also took a Chart of the Missouri from its mouth to the wanutaries nation, which following the windings of the river is little short of 1800 miles.[338]

On July 15, 1797, John Evans wrote to Samuel Jones, his Welsh friend in Philadelphia, giving him a lengthy report of his activities since leaving Philadelphia in 1793. All he ever had to say in writing about the Welsh Indians was contained in a very brief statement at the end of the letter:

> Thus having explored and charted the Missourie for 1800 miles and by Communications with the Indians this side of the Pacific Ocean from 35 to 49 Degrees of Latitude, I am able to inform you that there is no such People as the Welsh Indians, and you will be so kind as to satisfie my friends as to that doubtfull Question.[339]

In November, 1797 John Evans was appointed an assistant surveyor and assigned to the Cape Girardeau district, an American settlement in Spanish territory near the junction of the Mississippi and Ohio Rivers.[340] On June 18, 1798 Evans wrote to Zenon Trudeau, Lieutenant Governor at St. Louis. Trudeau had told Evans to look out for a piece of land for himself and Evans had found 50-55 acres at a beautiful elevation, on a creek well suited for a mill. Evans then became philosophical about his life and wrote to Trudeau—

> A reflection upon the shortness of life & the frowns of Delusive fortune—Convince me dayly of my duty to Live a retired life as soon as I can. For we can scarcly mount the stage of Life. Before we ought to prepare to leave it to make room for the next actors. In a solitary life a man is allowed a few moments to converse with himself—While in the crowd we are continualy Busy conversing with others untill unsensibly snatched away by death.

He reported he had heard General Wilkinson was expected shortly at Fort Massac on his way to either Natchez or Illinois and there were rumors America and France had declared war. [341]

Welsh historian Gwyn A. Williams writes that an American, Jabez Halliday, who was researching Welsh Indians in 1803, said he had known John Evans and that—

> In his cups, Halliday alleged, Evans used to boast that he knew more about Welsh Indians than anyone would ever know because he had been paid to keep his mouth shut. In any case, time and distance would soon remove any trace of their Welsh ancestry.[342]

Evans was known to be an alcholic and a troubled young man after he returned from the expedition. Author Richard Deacon says that Evans was offered a down payment of "the equivalent of two thousand dollars in cash" by the Spanish officials.[343]

Gayoso, the Governor-General of Louisiana, wrote to the Spanish court in 1797 about the Mandan question—

> It is in the interest of his Catholic Majesty that the reports of British Indians in the Mandan country be denied once and for all.[344]

By May, 1799, John Evans had come to live as a guest in Governor Gayoso's household in New Orleans. He would die there by the end of the month. In June, General Wilkinson came to stay with Gayoso on his way to the east coast. Wilkinson and Gayoso were close friends. Wilkinson's wife Ann was living in Gayoso's mansion in Natchez while the general traveled to New York to confer with Alexander Hamilton about plans for a cantonment on the Ohio.

The known dates are very close in time—within three weeks of each other. Wilkinson wrote to Andrew Ellicott from Gayoso's home in New Orleans on June 12, 1799.[345] Gayoso wrote to Mackay about John Evans on May 20, 1799:

> Poor Evans is very ill; between us I have perceived that he deranged himself when out of my sight, but I have perceived it too late; the strength of liquor has deranged his head; he has been out of his senses for several days, but with care he is doing better, and I hope he will get well enough to send him to his own country.[346]

John Evans would die by the end of the month of May, according to James Mackay, who described Evans in words almost identical to the words Thomas Jefferson would use in describing Meriwether Lewis after his death in 1809:

> Mr. Evans was a virtuous young man of promising talents, undaunted courage and perserverance.[347]

A young adventurer from Wales of "undaunted courage and perserverance,"arrived in America full of dreams about a 12th century Welsh prince and Welsh Indians. He wanted to visit the

Mandans and didn't care whether he traveled through British or Spanish territory. He obviously didn't understand the geopolitical significance of his mission, or the danger he was placing himself in by joining a Spanish expedition. The only explanation for why Spanish authorities allowed Evans to become part of Mackay's expedition is that a Welsh investigator could report "there is no such people as the Welsh Indians," and, as Gayosoa wrote, the British claim to the right of first discovery would be "denied once and for all."

When did Evans realize he could never talk about what he had learned at the Mandan villages or share his opinions? It was probably after he returned to St. Louis and met with Spanish authorities. He couldn't talk about it—not because he was paid to keep silent, but because his silence might keep him alive. Unfortunately, death was the only way to ensure he wouldn't talk. Evans's letter to Trudeau is a plea to allow him to retire quietly to the countryside where he will lead a solitary life:

> In a solitary life a man is allowed a few moments to converse with himself—While in the crowd we are continualy Busy conversing with others untill unsensibly snatched away by death.

Was John Evans another victim of an assassination organized by James Wilkinson? It seems quite possible. It would have been easy to add arsenic to his drinks. Arsenic was odorless and tasteless and in common use as a poison at the time. There were no means of detecting its presence in a body by forensic examination until the 1830's.

As Evans said, life was short and "we ought to prepare to leave it to make room for the next actors." Lewis and Clark would make use of Evans's discoveries on their expedition to the Pacific Coast in 1803–1806; they would also spend a winter at the Mandan villages, and they would be silent on the question of Welsh Indians.

Thomas Jefferson was fascinated by American geography and western expansion—subjects on which his father, Peter Jefferson, had been a leading authority. His father was a surveyor and partner in the Loyal Land Company, which in 1748 received a grant of 800,000 acres of land in Kentucky from the Virginia House of Burgesses. In 1751, Peter Jefferson collaborated with Joshua Fry in publishing a famous map of Virginia based on actual land surveys. The Loyal Land Company even planned to undertake an expedition to the Pacific in 1753, but the project was halted because of the coming war between France and Britain for control of the North American continent, the "French and Indian War" of 1754-63. All of his childhood, Thomas Jefferson was surrounded by maps and plans for western expansion.

Fifty years after Peter Jefferson and his partners planned an expedition to the Pacific, President Thomas Jefferson was going to make it happen. But at the same time, the United States was facing a potential French invasion of the Mississippi Valley which would lead to all-out war with France.

Jefferson was warned by Robert Livingston, the American minister to France, in December, 1801 that Napoleon was planning to invade Louisiana. In cipher, Livingston wrote:

> 12,000 of the men sent to st domingo are destined for louisianna in case tousaint makes no opposition.[348]

Toussaint L'Ouveture, called the "Black Napoleon," was the leader of the slave revolution in Saint Domingue. Slaves, free blacks and mulattoes, inspired by the French Revolution, had been fighting for their independence since 1791. It became the only successful slave revolt in history when in January, 1804 Saint Domingue became the nation of Haiti.

During the years of 1801–1803, Napoleon was determined to win back control of France's prize colony. Though only 1/3 of the island of Hispaniola (which Saint Domingue shared with the Spanish colony of Santo Domingo), it had produced 40% of the wealth of the French

Toussaint L'Ouveture, leader of the Saint Domingue revolution, and William Claiborne, Governor of Mississippi Territory in 1801-03. (Wikimedia Commons)

colonial empire before its slave revolution. The wealth came from the production of sugar and coffee through slave labor Napoleon was determined to regain control of Saint Domingue and then to proceed onto New Orleans and Louisiana. Louisiana would serve as a source of agricultural products for the French sugar islands in the Caribbean, whose lands were devoted to more valuable crops.

In February, 1802 France sent an invasion force of 46 ships carrying 40,000 troops to Saint Domingue. More ships and troops would follow. Four fifths of Napoleon's army perished in fighting the rebels and from yellow fever. Ultimately, as many as 60,000 French soldiers died in the attempt to retake Saint Domingue in 1802-03.[349]

France had acquired Louisiana from Spain in the secret Treaty of San Ildefonso signed on October 1, 1800. The French acquisition of Louisiana was rumored when Jefferson became president in March, 1801. He began at once to establish a state of military preparedness on the southwestern frontier. Jefferson appointed William C. C. Claiborne Governor of Mississippi Territory in 1801. Claiborne would go on to serve as Governor of the

Territory of New Orleans from 1804-1812, and then the State of Louisiana from 1812-16. For 16 years he played a vital role in securing the New Orleans area. He was the brother of Meriwether Lewis's friend, Ferdinand Claiborne.

Jefferson's appointment of Claiborne ensured that the New Orleans area was preparing for military action.[350] In 1801, General Wilkinson and Major Williams made plans for a military road in the Lake Erie area to defend against a British invasion, and then conducted training exercises at Cantonment Wilkinsonville on the Ohio River. After that, General Wilkinson spent two years in the Southwest negotiating Indian treaties and building military roads and forts, while Major Williams founded the West Point Military Academy.

Jefferson would write in the summer of 1803, after the Louisiana Purchase was accomplished:

> These grumblers too are very uneasy lest the administration should share some little credit for the acquisition, the whole of which they subscribe to the accident of war. They would be cruelly mortified could they see our files from April 1801, the first organization of the administration, but more especially from April, 1802. They would see that tho' we could not say when war would arise, yet we said with energy what would take place when it should arise.[351]

These were the years that Lewis was serving as his confidential aide. Preparations for war in the Mississippi Valley must have occupied much of their attention.

The American Ambassador to France, Robert Livingston, arrived in Paris in December, 1801. His departure for France had been delayed waiting for congress to ratify the terms of ending the Quasi-War with France. In April, 1802 Jefferson wrote to Livingston:

> The cession of Louisiana & the Floridas by Spain to France works most sorely on the US. . . . it compleatly reverses all the political relations of the US. and will form a new epoch in

> our political course. . . . there is on the globe one single spot, the possessor of which is our natural & habitual enemy. It is New Orleans, through which the produce of three eighths of our territory must pass to market, and from its fertility will ere long yield more than half of our whole produce and contain more than half of our inhabitants. .. the day that France takes possession of N. Orleans . . . from that moment we must marry outselves to the British fleet & nation.[352]

The next month in May, Secretary of State James Madison wrote to Livingston asking him to inquire if the cession of Spanish territory to France included New Orleans and the two Floridas. If they were included, then the United States wanted to know, what would be the price to purchase them?[353]

In July, 1802 Napoleon wrote to his Minister of Marine and Colonies, Denis Decrès:

> My intention, Citizen Minister, is that we take possession of Louisiana with the shortest possible delay, that this expedition be organized in the greatest secrecy, and that it have the appearance of being directed at St. Domingo. . . . I should also like to have made for me a map of the coast from St. Augustine Florida to Mexico . . .[354]

It wasn't until later in the year, however, on October 18, 1802, that Spain finally completed the cession of Louisiana to France. This procrastination cost France dearly, as Napoleon allowed it to delay his invasion of Louisiana.

At the very same time in October, a Spanish administrator in New Orleans closed the port of New Orleans to Americans, inciting a storm of outrage and demands for war from both westerners and Federalists. It was assumed that France was behind the order. In actuality, it was Carlos IV, the king of Spain, who ordered it in response to American smuggling of goods and their disregard for paying custom duties.[355]

Over the winter of 1802-03, Ambassador Livingston entered into talks with Joseph Bonaparte, Napoleon's brother. The United States was seeking to acquire New Orleans and at least West Florida, if not both Floridas. In January, Napoleon—upon learning of the death from yellow fever of his brother-in-law, General Charles Le Clerc, commander of the invasion forces—had exploded, "Damn sugar, damn coffee, damn colonies!"[356]

In March 1803, a colonial adminstrator, Pierre de Laussat, arrived in New Orleans to take possession of Louisiana for France, expecting that an initial military force of 3,000 troops would soon arrive. The French ships on the coast of Holland had been icebound over the winter. When the weather improved, the British set up a naval blockade of the Dutch coast.

By April 10th, Napoleon had decided to sell all of Louisiana to the United States. James Monroe arrived in Paris on April 12, as a special envoy of President Jefferson. War between France and England was about to resume, and Napoleon decided to cut his losses in the Caribbean and raise money for the coming war. Monroe's arrival coincided with news of Federalist demands for a war to take possession of New Orleans due to the closure of the Port of New Orleans. Monroe's instructions were to make an alliance with Great Britain if he couldn't receive satisfaction in France.[357]

The stage was set, and on April 11th, before Monroe arrived, Talleyrand asked Livingston if the United States "wished to have the whole of Louisiana." On April 12th, Napoleon sent orders to stop the sailing of ships to Louisiana. The order reached the fleet two days before they were to leave for New Orleans.[358] The treaty signed on May 2, 1803 was for the sale of Louisiana for $15 million. Adjusted for claims for American losses during the Quasi-War, the cost was $11.25 million. The exact amount of territory gained in this sale was unclear. It was defined as:

> the Same extent that it now has in the hand of Spain & that it had when France possessed it.[359]

The Louisiana Purchase is now understood to have encompassed

828,000 square miles, acquired at a cost of less than three cents an acre. The United States had been willing to pay up to $10 million for New Orleans and West Florida. Whatever the size and extent of its boundaries, the purchase approximately doubled the size of the American nation. When Livingston asked Talleyrand if he could clarify the boundaries, he replied:

> I can give you no direction. You have made a noble bargain for yourselves and I suppose you will make the most of it.[360]

In the summer of 1802 Jefferson and Lewis had begun making plans for Lewis to lead an expedition to the Pacific Coast. The expedition was fundamentally a military reconnaisance mission and a recognition of the need for haste in reinforcing the United States' claim to the Pacific Northwest. In 1802 they had no way of anticipating that Napoleon would end up selling Louisiana to the United States.

On January 18, 1803, Meriwether Lewis acted as Jefferson's courier in delivering a confidential message from the president to the senate and house of representatives. The message proposed a voyage of discovery, a "literary" expedition to explore the Missouri River in the interests of commerce and science, which would provide no offense to the nation [Spain] claiming possession of the territory. It would cost $2,500.

The message emphasized the great quantity of furs and peltries obtained by the British in Canada from the Indians on the Upper Missouri, and observed American traders would have a competitive advantage due to moderate climate and easy access. Jefferson wrote:

> An intelligent officer with ten or twelve chosen men, fit for the enterprize and willing to undertake it, taken from our posts, where they may be spared without inconvenenience might explore the whole line, even to the Western Ocean.

Harpers Ferry, 1803. Harpers Ferry National Historical Park
www.nps.gov/hafe/meriwether-lewis.htm

Jefferson attempted to keep the destination of the expedition secret, saying its mission was to explore the Mississippi. He had already informed Ambassador Irujo of the true destination, requesting a passport from the Spanish government. The Spanish passport was not forthcoming, but French and British officials provided passports.

By February Jefferson was writing to fellow members of the American Philosophical Society requesting their aid in training Captain Lewis, who would soon be coming to Philadelphia. He asked them to prepare notes for Lewis regarding inquiries he should pursue. To them, he revealed that the mission was to explore the Missouri and find a navigable water route to the Pacific Ocean. He wrote:

> Capt. Lewis is brave, prudent, habituated to the woods, & familiar with Indian manners & character. He is not regularly educated, but he possesses a great mass of accurate observations on all the subjects of nature which present themselves here, & will therefore readily select those only in his new route which shall be new.[361]

Andrew Ellicott's House, Lancaster, Pennsylvania
(Wikimedia Commons)

Lewis first went to Harpers Ferry, the U. S. Armory and Arsenal in West Virginia at the confluence of the Potamac and Shenandoah Rivers. He arrived there on March 16, 1803, expecting to spend a week. Instead, he spent a month working on a design for an iron boat frame which could be carried over the Rocky Mountains.The boat, called *The Experiment*, was built in sections and would weigh 176 pounds and carry a load of 8,000 pounds after being assembled. In addition Lewis placed orders for 15 rifles, extra gun parts, gunsmith repair tools, tomahawks and knives.[362]

He arrived at Lancaster, Pennsylvania on April 20th to begin training with Andrew Ellicott. The surveyor taught him how to make the astronomical observations used in determining the latitude and longitude of locations. When he arrived in Philadelphia in early May, mathematician Robert Patterson tutored him in the required mathematics, and he bought the scientific instruments needed for the observations.

At Philadelphia, Lewis met with Dr. Benjamin Rush, who advised him on medical matters. Dr. Rush's famous "bilious pills" for constipation and headaches were nicknamed "Rush's Thunderbolts." Lewis obtained more than thirty kinds of drugs for the expedition, including opium and laudanum (a diluted form of opium), and an assortment of medical instruments. He was knowledgeable about herbal remedies, having been trained by his mother.

He met with Dr. Benjamin Smith Barton who tutored him on the technical language of the natural sciences, purchased his newly-published *Elements of Botany,* and borrowed a book, du Pratz's *History of Louisiana*. He also visited the author of an anatomy book, Dr. Caspar Wistar, to discuss fossils.

Much of his time was spent shopping and placing orders with the Purveyor of Public Supplies, Israel Whelan, to whom he gave a draft of $1,000. Army goods were obtained from the Schuykill Armory. Lewis estimated the supplies and equipment would weigh 3,500 pounds and he hired a wagon with five horses and a driver to take the goods to Pittsburgh.

After spending a month in Philadelphia, Lewis returned to the President's House in early June.[363] It was decided then that Lewis needed a partner. Jefferson had been writing instructions in April. It was a question of handling the responsibilities of map making and scientific observations while on a dangerous expedition over unknown territory. A second leader would also ensure a backup plan in case Lewis lost his life.

On June 19th, Lewis wrote a long letter to William Clark at the Falls of the Ohio, describing his plans for the expedition:

> . . . it is to descend the Ohio in a keeled boat of about ten tons burthen, from Pittsburgh to its mouth, then up the Mississippi to the mouth of the Missourie, and up that river as far as its navigation is practicable with a boat of this discription, there to prepare canoes of bark or raw-hides, and proceed to it's source, and if practicable pass over to the waters of the Columbia or Origan river and by descending it, reach the Western Ocean; the mouth of this river lies about one hundred

William Clark by St. Memin, and later as Governor of Missouri Territory (Collection of State Capitol of Missouri)

and forty miles South of Nootka Sound; at which place there is a considerable European Tradeing establishment and from which it will be easy to obtain a passage to the United States by way of the East-Indies in some of the tradeing vessels that visit Nootka Sound annually, provided it should be thought more expedient to do so, than to return by the rout I had pursued in my outward bound journey.

. . . If therefore there is anything under those circumstances, in this enterprise, which would induce you to participate with me in it's fatiegues, it's dangers and it's honors, believe me there is no man on earth with whom I should feel equal pleasure in sharing them as with yourself; I make this communication to you with the privity of the President, who expresses an anxious wish that you would consent to join me in this enterprise; he has authorized me to say that in the event of your accepting this proposition he will grant you a Captain's commission, which of course will intitle you to the pay and emolulents of that office. . . . your situation if joined with me in this mission will in all respects be precisely such as my own.[364]

It was a generous offer. The expedition to the Pacific has become known as the "Lewis and Clark Expedition" because of it, but in truth it was an expedition under the command of Captain Meriwether Lewis, who bore the responsibility for it from beginning to end. It was not in President Jefferson's power to grant a captain's commission to William Clark. The War Office authorized a second lieutenant's commission, but members of the expedition were told he was a captain.

Jefferson instructions had been sent to Lewis and cabinet members for their comments in April. His instructions assumed Lewis would need a French passport to travel through Louisiana, and a British passport to deal with British traders.The Spanish passport had been refused on the grounds that the Upper Missouri was Spanish territory and not part of the cession of Louisiana to France.

The expedition would consist of 10–12 soldiers who would volunteer for the assignment and be recruited from western army posts. However, Lewis wrote to Clark that he was looking for civilians and wanted him to find:

> . . . some good hunters, stout, healthy, unmarried men, accustomed to the woods, and capable of bearing bodily fatigue in a pretty considerable degree. . .

The men would be told that it was an expedition to explore the Mississippi River to its source and that it would last about 18 months. Before they signed on they would be told the truth and Lewis expected that none would refuse. By 1805, when the expedition left Fort Mandan for the Pacific Coast, it had grown to 26 enlisted men in the army, Lewis and Clark, four civilians, a baby and a dog, but the instructions in 1803 were based on the original budget estimate of 10–12 soldiers.

One of the most important instructions was assigned to William Clark—making careful records, mile by mile, of the river courses and routes over land. He would use these measurements to create his famous map. It was an activity that required constant attention and note-taking, whether he did it himself or delegated the work. The crucial latitude-longitude readings would be handled by war office specialists after they returned.

Information about Indian tribes was needed—every kind of information. The natives were to be treated in "the most friendly & conciliatory manner" and told the U. S. wanted to engage in commerce with them in the fur trade. They were to record vocabularies in Indian languages.

They were to note soil, vegetation, animals, fossils, minerals, and climate. The dates of seasonal changes, when plants put forth or lost their flowers and leaves, fruits ripened, and the presence of birds and insects, were to be recorded. Anything new to the U. S. was especially important.

They were to be careful to not risk their lives, and to keep two duplicate sets of records. At least seven members of the expedition kept journals. Donald Jackson, who edited the *Letters of Lewis and Clark*, called them—

> the writingest explorers of their time. They wrote constantly and abundantly, afloat and ashore, legibly and illegibly, and always with an urgent sense of purpose.[365]

On July 2nd, Lewis wrote to his mother that he would start on his journey to the "Western Country" on July 4th. It would take 15–18 months, and was "by no means dangerous" as the Indian tribes were "perfectly friendly to the United States." He would be as safe as if he were to "remain at home for the same length of time" and he hoped she would not "indulge in any anxiety for my safety."[366]

However, on the evening of July 3rd, the most astonishing news reached the capitol—France had sold the entire territory of Louisiana to the United States! Lewis stayed in town one more day for what must have been one of most memorable Fourth of July celebrations in American history.

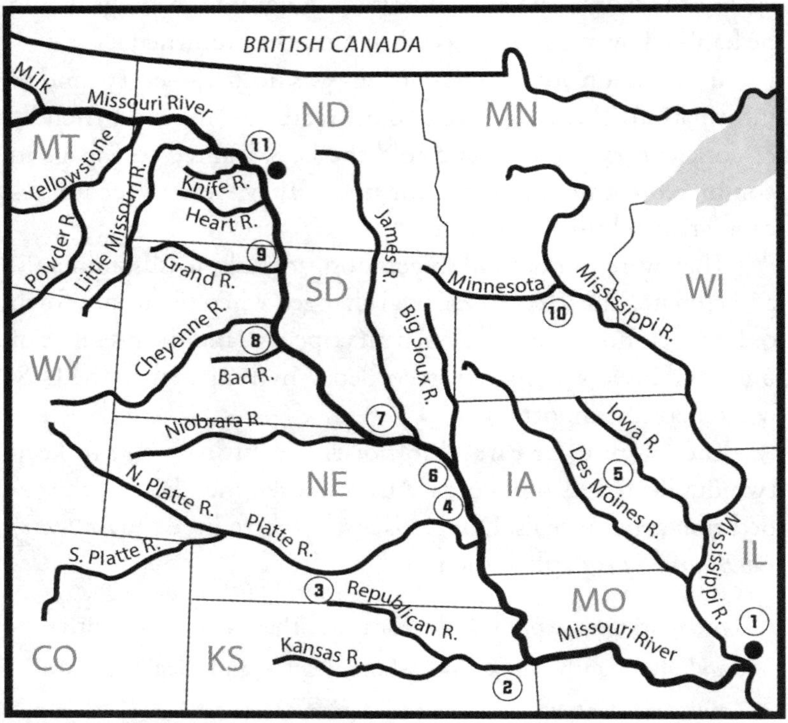

Expedition's Winter Camps:
① Wood River Camp (Dec. 12, 1803–May 14, 1804)
⑪ Fort Mandan (Nov. 2, 1804–April 7, 1805)

1. Wood River Camp and St. Louis
2. Kansas Indians
3. Pawnee village that stopped Spanish soldiers
4. Otoe and Missouria village at Platte & Elkhorn Rivers
5. Ioway Indians
6. Omaha Indians
7. Yankton Sioux
8. Teton Sioux (Brulé Sioux)
9. Arikara villages
10. Santee Sioux
11. Fort Mandan and Mandan & Hidatsa villages

Chapter 10

Journey to the Pacific: 1803–1806

Meriwether Lewis set off for the Pacific Coast on July 5th, 1803. He would return to the President's House on December 28th, 1806. Against all odds he had reached the mouth of the Columbia River on the Pacific Ocean, reinforcing the U.S. claim to the Pacific Northwest under the Doctrine of Discovery. It would establish the United States as a continental nation from "sea to shining sea."

History is a twice told tale and every generation tells its own tale. In honor of the 200th anniversary of the expedition in 2003-2006, its peaceful nature and contact with Indian nations was emphasized. The story of Lewis's supposed suicide in 1809 was presented as a fact, which distorted all interpretations of his character. That his death was immediately labeled a suicide has unintentially colored the assessment of who he was and what he accomplished for over two centuries.

He celebrated his 29th birthday at Pittsburgh on August 18, 1803, while he was living in the officer quarters at Fort Fayette, (Fort Lafayette) waiting for his boat to be built by a drunken contractor. Lewis himself had arrived at Pittsburgh on July 15th, only ten days after leaving the capitol. If William Clark was unable to go, Lieutenant Moses Hooke, commander of Fort Fayette, was going to be his replacement. Hooke would serve as second in command, not as co-commander. But by the 29th of July, Lewis had received a reply from Clark:

> I will cheerfully join you in an "official Charrector" as mentioned in your letter, and partake of the dangers, difficulties, and fatigues, and I anticipate the honors & rewards of the re-

> sult of such an enterprise, should we be successful in accomplishing it. This is an undertaking fraited with many difeculties, but My friend I do assure you that no man lives whith whome I would perfur to undertake Such a Trip &c. as yourself, and I shall arrange my matters as well as I can against your arrival here.[367]

On July 22nd, Lewis wrote to Jefferson that the goods had safely arrived from Harpers Ferry and the army volunteers who would take the boat down the Ohio had arrived from Carlisle. The boat would be ready by August 5th. The water on the Ohio River was exceedingly low and getting lower. No matter how low the river got he was—

> determined to go forward though I should not be able to make a greater distance than a mile pr. day.[368]

By the time he wrote to William Clark on August 3rd, he was going to do it, even if he could make no more than a "boat's length pr. day." He wrote the boat would be ready by mid August and he would be at the Falls by the end of the month.

He cautioned Clark he was pleased to hear he had engaged some men for the expedition, but they should be told it was conditional on Lewis's approval. He was "well pleased" that Clark had rejected the young gentlemen who were applying, as "they will not answer our purposes." They needed hunters, who were willing to share in the general duties of the party.[369]

Lewis had successfully managed things so far. The purveyor's summary of his purchases for the journey listed the following categories and took many pages to describe in detail—mathematical instruments; arms, ammunition and accoutrements; medicine &c; clothing; provisions &c; Indian presents; and camp equipage. Altogether his purchases totaled $2,160.14, and were based on the estimate of 12-15 soldiers making up the party.[370]

The expedition was expanding. His original plan had been to use volunteers from Southwest Point, an army post in Tennessee, and to purchase a boat at Nashville. However, Lewis

A portrait of the original breed
of Newfoundland dogs, circa 1790.

learned there were not enough good men stationed at Southwest Point for his purposes and no boat was available at Nashville. Building a boat at Pittsburgh was a substitute plan. The use of recruits from Carlisle to take the boat down the Ohio was also improvised. The soldiers would be traveling to their first assignment at Fort Adams in Mississippi Territory, and would leave the boat once it reached Fort Massac near the Mississippi River.[371]

On July 2nd, Secretary of War Henry Dearborn authorized the commander at Kaskaskia in Illinois Territory to supply eight men and a sergeant who understood how to row a boat and to provide the best boat at the post to accompany Lewis up the Missouri. The soldiers were expected to return to Illinois before the river iced over.[372]

Dearborn also instructed Captain Lewis that he should have no more than 12 non-commissioned officers and privates under his command. Another officer (Clark) and a hired interpreter, would bring the total to 15.[373] Lewis interpreted these orders to mean that he could hire as many civilians as he needed when he wrote to Clark.

Lewis made a decision that was important for the safety of the party, and certainly for its morale. He purchased a Newfoundland dog for $20 at Pittsburgh which he named Seaman. The big dogs originally came from Newfoundland and were famous for their swimming ability. Newfoundlands were working dogs used by fishermen to retrieve lines and nets of fish. They had short, thick coats and were very strong.[374]

Keelboat at Lewis and Clark State Park,
Onawa, Iowa. (Photo by Kira Gale)

Finally the boat was completed. Lewis sent two wagons loaded with goods down to Wheeling on the Virginia side of the Ohio River. The Ohio was the lowest it had ever been in the memory of the "oldest settler in this country." The boat had to be unloaded and lifted by hand over bars of stones and driftwood in the river, or dragged by horses and oxen over the bars. The water level at many bars was six inches or less. Lewis left Pittsburgh on August 31, 1803. When he reached Wheeling on September 8th, he wrote Jefferson that he had been "most shamefully detained by the unpardonable negligence of my boat-builder," and that he had averaged about 12 miles a day to Wheeling, but some days it was 5 miles or less.[375]

At Wheeling he purchased the red pirogue—a large row boat manned by seven oars and equipped with two sails—to haul some of the goods brought by the wagons. The water level was deeper from Wheeling on down the Ohio to its mouth on the Mississippi River.

One of Lewis's prize possessions was his newly purchased airgun. When the boat left Pittsburgh he stopped at Brunot's Island, five miles down river, to say goodbye to his fellow Masons who had gathered to bid him farewell. He demonstrated the airgun, firing it seven times on one charge before giving it to a friend to try out. The friend accidentally shot a woman bystander, standing about 40 yards away. The .46 caliber (nearly half an inch) lead ball passed through her hat and cut her temple. She fell to the ground, bleeding. Lewis wrote in the first day's entry in his journal:

> . . . we were all in the greatest consternation supposed she was dead [but] by a minute she revived to our enespressable satisfaction, and by examination we found the wound by no means mortal or even dangerous; called the hands aboard and proceeded. & [376]

The next account of the airgun comes from the journal of another traveler, Thomas Rodney, who was going down the Ohio to become a judge and land commissioner in Mississippi Territory. Rodney had traveled by horseback to Wheeling and was going by flatboat the rest of the way. On September 8th he wrote:

> Visited Captain Lewess barge. He shewed us his air gun which fired 22 times at one charge. . . . It is a curious peice of workmanship not easily discribed and therefore I omit attempting it.[377]

Rodney did not actually witness the gun fire 22 times on one charge. Lewis was having trouble with the firing mechanism, but he could manage to fire it seven times on a charge. If you were 50 yards away, the airgun would appear to fire silently. Closer to the firing, a "whack" sound was heard. The gun was a Girandoni airgun, a military weapon made in Austria and was one of only one or two repeating airguns in the U. S.[378]

The Grave Creek Mound, from an 1847 engraving.[379]

After leaving Wheeling, Lewis visited the Grave Creek burial mound at Moundsville, West Virginia. He estimated the "remarkable mound of earth" to be 65 feet high, ending in a blunt point of 30 feet in diameter. He wrote:

> near the summet of this mound grows a white oak tree whose girth is 13 $^{1}/_{2}$ feet, from the aged appeance of this tree I think it's age might resonably be calculated at 300 years.[380]

The Ohio Valley was covered with Indian burial mounds and ceremonial and astronomical observation earthworks, built by prehistoric Indians who moved countless tons of earth in constructing them. Lewis wrote that some brass beads found near two skeletons excavated from the site had been sent to Mr. Peale's museum in Philadelphia. The Grave Creek Mound dates back to 250–150 B. C. and is now a National Historic Landmark.

Two days later he arrived at Marietta, Ohio, established by veterans of the American Revolution as the first settlement in Northwest Territory. Located at the junction of the Muskingum and Ohio Rivers, Marietta is the site of an ancient complex of platform mounds. There he visited with the postmaster, an "excalent republican," and wrote another letter to Jefferson.

Seaman was having a great time, catching black squirrels who were swimming across the Ohio, from the west to the east shore. Lewis wrote the squirrels were fat and he found them "when fryed a pleasant food."

Lewis passed Blennerhassett Island without comment in his journal. The island was the home of one of the largest mansions in the United States, built by a wealthy and eccentric Irish couple, Harmon and Margaret Blennerhassett, who emigrated to America in the 1790's. Blennerhassett Island would become an important part of the Aaron Burr treason trial in 1807.

After Lewis reached Letart Falls on September 18th, he quit writing in his journal for almost two months. The Falls were near the bend of the Ohio, where the river heads west to join the Mississippi. During this time he visited Cincinnati and Big Bone Lick to collect bones for Jefferson, and then spent almost two weeks at the Falls of the Ohio.

When he reached Cincinnati on September 28th he wrote to Clark that he had conditionally engaged two young men. They were George Shannon and John Colter. Shannon would become the youngest member of expedition at age 16. Shannon is often said to have been 18 years old in 1803, but it seems more likely he was two years younger. Perhaps Lewis was remembering how disappointed he had been when he wasn't allowed to join Michaux's expedition at the age of 18.[381]

In his letter to Jefferson on October 3rd, Lewis described mastodon bones (which he called "mammoth" bones) that he examined in the collection of Dr. William Goforth of Cincinnati. He then went to Big Bone Lick, 17 miles southwest of Cincinnati, to collect bones for the president—which unfortunately were lost in a boat accident on their way to Monticello. Mastodons roamed North America until 10,000 to 11,000 years ago. The salt licks and a sulphur springs at Big Bone Lick had supplied salt and minerals to animals living in the Pleistocene era and for many years afterwards. The mastodons were the animals, the "American Incognitum," which they hoped to find were still living in the unexplored regions of the West.

Lewis included some alarming news in his letter. He had decided to explore the country south of the Missouri on horseback over the winter, perhaps going by the route of the Kansas River to Sante Fe. He added that if Clark could be spared from other duties, he could explore the area in another direction. Jefferson wrote back to Lewis on November 16th—

> We are strongly of the opinion here that you had better not enter the Missouri till the spring.... One thing however we are decided in: you must not take the winter excursion which you propose ... Such an expedition will be more dangerous than the main expedition up the Missouri & would by an accident to you, hazard our main object, which since the acquisition of Louisiana interests every body in the highest degree..... The object of your mission is single, the direct water communication from sea to sea formed by the bed of the Missouri & perhaps the Oregon [Columbia]. By having Mr. Clarke with you we consider the expedition double manned & therefore the less liable to failure, for which reason neither of you should be exposed to risques by going off of your line.[382]

Jefferson also advised him to spend the winter in Illinois Territory at Kaskaskia or Cahokia in order to avoid all danger of Spanish opposition. He gave Lewis the news that he expected congress would authorize $10-12,000 in additional funding for three more expeditions to explore the principal waters of the Mississippi and Missouri. Lewis received this letter after he arrived at Cahokia in December.

William Clark was waiting for Lewis when he reached the Falls of the Ohio on October 14th. The boat was piloted through the Falls on the 15th and moored at Mill Creek near George Rogers Clark's home on Clark's Point overlooking the falls. William had moved across the river from Louisville in March to live with his brother in Clarksville, Indiana. His brother's home was a two room log cabin with an upstairs sleeping loft.

Due to his brother's financial entanglements, William owned the land adjacent to Mill Creek, and had entered into an agreement with the town trustees in March to build a boat canal and water-works on the land.[383]

Lewis and the two Clark brothers must have immediately agreed that the seven men William Clark had recruited, and the two recruits with Lewis, needed to be enlisted in the army rather than hired as civilians. These became the expedition members known as "The Nine Young Men from Kentucky." In addition, Clark informed Lewis he was bringing his slave York along on the journey, making it a total of ten men who joined at the Falls.

William Clark and the two Field brothers, Reuben and Joseph, have the date of August 1, 1803 for their army enlistments. Clark must have recruited them and secured their commitment very soon after receiving Lewis's letter. At the end of the expedition, Lewis would recommend bonus pay for the Field brothers, describing them as—

> . . . two of the most active and enterprising young men who accompanied us. It was their peculiar fate to have been engaged in all the most dangerous and difficult scenes of the voyage, in which they uniformly acquitted themselves with much honor.

The other men enlisted between October 15–20th. Charles Floyd and Nathaniel Pryor, who were cousins, were appointed sergeants by the captains after the expedition set out. Another of Clark's recruits, John Shields, was praised by Lewis who wrote at the end of the expedition—

> Nothing was more peculiarly useful to us, in various situations, than the skill and ingenuity of this man as an artist, in repairing our guns, accoutrements, &c.

Lewis had written to Clark that "much must depend upon the judicious scelection of our men" and of the seven men chosen by Clark five received special recognition. His ability to recruit the right men was essential to the expedition's success.

The other two men who joined the party were William Bratton and George Gibson who also became members of the permanent party. Bratton was a blacksmith and Gibson was a fiddle player among their other accomplishments. The expedition left the Falls on October 26th.

Lewis resumed writing in his journal on November 11, when they reached Fort Massac, where he hired the services of a civilian, George Drouillard (pronounced "Drewyer"), a sign language interpreter and hunter of mixed Shawnee and French Canadian descent. Lewis urged additional pay for "Drulyard" at the end of the expedition, writing—

> A man of much merit; he has been peculiarly usefull from his knowledge of the common language of gesticulation, and his uncommon skill as a hunter and woodsman.... It was his fate also to have encountered on various occasions, with either Captain Clark or myself, all of the most dangerous and trying scenes of the voyage, in which he uniformly acquited himself with honor.[384]

The soldiers who had manned the boat from Pittsburgh were dropped off at Fort Massac. They passed Wilkinsonville on November 14th with just a brief mention in Lewis's journal. Lewis had been stricken with malarial fevers the day before, and Clark was ill with violent stomach and bowel pains from November 16th until reaching Kaskaskia on November 27th.

Seven of the soldiers who volunteered at Kaskaskia became members of the permanent party—John Collins, Patrick Gass, John Ordway, Peter Weiser, Joseph Whitehouse, Alexander Willard and Richard Windsor. The others were part of the military escort who brought the barge back to St. Louis. The white pirogue, the larger of the two rowboats with sails, was acquired at Kaskaskia.

Lewis left Kaskaskia on November 5th, traveling on horseback to Cahokia. There he met with postmaster John Hay and French fur trader and land speculator Nicholas Jarrot, who agreed to serve as translators at his meeting with the Spanish Lieutenant Governor of Louisiana, the Marqués Carlos Delassus in St. Louis. At the meeting, Lewis informed Delassus that twenty five men would be accompanying him on the expedition to the Pacific, and that Louisiana would soon be formally acquired by the United States. On December 9th, Delassus wrote to the military governor of Louisiana, the Marqués de Casa Calvo, and to Governor Juan Manuel de Salcedo, that—

> Mr. Merryweather Lewis, Captain of the United States army and former secretary of the President of them has presented himself at this post. . . . I should inform Your Excellencies that according to advices, I believe that his mission has no other object than to discover the Pacific Ocean, following the Missouri, and to make intelligent observations, because he has the reputation of being a very well educated man and of many talents.[386]

On November 30th in New Orleans, Casa de Calvo and Manuel Salcedo had presided over the transfer of Spanish Louisiana to France. Twenty days later, Pierre de Laussat—the bureaucrat sent by Napoleon to govern Louisiana after the planned invasion by the French army—unhappily presided over the peaceful surrender of Louisiana to the United States.

General Wilkinson was prepared to enforce the transfer of power from France to the U. S. with about 350 regular troops and volunteer militia who accompanied him to New Orleans. The general and the new Governor of Louisiana Territory, William Claiborne, were given the keys to the city at the ceremonies on December 20th and the American flag rose over the Cabildo and Place d'Armes on the city's waterfront. The eight to ten thousand citizens of New Orleans—who almost all spoke French as their native language—were not happy. Most of their quarrels started over French or American style of dancing at public balls, and the songs and music played at them. Wilkinson, with the small number of troops under his command, managed to maintain order during this time of government and cultural transition.

In February, 1804, the general met in secret with Vincent Folch, Governor of Spanish Florida, and the Marqués de Casa Calvo, the newly appointed Spanish boundary commissioner. Casa de Calvo was in charge of determining the extent of the land Spain had ceded to France. Wilkinson wanted to resume his career as Agent Number 13 at a pay scale of $4,000 per year; to receive $20,000 in back pay; and to sell 16,000 barrels of flour in Havana. In return he would write a report, "Reflections on Louisiana." After negotiations he was paid $12,000, which the general invested in sugar to be shipped to the east coast.

Over the next month Wilkinson wrote a long report, which Governor Folch translated into Spanish and sent under his own name to his superior in Cuba. The report urged that the Lewis and Clark expedition be arrested or forced to retire from the Upper Missouri country. On March 5th, Casa de Calvo and Manuel Salcedo wrote a letter to Salcedo's brother, Nemesio Salcedo, Governor of Spanish Texas, advising him that it was "absolutely necessary for reasons of state" to carry out the arrest of Captain Merry Weather and his party.[387]

After writing his "Reflections," Wilkinson rewrote it for an American audience and brought it to President Jefferson. This version described the country between the Mississippi and the Rio Grande and was accompanied by 28 maps made by the late Philip Nolan. Wilkinson did not provide any maps to the Spanish government. Essentially, his report on the western frontier served at least two purposes for the general, and he was paid handsomely for it by the Spanish government.[388]

Wilkinson's Spanish report explains why Lewis and Clark's original mission—to find a direct water route from the Missouri to the Pacific Ocean—was expanded to include discovering the source of the Missouri. Finding the source of the Missouri River would establish America's claim for the northern boundary of the Louisiana Purchase. The Spanish insisted the northern boundary was based on the boundary between French Louisiana and New Spain in the 1700's, when the *Arroyo Hondo* (the Calcesieu River in today's state of Louisiana) served as the boundary line between the two colonial empires. The United States's position was that the Louisiana Purchase included the mouth of the Missouri River at St. Louis, and all of the land whose waters drained into the Missouri from its source in the western mountains.

Wilkinson's report, signed by Folch, stated:

> If anyone should doubt the dispositions of the American government with respect to the western boundaries of Louisiana it will be necessary to refer those incredulous ones to the doctrines uttered in full Congress and published in the ministerial gazettes of the city of Washington, and those of this capitol [Pensacola,West Florida]. To these proofs may be added the following interesting facts: that the president of the United States has commissioned his astronomer, who is at this very moment in the province of Louisiana, with the duty of determining by all practical investigations the relative positions of the mouth of the Rio Bravo or Rio Grande, and the source of the Missouri River; that, as an aid to this end, orders have been given to the presidents's private secretary and the Infantry Captain Lewis to ascend the Missouri River with a military command, and supplied with the articles necessary to make the proper observations at different points, and if circumstances favor the extension of the enterprise, they are to proceed as far as the Pacific Ocean; while wagers of consideration have been laid that the United States will have a seaport on said ocean before five years roll by.[389]

Wilkinson accompanied his sugar cargo back to the east coast, amidst much speculation in New Orleans about the source of his new found wealth. Upon his arrival in New York City in May of 1804, the general met in secret with Aaron Burr, which will be the subject of the next chapter.[390]

The Spanish authorites did send four expeditions out to capture "Captain Merry" as they called him, and in their last attempt they almost succeeded. It happened in September, 1806 as the Lewis and Clark expedition was returning to St. Louis on the

Missouri River. About 300 Spanish soldiers were 140 miles west of the expedition when it passed the mouth of the Platte River on September 9th. The soldiers, commanded by Lieutenant Facundo Melgares, were being detained at the Pawnee Indian village on the Republican River near Red Cloud, Nebraska. The Pawnee chief had refused to allow the soldiers to continue on their march to the Missouri, where they almost certainly would have caught Captain Merry and his men.

It seems quite likely the detention of the Spanish force had been prearranged by General Wilkinson with the Pawnee chief. It would have been a major international incident if the Spanish had arrested Lewis and Clark. Only a few weeks later, in October, 1806, Captain Zebulon Pike and his men arrived at the same village and were allowed to continue on a reconnaissance mission into Spanish Texas. Wilkinson's son, Lieutenant James Biddle Wilkinson, was a member of Pike's expedition. After leaving the Pawnee village, Pike and Wilkinson split up and Wilkinson's son and his exploring party returned safely to St. Louis before Pike's party was captured by the Spanish.[391]

The general organized three exploring expeditions into the former Spanish Louisiana during this time—the Pike and Wilkinson, Dunbar and Allis, and Freeman and Custis expeditons. These were the expeditions for which the president requested funding of $10–12,000 from Congress.

The expedition wintered at Wood River in Illinois Territory at a place they called Camp Dubois. ("Bois," pronounced bwa, means wood in French.) Wood River was 18 miles north of St. Louis, northeast of the junction of the Mississippi and Missouri Rivers. Clark was in charge of the camp while Lewis spent much of his time at Cahokia and St. Louis. The men arrived at Wood River on December 12th and the next day began clearing land and cutting logs for their huts. Clark's cabin was ready by December 30th. The weather was sleety with mixed rain and snow and the Mississippi was covered with running ice.

The reconstructed Camp Dubois, an Illinois State Historic Site in Hartford, IL. Although the historic site is enclosed like a fort, the name "Camp" indicates the original site had no walls. www.campdubois.org (Photo by Rick Blond)

George Drouillard arrived with eight soldiers from the Tennessee army post on the 16th. Some were unsuitable, but Hugh Hall, Thomas Howard, and John Potts became members of the permanent party. Drouillard was satisfied by what he saw at the camp, and on Christmas Day announced he would return to Fort Massac to settle his affairs and join the expedition.

The days were a routine of completing the living quarters and hunting for deer and turkeys. Clark had to deal with drunkenness, fighting, insubordination and going AWOL (absent without leave). Matters of discipline would occupy the captains until the men settled down into the tight knit group that became the permanent party going to the Pacific Coast.

Storage lockers were built for keelboat, lining both sides of the boat. Their lids could be raised to provide a shield in case of attack. The men practiced target shooting, competing against local men in the neighborhood. At first the locals won the matches, but by April 27th, Clark was able to report:

> Several Country men Came to win my mens money, in doing So lost all they had, with them.[393]

Meanwhile, Lewis was busy getting to know the St. Louis leaders. The Louisiana Purchase would soon take effect, and the United States needed information about the people living there. After unsuccessful attempts to obtain census data and maps from Spanish officials, Lewis wrote to Jefferson that the government officials and residents of St. Louis feared Colonel Delaussus more than God because—

> . . . he has for very slight offences put some of the most wealthy among them into the Carraboose; this has produced a general dread of him among all classes of the people.

He had no doubt that once Americans came into power that many of the best informed people would be coming forward with information, but until then "every thing must be obtained by stealth." Nevertheless Lewis had already managed to learn a great deal about the country and its inhabitants which he passed along to Jefferson.[394]

Lewis stayed in Cahokia as a guest of John Hay, the Cahokia postmaster. Hay, who served in a variety of government posts, was called "the generalissimo of the pen" for his ability to help people with contracts and deeds. He had come to Cahokia as a merchant representing the British Canadian trading company of Todd & Hay, and he advised Lewis and Clark on their Indian goods—what additional goods to purchase, and how to select the gifts for the individual tribes.

Hay was the son of Jehu Hay, second in command to the "Hair Buyer" General Henry Hamilton who had surrendered to George Rogers Clark at Vincennes in 1789. Both men had been imprisoned in Virginia by Thomas Jefferson. After his release in 1781, Jehu Hay served as Lieutenant Governor of Detroit. In 1785, when John Hay was 16 years old, his father died.[395]

How would John Hay deal with Clark, the brother of the man who had captured his father, and Lewis, who worked for Jefferson, who imprisoned him? The answer is treacherously, as historian Thomas Danisi has documented. Lewis, who became Governor of Louisiana Territory in 1807 after his return from the expedition, had great difficulties with the postal service.[396] This was the world of Louisiana and Illinois Territories—the loyalty of its residents to the new American government could not be taken for granted.

On March 9th, Meriwether Lewis and William Clark attended the ceremonies in St. Louis in which Louisiana was transferred from Spain to France. The flag of France flew over St. Louis for one day. The next day, Upper Louisiana was transferred to the United States. Captain Amos Stoddard, the Kaskaskia commander, was named military commandant for Upper Louisiana and acting governor. A year later in March, 1805, General Wilkinson was named the first governor of Louisiana Territory, while retaining his position as commanding general of the United States Army. Three years later, in March, 1807, Meriwether Lewis would become the second governor.

On the 26th of March, the Secretary of War wrote to Lewis notifying him that William Clark could receive no appointment higher than second lieutenant in the Corps of Artillerists, and enclosed his commission. He assured Lewis that Clark's military grade would have no effect on his compensation, which would be as Lewis requested.

After the expedition, Nicholas Biddle, the editor of the journals, asked Clark to clarify the matter of his rank, and Clark wrote that when he received his appointment as second lieutenant instead of the captain's rank he was promised—

> My feelings on this Occasion was as might be expected. I wished the expidetion suckcess, and from the assurence of Capt. Lewis, that in every respect my situation Command &c. &c. should be equal to his viewing the Commission as mearly Calculated to autherise punishment to the soldiers, if necessary, I proceeded.... Be so good as to place me on equal footing with Cap. Lewis in every point of view without exposeing any thing...[397]

Clark was extremely hurt by this action, and viewed his commission as simply authorizing his ability to discipline the men. He returned the commission as soon as the expedition ended. Lewis and Clark never revealed that Clark had not received a captain's commission. In 2001, President William Jefferson Clinton promoted William Clark to the rank of captain posthumously.

The expedition set out from Wood River on May 14th, going only as far as St. Charles, where they waited for Lewis to join them from St. Louis. At St. Charles, a small riverfront village on the Missouri River, they were joined by two experienced boatmen, Pierre Cruzatte and François Labiche, who had earlier been enrolled as privates in the U. S. Army. Cruzatte was half French-half Omaha Indian. Labiche was half French-half African American, or "mulatto."[398]

Cruzatte was a fiddle player, as most bowmen were. ("Bow" is pronounced like "cow.") The bowman stayed at the bow, or front of the boat, and directed the course of the boat through the water, paddling with the bow oar. With his fiddle playing and singing, he set the pace for the men rowing with oars or pushing the boat along with iron-tipped poles. Cruzatte would have taken the lead in call and response songs.

In today's world we can scarcely imagine the importance of singing in performing manual labor—particularly timed and coordinated labor. So as we imagine the three boats going up the Missouri, we should include their music. In the evening the men often danced. They would have danced contra dances (similiar to square dancing) and jigs and reels. It was a chance to get moving after a day in the cramped conditions of the boats.

Labiche served as an interpreter for the expedition, as he spoke both French and English. Like Cruzatte, he was an experienced river man and Indian trader. Labiche and Cruzatte alternated between the larboard oar (left side of the ship, or port) and the bow oar in their duties. Cruzatte, however, was the better navigator and was called upon for the more challenging navigation problems.

The boat was a keelboat, but of an unusual design. It was similiar to the Spanish war vessels that Clark had observed on the

Keelboat replica built by Butch Bouvier at Lewis and Clark State Park, Onawa, Iowa. The rudder tiller is seen at the rear, on top of the cabin. The men rowed with single oars standing up, in the center deck. There would have been oarlocks to hold the oars in position. It was a 22 oar boat, 11 men on a side. The boat measured 55 x 8 feet. The cabin and foredeck were 10 feet each; the main deck was 35 feet. It drew three feet of water.[399]
(Photo by Kira Gale)

Mississippi. The keel is a piece of sturdy wood running underneath the entire length of the boat, to which the rest of the boat frame is attached. Lewis and Clark usually referred to their boat as a barge, rather than a keelboat. A keelboat was generally a long narrow boat with a cabin the middle, used for hauling goods. With a keel, you could go up river against the current. Flat boats, without a keel, could only travel downriver. Their barge had a captains' cabin at the stern.

The three sergeants, Charles Floyd, John Ordway, and Nathaniel Pryor, were assigned to the front, center, and rear of the boat. The bow sergeant used a setting pole to help the bowman navigate. He kept a look out for all danger, enemies, Indian camps, and obstructions and notified the sergeant at the center, who notified the commanders. The sergeant at the center managed the sail and all matters relating to the crew—rowing, break times, spiritous liquors, and guard duties at the campsites. He was to keep a look out for all notable places, such as rivers, creeks and islands, and notify the captains immediately. The sergeant at the rear, or helmsman, steered the boat, and saw that all baggage on the quarterdeck was stowed away with no loose articles. He was also to "attend to the compass when necessary."

The sergeants were in charge of the messes, in which the privates were divided into three squads of 8–9 men each. Each mess camped together and had its own cook. The French engagés in the red pirogue and the American soldiers on loan from Kaskaskia in the white pirogue formed their own messes. Altogether there were at least 46 men on the expedition going up river to the Mandan villages in today's North Dakota. The exact number is unknown as the engages also went by their "dit" names, or nicknames. Pierre Cruzatte, for example, was called "St. Peter."[400]

William Clark had the responsibility of map-making and maintaining course and distance records. He most likely used a slate during the day and transferred his notes to his field journal and maps in the evening. He recorded compass directions, distance between points, landmarks, and miles traveled each day. The notes appeared like this for June 4th:

N. 30^0 W.	4	ms. a pt. on S. Sd. psd. a C & 2 Isd.
N. 25^0 W.	3	ms. to a pt. on S. Sd. psd. Seeder C.
N. 58^0 W.	7½	ms. to pt. on L.S. a Creek on L. S.
N. 75^0 W.	3	Ms. to a pt. on S. Sd. opsd. Mine Hill
	17½	

S. Sd. was Starboard Side, or right. L. S. was Larboard or left side, looking towards the front of the boat. The miles would be measured from landmarks or "points." Geographer John Logan Allen has written that Clark's—

> coordinate positions are accurate to within 5%, an accuracy level that would not be matched by many cartographers until the advent of mapping aided by aerial photography in the early 20th centry.

Clark calculated the miles from the mouth of the Missouri River to the mouth of the Columbia River at 4,142 miles. Today's National Historic Lewis and Clark Trail is 3,700 miles long.

The Missouri has been channelized by the Corps of Engineers, so it is no longer the river they experienced. Lewis wrote to his mother—when he was 1609 miles up river at Fort Mandan—that the Missouri River was more dangerous than any Indians they encountered. He described the force of the river as being so strong from the mouth of the great river Platte to the mouth of the Missouri at St. Louis, that they couldn't use oars or poles to travel in the center's fast current and had to stay close to the river banks. The banks were often collapsing and bringing trees down with them. Floating trees were hazards, but others anchored in the river bed were hidden from sight in the muddy waters. Sawyers were trees which bobbed up and down and planters were firmly fixed in the river bed. Both could destroy a boat. Sandbars and quick sands could overturn a boat.

On June 9th, William Clark wrote that—

> the Sturn of the boat Struck a log which was not proceiveable the Curt. [current] Struck her bow and turn the boat against some drift & Snags which [were] below with great force; This was a disagreeable and Dangerous Situation, particularly as immense large trees were Drifting down and we lay immediately in their Course,—Some of our *men* being prepared for all Situations leaped into the water Swam ashore with a roap, and fixed themselves in Such Situations, that the boat was off in a few minits, I can Say with Confidence that our party is not inferior to any that was ever on the waters of the Missopie

On June 15th, he described the "worst part" of the river—

> we wheeled on a sawyer which was near injuring us Verry much ... passed between two Islands a verry bad place, Moveing Sands, we were nearly being Swallowed up by the roleing Sands over which the Current was so Strong that we Could not Stem it with our Sales under a Stiff breese in addition to our ores, we were compelled to pass under a bank which was falling in, and use the Toe rope occasionally.[401]

They stopped to rest for a couple of days and to make oars and a tow rope from cable Lewis had purchased in Pittsburgh. Two thirds of the men were suffering from boils and ulcers, caused by bacterial infections from the dirty water. Some had 8 or 10 of these "tumers." The "Mesquetors verry bad." Despite this, the men were in "Spirits." The journals are filled with interesting descriptions of their daily lives. Always some men were out hunting. George Drouillard was on permanent duty as a hunter, and exempt from camp duties. He brought in a bear, deer, and turkeys. [402]

The sergeants had been ordered to keep journals. Ordway's and Floyd's journals are available, but Pryor's journal hasn't been found. Ordway was the most faithful journal keeper. He made entries for all 863 days of the expedition from St. Louis to the coast and back (May 14, 1803 to September 23, 1806). Four privates also kept journals. Robert Frazier's journal is lost, but Patrick Gass's and Joseph Whitehouse's journals are published. Gass's journal—the first to be published in 1807—was edited and his original journal entries are lost. The fourth writer is unknown.[403]

How did these journal keepers see the dramatic events of June 9th? Ordway wrote: "we got fast on a log Detained us half a hour." Floyd wrote: "this is a butifull Contry of Land the River at this place is 300 yards. wide the current Strong." Gass wrote: "This day going round some drift wood the stern of the boat became fast, when she immediately swung round, and was in great danger; but we got her off without much injury." Whitehouse wrote: "The day proving stormy, we were obliged to waite for our hunters who were on the opposite side of the River, and it being unsafe to venture across in the Pettiauger [pirogue] for them." Their journals are online at the University of Nebraska Press, with daily journal entries grouped together for readers who like to read original sources. (www.lewisandclarkjournals.unl.edu).

Lewis wrote he was sending journals by separate routes—

> in order as much as possible to multiply the chances of saving something. We have encouraged our men to keep journals, and seven of them do so, to whom in this respect we give every assistance in our power.[404]

When the men sat around the camp at night writing up their accounts of the day's happenings, was Lewis—the president's former secretary—also writing his daily account, or was he working on his scientific observations?

There are no daily journal entries by Lewis for the journey from St. Louis to the Mandan villages, except for fragments, dated May 15 and May 20 and September 13 &14, 16 & 17, 1804. They were pages torn from a journal that were found fastened to a later journal with sealing wax. Historian Donald Jackson described the paper as being "worn and discolored." Jackson, the editor of *Letters of the Lewis and Clark Expedition*, has suggested that Lewis's journal was lost in a boating accident on May 14th, 1805, five weeks after they left Fort Mandan. By then the keelboat was on its way back to St. Louis and the captains were traveling in the white pirogue westward towards the mountains.[405]

Both captains were on shore hunting buffalo when a violent gust of wind turned the pirogue on its side, filling it with water. Only the boat's awning and sail prevented it from turning over. Sacagawea, who was in the rear of the boat, managed to retrieve "nearly all" of the articles as they floated by. Lewis wrote:

> in this perigue were embarked, our papers, Instruments, books medicine, a great part of our merchandize and in short almost every article indispensibly necessary to further the views, or insure the success of the enterprize . . .

He was preparing to jump in the water and swim to the boat when he realized he would almost certainly lose his life in the attempt. The boat was 300 yards distant, the waves exceedingly high and the water very cold. Private Whitehouse wrote:

> We found that most part of her loading was wet, the Medicine damaj'd, & part of it Spoiled—We also found that some of the papers and books had got wet, but not so much as to be spoiled.

Gass wrote: "we found a great part of the medicine, and other articles spoiled." Ordway wrote: "we opened the goods &c. to get them dry before we packed them up." Clark simply copied Lewis's account as he often did. If Lewis's journal was spoiled in the boat accident, no one was going to reveal it.

When they left Fort Mandan on April 7, 1805 Lewis wrote a letter to Jefferson, sending it back with the keelboat. The keelboat was starting on its return journey to St. Louis bearing articles and specimens for Jefferson and the American Philosophical Society, including live magpies—large and noisy birds—and a live "barking squirrel" (prairie dog). Both arrived safely at the President's House and then were sent to Philadelphia where they were put on exhibit at Peale's museum.[406]

Lewis wrote that he was sending Clark's journal to Jefferson, which would "serve to give you the daily detales of our progress," but before copies were distributed to others, Clark wanted his spelling and grammer errors corrected. Lewis wrote that he planned to send a canoe back from the "extreem navigable point of the Missouri" with three or four men, and at that time would send his own journal back with one or two of the best journals being kept by the men.[407]

Since this letter predates the boat accident, it appears to confirm that Lewis kept a journal en route to Fort Mandan—certainly the pages for six days of entries support this idea. Donald Jackson offers one final bit of evidence in support of his theory. After the boat accident, Clark, who had been simplifying Lewis's scientific descriptions of plants and animals, began copying the descriptions word for word in his own journal.

The captains split the science duties. Clark had the task of recording course and distance so he could make his maps, which required his steady attention throughout the day. Lewis was responsible for daily weather records and writing descriptions of noteworthy features of the landscape—mineralogy, geology, soil, seasonal changes, etc. He collected plants new to the United States and described them using a botanical text for reference. He did the same for animals and preserved specimens to be sent back to Peale's Museum. He created a chart of rivers and creeks on the Missouri. They had a library of travel journals, scientific references, and maps in their boat cabin, including the Mackay-Evans map and Evans's Indian notes.

The captains collaborated on the task of making celestial observations in order to determine the latitude and longitude of river junctions and other points. The position of the North Star at night was used to determine their latitude by measuring its angle to the horizon with a sextant. The position of the sun at its high point was used to compare the local noon hour with the time on the chronometer clock to find their longitude. The chronometer's time had been set at Pittsburgh. Their latitude readings were accurate, but when their chronometer clock ran down and had to be recalibrated by guesswork, errors crept it. These observations required the mathematical calculations that Lewis had been trained in before leaving on the expedition.

As they encountered Indian tribes they sought answers to the questions proposed by Jefferson and others—the tribe's life style, population, geography, number of warriors, enemies, cultural traditions, health practices, and trade networks. The commercial fur trade was a major focus. What kinds of furs were available, what were the best places to build trading forts, and what were the most desirable articles to exchange for furs? During the winter at Fort Mandan, Lewis and Clark prepared an elaborate chart, which they called "Estimate of Eastern Indians." The chart provided data on nearly 50 tribes and bands from information they had obtained either first hand, or from reports by other Indians and fur traders.

Jefferson and his fellow American Philosophical Society members were fascinated with comparing Indian languages to European and Asian languages, believing it might hold the key to discovering common ancestries. Lewis recorded the vocabularies of twenty three different Indian tribes on blank vocabulary sheets with 250 English words printed on them. Lewis intended to publish his scientific information and the Indian vocabularies in the third volume of his planned three volume account of the expedition. Unfortunately, these vocabulary sheets were lost after his death and the third volume was never published.[408]

First Council with the Otoe and Missouria, woodcut from Patrick Gass's published journal.

Their first council with Indians was held with the Otoe and Missouria Indians at Council Bluffs on August 3rd, 1804. The Otoe lived near the junction of the Platte and Elkhorn Rivers in an earth lodge village. Their kinfolk, the Missouria had joined them there, after warfare and disease had driven them from their home on the Grand River in Missouri. The two tribes had gone off on their annual summer buffalo hunt, leaving the village deserted.

Cruzatte and Drouillard had been sent to invite them to come in for a council. Cruzatte was half Omaha and knew the area well. He had operated a trading post for the Omaha near Council Bluffs for two years. It seems likely that he had accompanied MacKay's expedition up river in 1796–97 and decided to remain. Some days after finding the village deserted, Drouillard heard gun shots and encountered a group of Missouria out hunting elk. Messengers were sent to find the Otoe and bring them in for the council.

The Otoe chiefs arrived bringing gifts of "water millions." Speeches were made and medals, a flag, and presents given out. The French trader who came with them said the Otoe traded with the Spanish at Sante Fe, a journey of 25 days. The captains recommended that a trading post be built at Council Bluffs and Fort Atkinson was established there in 1819.

The expedition traveled much more easily on the Missouri River after passing the mouth of the Platte. The river no longer rushed downstream at a rate of 5-6 miles per hour, but moved along at a leisurely 3 miles per hour. However, the bends of the river could make it a frustrating day's travel. One bend measured a little more than half a mile (973 yards) by land, but it was a day's journey, 18 3/4 miles, by water.

The Otoe chiefs had asked the captains for help in arranging a peace between themselves and the Omaha, and it was agreed they would come up for another council at the Omaha village, some eighty miles up the Missouri. As they left Council Bluff, Private Moses Reed deserted the expedition, and Joseph Barter, "La Liberte," a French engagé, also deserted. Drouillard was sent with three men to find them, with orders to kill Reed if he didn't come in peaceably and to bring back La Liberte from the Otoe village, where it was presumed he had gone.

As they neared the Omaha village they stopped to climb the hill where Omaha Chief Blackbird was buried, and planted a white flag at his gravesite. Blackbird and 400 members of his tribe had perished in a small pox epidemic four years earlier. The next day they passed the location of MacKay's old trading post, Fort Charles, and formed a camp 5 miles further up river. They called it "Fish Camp" because they caught about 1,100 fish, using a brush barrier dam on a creek, while waiting there.

The Omaha village, like the Otoe village, was deserted. The Omaha were away on their summer buffalo hunt. When the three Otoe chiefs, accompanied by 8 lesser chiefs and warriors, arrived with Drouillard at the camp on August 18th, there was no chance to arrange a peace with the Omaha. Drouillard and his party had found both Moses Reed and La Liberte. However, en route to the Omaha village, La Liberte had escaped again.

August 18th was a memorable day. It was Meriwether Lewis's 30th birthday. William Clark had celebrated his 34th birthday at Council Bluffs on August 1st with a feast of game and ripe fruit. Lewis's birthday, however, was definitely mixed. The deserter Reed was ordered to run the gauntlet four times, and received about 500 lashes from the men. Clark wrote that they were "as favorable to him as they could be, consistent with their oaths."[409] Before his punishment, the Otoe chiefs requested his pardon. After the captains "explained the customs of their country," the chiefs were satisfied. Reed was dismissed from the permanent party, and would return with the soldiers from Kaskaskia to St. Louis in the spring. After the court martial and punishment, they held a dance. The dance, celebrating Lewis's birthday, lasted until 11 that night.

The next day it became obvious that Sergeant Charles Floyd was seriously ill. The captains were still meeting with the chiefs counseling with them about their altercations with the Omaha and Pawnee. Clark wrote:

> Sergt. Floyd was taken violently bad with the Beliose Cholick and is dangerously ill . . . nature appear exosting fast in him every man is attentive to him (york prlly) [York, primarily].[410]

Floyd, apparently, was suffering from a ruptured appendix and peritonitis. He had not been feeling well for more than a month. He wrote on July 31st:

> I am verry sick and Has ben for Somtime but have Recovered my helth again.[411]

If he was treated with Dr. Rush's Bilious Pills—laxatives called "Thunderclappers"—it could have irritated an already inflamed appendix in the days before his death. They may have bled him, another treatment promoted by Dr. Rush, who urged removal of blood, from a few ounces to 80 ounces (about half the blood in the body), as a cure for illness.[412]

Sergeant Floyd was only 21 years old when he died. As they proceeded up river the next day, they stopped to make a warm bath for him. But, before they could get him into the bath, he died. Clark wrote:

> Serj. Floyd Died with a great deel of Composure, before his death he Said to me, "I am going away" I want you to write me

The Sergeant Floyd Monument at Sioux City, Iowa on Sergeant Floyd's Bluff. It became America's first National Historic Landmark in 1960. (photo by Kira Gale)

> a letter"—We buried him on top of the bluff ½ Miles below a Small river without a name to which we Gave his name, he was buried with the honors of War much lamented; a Seeder post with the name Sergt. C. Floyd died here 20th of August 1804 was fixed at the head of his grave— This man at all times gave us proofs of his firmness and Deturmined resolution to doe Service to his Countrey and honor to himself after paying all the honor to our Decesed brother we Camped in the mouth of floyds river about 30 yards wide, a butifull evening.—[413]

Captain Lewis read the funeral service, and the bluff was named Sergeant Floyd's Bluff. He was the only man to die on the expedition. On August 26th, by a vote of a large majority of the men, Patrick Gass was chosen to replace Floyd. In his orders appointing Sergeant Gass, Lewis used the term "the corps of volunteers for North Western Discovery" for the first time in the journals.

On the very same day, August 26th, a new crisis was about to develop. Drouillard and George Shannon were sent out to look for lost horses. The expedition had a small number of horses accompanying them, used by the men in hunting game. There were now ripened grapes, and three kinds of plums of "delicious flavor" in the fields ready to eat. They had been checking out the fruit as hungrily as birds for weeks.

It was fortunate for George Shannon that the fruit was ripe, because he became lost after separating from Drouillard. Shannon believed he was behind the boats and hurried to catch up with them—but, instead, he was in front of the boats. Shannon was lost for 16 days until the boats finally caught up with him on September 11th. He had eaten only grapes and one rabbit—which he shot with a hard stick from his gun—for twelve days. He still had his horse, which he would have killed and eaten as a last resort.They saw from his horse tracks that he was ahead of the boats, but the men sent after him couldn't overtake him. During Shannon's absence he missed some big events—the election of Patrick Gass, the council with the Yankton Sioux, and catching a live prairie dog.

The council with the Yankton Sioux took place at Calumet Buffs on August 30-31. Coming up river they had acquired the services of Pierre Dorion, Sr. a long time trader with the Yankton Sioux. They realized they needed someone who could interpret for them with the Sioux. However, once they met in council with the Yankton, they decided Dorion could be more usefully employed in escorting a delegation of Sioux chiefs to Washington in the spring of 1805. Dorion remained with the Yankton, and it meant they would be without the services of a skilled interpreter when they encountered the more hostile Teton Sioux up river.

They spent most of September 7th trying to catch a prairie dog. The village of underground tunnels of the "barking squirrels" covered three acres. First they dug through six feet of hard clay in an attempt to reach their lodges. Then they resorted to a bucket brigade, pouring gallons of water down a hole, to flush one out. Seaman, the dog, must have been eager to help. The little creature was put in a cage and lived with the captains for months before it was sent to Washington.

They were seeing a great many buffalo by now. Clark killed a male pronghorn, which they called an antelope or wild goat, on September 14th. On the same day, John Shields killed a prairie hare, or jack rabbit. Lewis measured the leap of one hare at 21 feet, noting "the ground was a little descending." They stuffed the pronghorn and hare to send back to Peale's Museum. Among the pages of Lewis's notes that were found for September 13 & 14, 16&17, was this observation by Lewis on September 16th—

> . . . these extensive planes had lately been birnt and the grass had sprung up and was about three inches high. vast herds of Buffaloe deer Elk and Antilopes were seen feeding in every direction as far as the eye of the observer could reach.[414]

The Indians practiced fire burning in the spring to create new grass for the animal herds. They also set the prairies on fire as signals—to bring in other tribes, to warn of danger, or to announce the presence of a buffalo herd. The prairies were not naturally treeless—regular burning destroyed all trees except those near creeks and rivers.[415]

On the 17th, Lewis killed a female pronghorn to send back. These animals were new to science. The pronghorn is the fastest quadruped in America; it can run up to 60 miles per hour and Lewis said it could smell a hunter at a distance of three miles. He estimated he saw 3,000 buffalo in one view that day.

Naturalist Paul Cutright wrote that zoologically this area was the "most important and most exciting of the entire trip." From the Niobrara River in Nebraska to the Bad River in South Dakota, a distance of 263 miles, they discovered and named nine animals for science—the pronghorn, coyote, plains gray wolf, prairie dog, mule deer, white-tailed jack rabbit, black-billed magpie, sharp-tailed grouse and desert cottontail.[416]

On September 20th, they arrived at the Big Bend of the Missouri River—30 miles by water and $1^1/_4$ miles by land. The next day, the bank of the sand bar on which they were camping began to give way during the night and would have swallowed their boats within minutes. They rushed to the boats and the bank caved in before they reached the opposite shore. They passed Loisel's trading post on Cedar Island with a number of Indian camps around it on the 22nd. No one was there.

On the 23rd, three Sioux boys swam the river and informed them the Brulé Sioux (Teton or Lakota Sioux) were camped a short distance above. Some Sioux stole Colter's horse while he was out hunting on the 24th. Two thirds of their party stayed on board the boats that night, and the rest stayed with the guard on shore. On the morning of the 25th they raised a flag on a staff, and set up the sail for a shade awning to council with the Brulé chiefs.

They gave out medals and presents and invited the 3 chiefs and a warrior to come on board the keelboat to view the "many Curiossites." After a small drink of whiskey the 2nd chief, Partisan, became troublesome, pretending he was drunk. He was a rival to the 1st chief, Black Buffalo, and resented the more splendid presents he had been given. Clark got them into a pirogue to take them back to shore. At the shore they were met by three young Sioux, who seized the cable rope of the pirogue, while one warrior held fast to the boat mast. Partisan was insulting and threatening, saying he wouldn't get off the boat until he received more presents. Clark wrote: "I felt my Self warm & Spoke in verry positive terms."

Clark drew his sword and the men in the pirogue took up arms, while Captain Lewis and his men took up arms and loaded the 3 swivel guns on the boats for action. The Indians on shore had their bows and arrows and guns ready to fire. At this point, the 1st chief of the Brulé Sioux, Black Buffalo, took hold of the rope from the young men and defused the situation. After some time on shore, Black Buffalo, two of his warriors, and the 3rd chief, came on board the keelboat to escort them to a new campsite, which the captains named "bad humered Island as we were in a bad humer." This is how the Bad River in South Dakota got its name.

The next day Black Buffalo and his entourage stayed with the keelboat as it proceeded five miles up river to his own camp, with Indians lining the shore to observe the boats progress. Black Buffalo's camp had about 100 tipis, all made of white colored buffalo hides, arranged in a circle around a large council tipi. Captains Lewis and Clark were carried into the village on white buffalo robes, a mark of great esteem. There was feasting and a dance that evening, with food sent out to the men in the boats. A council was held in the central lodge with about 70 elders and warriors. An elder and Black Buffalo "spoke approvingly of what we had done." That evening Black Buffalo slept on the boat with them.

The Sioux had lately been at war with the Omaha and had killed 75 men. They had also captured 48 women and boys, held captive in two villages. Pierre Cruzatte, who spoke Omaha, learned from them that the Sioux intended to stop the expedition. At the council, the chiefs had promised to deliver the Omaha captives to Pierre Dorion, the Yankton trader.

The Indians wanted the expedition to stay for several more days, as they expected the arrival of more Sioux to see the Americans. After another night of feasting and dancing, Partisan and one principal man were accompanying William Clark in the pirogue, going back to the keelboat. Through bad steering, the pirogue cut the cable holding the keelboat anchor. The keelboat anchor sunk and the pirogue sprung a leak. All of this alarmed the Indians. Black Buffalo and about 200 warriors came down to the river bank, armed for action. Seeing that it was a false alarm, they left, but about 20 warriors remained on shore and Black Buffalo came on board again.

The next day, after searching in vain for the anchor, the captains decided it was time to leave.There were theatrics. The 2nd chief, Partisan, wanted tobacco and a flag, which they refused to give him, and some young men held the keelboat cable, demanding tobacco. The captains reluctantly gave Black Buffalo two twists of tobacco, which he gave to the men, and took the cable from them. It was a face saving maneuver. Tempers were frayed on both sides. As they continued up river, Black Buffalo stayed with the boat for two more days until announcing that all things were clear for them and they would not see any more Tetons.

The Sioux were a roving tribe, the warriors of the plains. They were the tribe that Jefferson most wanted to conciliate and establish trade relationships with. They had a long connection with British traders. Jefferson had enclosed excerpts from the journal of Jean-Batiste Truteau in a letter to Lewis. Truteau, who traded with the Upper Missouri tribes, had written:

> The Sioux inhabit the Northern part of the Mississipi, and are hostile to the Ricaras, Mendanes, big-bellies and others. Others of them live on the river St. Pierre. They have from 30 to 60,000 (?) men and abound in fire-arms. They are the greatest beaver hunters; and could furnish more beavers than all the other nations besides, and could bring them to a depot on the Missouri, rather than to St. Pierre, or any other place. Their beaver is worth double of the Canadian for its fineness of it's fur and parchment.[417]

The rivalry between the two chiefs had complicated matters, but it allowed Black Buffalo to establish a relationship with the Americans which would later have a great impact on the War of 1812. In 1813, after Lewis's death, William Clark was serving as both the Governor of Missouri Territory and Superintendent of Indian Affairs. At the start of the war, he appointed Manuel Lisa a sub-agent for Indian Affairs, and provided trade goods to send up to Lisa's new post on Cedar Island to keep the Brulé Sioux out of the war.

Black Buffalo sent a messenger to the Santee Sioux of the Minnesota River (St. Pierre River) to stop fighting for the British in Michigan and come home, or he would burn their villages. The Santee had been recruited by the British trader Robert Dickson. After the war, at a British court of inquiry, Dickson blamed the Sioux defection for the American victory on Lake Erie, the loss of Amherst and Detroit, and the subseqent capture of General Proctor's army. About 2,000 warriors returned to their homes after the Siege of Fort Meigs, reducing Tecumseh's Indian army to about half its size.[418]

Lewis and Clark's diplomacy with the Indians of the Missouri was continued by Manuel Lisa, who employed former members of the Lewis and Clark expedition in his fur trading empire. Lisa observed at the end of the war:

> The Indians of the Missouri are of those to the Upper Mississippi as four to one. Their weight would be great, if thrown into the scale against us. They did not arm against the Republic; on the contrary, they armed against Great Britain, and struck the Ioways, the allies of that power.

Lisa arranged for both Yankton Sioux and Omaha war parties to attack the Ioway. The Ioway didn't care about a peace treaty ending the war and were continuing to attack settlers.

At the end of the War of 1812 in the summer of 1815, Manuel Lisa brought 33 chiefs and important men of the Missouri tribes to William Clark's treaties council at Portage des Sioux at the junction of the Mississippi and Missouri Rivers near St. Louis. Over 2,000 Indians attended and at least 19 tribes signed peace treaties with the United States.

Both Partisan and Black Buffalo came with Lisa's delegation. Black Buffalo died suddenly of an unknown cause at the council. Was he assassinated? He was surrounded by Indians who had fought for the British; he was a known collaborator; and Partisan was his enemy. His death was marked by great ceremonies.

Omaha Chief Big Elk delivered the funeral oration. Big Elk's daughter was married to Manuel Lisa. The English translation of his speech, "Big Elk's Funeral Oration on the Death of Black Buffalo," was immediately published in books by John Bradbury and Henry Brackenridge and it remains a classic of Indian oratory. It begins—"Do not grieve—misfortunes will happen to the wisest and best men . . ." and eloquently described the ceremonies honoring Black Buffalo: "a noble grave with a flag raised over it, a grand procession, rolling music, and a thundering cannon—".[419]

On October 1st, near the Cheyenne River, they visited with Jean Vallé at his trading post for the Sioux. He came on board and stayed with them for a day. It was turning cold and the geese were flying south in large flocks. Small groups of Teton

Sioux were still following them, begging for tobacco. On October 6th, near the Grand River, they saw a deserted Arikara village of about 80 lodges made of earth, each lodge 20-60 feet in diameter. They were packed closely together and surrounded by a picketed stockade fence. Squashes of three different kinds were growing at the village. On the 8th they passed another deserted village of about 60 lodges, which apparently had been used the last winter.

Finally they came to an inhabited Arikara village on an island three miles long, where they were growing corn, beans, squash and tobacco. They met the trader, Joseph Gravelines, who had lived with the Arikara for over 20 years, and Pierre Tabeau, a trader from St. Louis. They saw their first bull boats in the water—bowl-shaped boats made from buffalo skins. Clark remarked:

> I saw at Several times to day 3 squars [women] in single Buffalow Skin Canoes loaded with meat Cross the River, at times the waves were as high as I ever Saw them in the Missouri—

On October 10th, they raised the flag and set up the awning to hold a council with the Arikara chiefs. They designated three chiefs, one for each village. and gave medals to them. A Sioux attended the council and they were told he had come to the Arikara to persuade them to stop the expedition.

York was the first black man the Arikara had ever seen. They called him "the big medison." York took great delight in his new celebrity. Clark wrote:

> the Inds. much astonished at my black Servent, who made himself more turrible in thier view that I wished him to Doe as I am told telling them that before I cought him he was wild & lived upon people, young children was verry good eating Showed them his Strength &c. &c.

The next day they took on board the two chiefs whose villages were up river from the island. These villages—surrounded by picketed stockade fences and located across the river from each other—would prove the downfall of the Missouri River fur trade in the years to come. Any expedition attempting to come up river was completely vulnerable to an Arikara attack.

Three years later in 1807, some of Lewis and Clark's men, who were part of a new expedition led by Nathaniel Pryor, were attacked by the Arikara. Two were wounded and one most likely lost his life. George Shannon lost a leg, George Gibson was wounded, and Joseph Fields was probably killed, as William Clark listed him as "killed" in 1807. In 1823, another battle took place between the Arikara and the Ashley-Henry expedition. After that, the fur trade avoided the Missouri River and switched to the rendezvous system, hauling goods back and forth by wagon along the Platte River to the mountain men who remained in the West year round.

Tabeau reported that the Arikara once lived in 18 villages along the Missouri, but their population had been drastically reduced by small pox epidemics and war with the Sioux. In the three villages, there were at least ten different bands of Indians and more than 42 chiefs. The bands did not all speak the same language, and intense rivalries and jealousies hindered any cooperation. They relied on trading their agricultural products with the hunting tribes of the plains for meat and hides. They also acted as middlemen—trading in horses, obtained from Indians of the Southern Plains, and trading in guns, ammunition, and manufactured goods, obtained from the Assiniboins and Sioux who got them from British traders.

The Arikara had 500 warriors and a great quantity of arms. The Sioux, while trading with them, took what they wanted, set prices, stole horses, and sometimes murdered them. The Arikara had been at war with both the Sioux and the Mandan. One of Lewis and Clark's diplomatic goals was to arrange a peace between the Arikara and their neighbors, in order to "clear an open road" for American traders. An Arikara chief joined the expedition and accompanied them up to the Mandan villages, hoping to arrange a peace which would allow them to present a united front against the Sioux. But because of internal rivalries and the military power of the Sioux, the alliance proved unworkable.[420]

The Arikara, like the Sioux and Mandan, offered their women to white men as sex partners. They encouraged sex as a way of extending hospitality, receiving presents, and building business relationships. They believed that spiritual power was transferred in the sexual act, and this power would be passed onto the husbands of the wives who engaged in it. Indian women followed the boats as they left the Sioux, and again as they left the Arikara. Patrick Gass wrote approvingly of "a great number of smart and handsome women and children" among the Arikara. As they left the Arikara, they took on board two women. Clark wrote:

> a curious Custom with the Souix as well as the receres is to give handsom Sqars to those whome they to Show Some acknowledgements to—the Seauix we got Clare of without taking their Squars, they followed us with Squars 13th two days. The Rickres we put off dureing the time we were at the Towns but 2 Handsom young Squars were Sent by a man to follow us, they Came up this evening and persisted in their Civilities.[421]

Clark spoke more honestly about these matters in talking with Nicholas Biddle, the editor of the journals, and told him that "the men by means of interpeters found no difficulty in getting women" at the Arikara villages.[422]

The two young women came on board on October 12th. On October 13th Private John Newman was "Confined for mutinous expression" and a court martial was held the same day. He was charged with "having uttered repeated expressions of a highly criminal and mutinous nature." A jury of his peers, two segeants and seven privates, unanimously judged him guilty. Two-thirds of them concurred with his sentence of 75 lashes on his bare back. It may, or may not, be relevant that at least one of the young women stayed on board the boat until October 15th.

He was discharged from the permanent party and ordered to labor on the red pirogue with the Frenchmen. Both John Newman and the deserter Moses Reed returned to St. Louis with the keelboat in the spring. The Arikara chief cried out loud as he witnessed Newman's whipping on October 14th. Clark wrote:

> I explained the Cause of the punishment and the necessity He thought examples were also necessary & he himself had made them by Death, his nation never whiped even their children, from their birth.

Later, Lewis wrote to the Secretary of War stating that Newman had performed valuable services at Fort Mandan and saved the keelboat from harm on several occasions on the return journey. He urged that Newman receive his full pay, and wrote that Newman had begged to accompany them on the journey to the coast, but he thought it "impolitic to relax from the sentence altho' he stood acquitted in my mind."[423]

The captains had devised their own, irregular, form of a court martial, as they had no access to a court composed of the required minimum of three commissioned officers. They could have administered punishments themselves, but instead chose to use a jury of enlisted men in every case except one— the captains sentenced Alexander Willard to 100 lashes for sleeping on duty on July 12, 1804. John Collins was sentenced twice to receive whippings, once for disrespect to an officer and going AWOL at St. Charles on May 17th (50 lashes); the other time for drunkeness and allowing Hugh Hall to open a barrel of whiskey on June 29th (100 lashes). Hall received 50 lashes. Clark wrote that the men were always "found verry ready to punish such crimes." On February 9th, 1805 Thomas Howard was punished for violating security at Fort Mandan. He received a sentence of 50 lashes for climbing over the stockade fence, which was not carried out.

There were no more corporal punishments and they worked together as a team under extraordinary circumstances. At the end of the expedition Whitehouse wrote —

> . . . the manly and soldier-like behavior; and enterprizing abilities; of both Captain Lewis and Captain Clark . . . and the humanity shown at all times by them, to those under their command, on this perilous and important Voyage of discovery.[424]

Robert Hunt, writing about crime and punishment on the expedition, raised an interesting question about the captains asking the men to decide matters by a vote. Is Patrick Gass the only sergeant in the history of the United States Army to have been elected by a vote of the men? [425]

Immense herds of pronghorns were now swimming across the Missouri heading west to their winter grounds in the Black Hills. Near the mouth of the Cannonball River—named for the round stones in its waters—they met two French trappers who complained about being robbed of their beaver traps by the Mandans. The trappers joined the boat flotilla, hoping to get their traps back.

On October 19th Clark counted 52 "gangues" of buffalo and 3 of elk as he walked on shore. He also saw the ruins of an old Mandan village with a "high strong watchtower" on top of a 90 foot high hill. The Arikara chief said they would see several old Mandan villages ahead near the Heart River, which the "Troubleson Seauex" had forced them to leave.

They had their first encounter with a grizzly bear, recorded by Lewis in his natural history notes for October 20th. It was the first of many encounters to come. Lewis wrote:

> Peter Crusat this day shot at a white bear he wounded him, but being alarmed at the formidable appearance of the bear he left his tomahawk and gun; but shortly after returned and found the bear had taken the opposit rout.—

Clark wrote, "I saw Several fresh tracks of that animal double the Sise of the largest track I ever saw." [426]

On the 21st, Clark saw a sacred stone boulder, the "Medicine Rock" of the Indians, marked with petroglyphs and a ceremonial oak tree on the open prairie. The chief told Clark that Mandan warriors suspended themselves from the tree, by piercing the skin of their necks and threading cords through the holes, "to make them brave." They saw a war party of 12 naked Sioux on their way to attack the Mandans.

On October 24th they met a grand chief of the Mandans camping on an island with a hunting party. Lewis went with an interpreter and the chief to his village, located a mile up the river. The next day, the British trader Hugh McCracken greeted them and on the following day they met René Jessaume, the long time resident trader, whom they hired as an interpreter. Jessaume was the trader who had tried to kill John Evans.

The expedition remained with the Mandan and Hidatsa for over five months. They left on April 7th, 1805. The two tribes lived in five villages. There were two Mandan villages along the Missouri near the Knife River, and three Hidatsa villages along the Knife. After the Mandans were decimated by small pox and attacks by the Sioux, they had moved from their villages on the Heart River, where they had lived for two hundred years, to join forces with the Hidatsas in the 1780's. The Hidatsa were also known as the Minatarees or the Gros Ventre of the Missouri. They were three related tribes who spoke different dialects. The Hidatsas roamed more widely than the Mandans, although both were agricultural people.

The five villages had a combined population of 3,000 to 4,000. (Washington D. C.'s population was 3,210 and St. Louis's, 795.) Clark in his "Estimate of Eastern Indians" reported the Mandans numbered between 1,250–1,500; the Awaxawis at the first village of the Hidatsa, 200-300; and the Minetarees of the upper two villages, 2,500. The report described the Mandans as "the most friendly, well disposed Indians inhabiting the Missouri. They are brave, humane, and hospitable."[427]

The villages were the trading center of the Northern Plains, with a network of trade stretching thousands of miles. They traded their corn, beans, squash and tobacco with the Cree and Assiniboin of Canada, who provided manufactured goods obtained from British and Canadian traders. The nomadic tribes of the plains, the Cheyenne, Crow, and Arapahoe, brought buffalo meat and hides to the market. The Cheyenne brought Spanish horses and mules and fancy leather clothing. The most prized trade goods were English trade guns and ammunition.[428]

North Dakota Lewis & Clark Interpretive Center, Fort Mandan; Knife River Indian Villages National Historic Site, Knife River and Hidatsa earth lodge at Sakakawea village site (Photos by Rick Blond) www.fortmandan.com and www.nps.gov/knri

On the 29th of October, they held a council under the sail awning and gave out medals to the village chiefs. The head chiefs received medals with a likeness of Jefferson and presents of coats, hats and flags. The lesser chiefs received medals with images of weaving, domestic animals, and a man sowing wheat. The ceremonies ended with the airgun demonstration. Its ability to fire without making a loud sound always astonished the Indians.

Big White, the chief of the first village, had been out hunting on the council day. The next day they met with him and gave him a medal. Big White, whose Mandan name was Sheheke, or White Coyote, was tall, fat, and white-skinned. They would come to know Big White well in the years after the expedition, because he went to Washington with them to meet the president and then lived in St. Louis while awaiting his return to his village.

The Mandan chief of the second village, Black Cat, was the main chief. They met with him the following day, and he gave them back the two traps stolen from the French trappers, a robe, and 12 bushels of corn. Black Cat spoke in favor of the peace negotiations they had promoted at the council, and said he believed the road was open and that peace would give him and his people great satisfaction, because—

> they Could hunt without fear, & ther womin Could work in the fields without looking everry moment for the Enemey, and put off their mockersons at night.[429]

He offered to send some brave men to accompany the Arikara chief back to his village and smoke the pipe with them.

Big White came the next day and said he would send men and "we will make a good peace." He promised, "if we eat you Shall eat, if we Starve. you must starve also." If they had accepted his offer to live at his village, he would have fed them as his guests, but since they were building their own living quarters, it was a very generous offer.[430]

On November 2nd, Clark found a woodland, three miles below Big White's town, and the men began cutting down cottonwood trees to build Fort Mandan.The Arikara chief departed for his village, accompanied by several Mandans and Minatarees. On November 3rd, they discharged the French hands. Biddle's notes say Clark told him that 9 engagés built a pirogue and returned to St. Louis, but 2 or 3 stayed with them until they left in the spring. Five engagés received pay in St. Louis in 1805.[431]

René Jessaume and his wife and children moved into the camp, so he could serve as their interpeter. The next day they were visited by Toussaint Charbonneau ("Shar-bone-oh"), a trader who lived with the Hidatsa and had two young Shoshone wives. They hired him to accompany them and bring one of his wives to serve as an interpreter when they left in the spring. They were relying on the Shoshone to supply the horses they needed to cross the Rocky Mountains. The Shoshone wives had been captured by Hidatsa warriors on raiding parties and sold to Charbonneau. One of them was Sacagawea, who was pregnant with their first child. She was 14–15 years old. Charbonneau brought his wives to live at the camp on November 11th.

Big White came to visit with his wife who carried about 100 pounds of meat on her back for them. They gave her a small axe as a present, which "much pleased" her. The swans were flying south and the weather was turning very cold. Big White visited with them, relating Mandan history. Clark observed that the Mandans spoke "a language peculial to themselves (verry much)." Clark wrote:

> The Mandans are at War with all who make war on them, at present with the Seauex only, and wish to be at peace with *all* nations. Seldom the agressors—[432]

They began to move into their rooms on the 13th. Black Cat brought a delegation of Assiniboin to meet them. Clark called them a "bad disposed set" who could raise about 1,000 warriors. It snowed all day. About 70 lodges of Assiniboin and Cree had come to trade and warned the locals they couldn't rely on the

Americans to supply guns and ammunition. John Evans had promised to do so, but it hadn't happened. On the 20th, Lewis and Clark moved into their rooms at the fort.

A quarrel involving an Indian woman took place. She had sex with Sergeant Ordway with the consent of her husband. But then her husband beat and stabbed her, so she came to stay with the interpreters's wives. Lewis told Ordway to give the husband gifts and Black Cat came and lectured the husband, who took his wife home. Ordway kept silent about it in his journal.

A new trouble came when seven traders arrived from the British Northwest Company. Their interpreter spread the rumor the Americans were going to join forces with the Sioux and attack the Minatarees. Two Minataree chiefs came to see for themselves. Clark reported they were much pleased by the treatment they received and the men had danced to amuse them. Ordway wrote that François Rivet danced on his head. The captains warned the Northwest trader François Larocque not to give out any British flags or medals.

Some Sioux and Pawnee attacked a Mandan hunting party. A young chief was killed, two men were wounded, and nine horses were stolen. Clark took 23 men and went to Big White's village to collect men from the villages to fight. The chiefs were impressed to find the Americans were speaking the truth and would fight for them, but said it was too cold and the snow was too deep for the horses to go after the Sioux. They would send out a war party in the spring. The captains replied they could only help while they were in residence, but when the Mandans and Hidatsas had arms and ammunition from American traders they wouldn't have to take any insults. Clark reported the chiefs "frequently thanked" him for coming to their aid, and "the whole village appeared thankful." They sent a message to the traders at the Arikara villages to tell them of the attack and discover what they knew about it. The father of the young man killed brought them a present of dried pumpkins.

Lewis went out with a hunting party. They killed eleven buffalo, but the wolves got most of them, leaving only five. The thermometer was at 12 degrees below zero and several of the men were bady frostbitten. By mid December the temperature was 21 degrees below zero and the snow was 10 inches deep. Two chiefs brought in loads of meat as gifts. They had been accepted as friends and hunting partners by the Indians.

The fort had many visitors, most bringing corn to trade. The Indian who wanted to murder his wife came with his two wives to apologize to Ordway. Lewis treated an Indian child for an abcess on his back and was paid in corn. Indian women and some Indian men dressed in women's clothing came to sell corn. The Mandan chiefs liked to stay at the fort and sleep overnight. A chief's wife made a kettle of boiled pumpkins, beans, corn and choke cherries as a treat.

On Christmas day they were awakened before dawn by the men firing three volleys. The men were given taffia, a rum drink. The three cannons were fired and the flag raised. Clark wrote:

> Some men went out to hunt & the Others to Danceing and Continued until 9 oClock P.M, when the frolic ended &tc.[433]

Etcetera (&tc) appears to have been used sometimes by Clark as a shorthand for unmentionables.

On the 27th of December, the Indians made an important discovery—they saw John Shields at work at his forge, and learned that "sundery articles of iron" could be made by the expedition's blacksmith. More Indians soon came to observe even though the temperature was 20 degrees below zero.

On New Years Day, 1805 they fired off two cannons, and sixteen dancers and musicians went over to Big White's village to give a dance exhibition. Clark went with them and wrote:

> I ordered my black Servent to Dance which amused the Croud verry much, and some what astonished them, that So large a man Should be active &c. &c.[444]

As Clark was about to leave, two chiefs came in with the news that 150 Gros Ventres ("Big Bellies") camped above the villages were demanding the return of a Gros Ventre girl who had been stolen by a Hidatsa chief. The girl had sought refuge at the fort. On January 3rd, her husband came for his wife, "who had been much abused, & come here for protection." The Gros Ventre lived in northern Montana and were called the Gros Ventre of the Prairies. They did not have "big bellies." It was a misinterpretation of sign language. They had lived on the southern branch of the Saskatchewan River, which was full of water falls, and they passed their hands down their body like water falling. The Hidatsas were known as the Gros Ventre of the Missouri.[445]

On January 4th they sent a man to the Medicine Dance, who had sex with four girls, in order to ensure the buffalo would return and come near so they could kill them.The medicine dance was held at Big White's village for three nights. The young men offered their wives, naked, except for a buffalo robe to old men. Very often the old men could scarcely walk, let alone perform. They were also offered to white men. An unknown member of the expedition volunteered his services.

A boy of about 13 years old came to the fort with his feet frozen. He had been out all night without fire, and only a small robe. It was 40 degrees below zero. Another Indian, barely clothed and also without fire, had stayed out all night without being frostbitten. Lewis was eventually forced to amputate the boy's toes. Clark wrote that the Indians were able to bear more cold than he thought it was possible to endure.

Several men now had venereal disease, caught from the Mandan women. One man had syphilis, or the "pox." By the end of their stay, Clark wrote on March 30th:

> All the party in high Spirits they pass but fiew nights without amuseing themselves dancing possessing perfect harmony and good understanding towards each other Generally healthy except venerial complains which is verry Common amongst the natives and the men Catch it from them.

The men with gonorrhea were treated with copaiba oil taken orally, or sugar of lead mixed with water injected into the urethra with a penile syringe. The men with syphilis were treated with mercury ointment, calomel, and the universal remedy, Dr. Rush's

"Thunderbolts." As an experienced army officer, Captain Lewis had laid in a good supply of medicines, with 15% of his total purchases for venereal disease.[446]

The next incident came about when Clark had York give Sacagawea, who was sick and eight months pregnant, some stewed fruit and tea. This made the interpreter René Jessaume's wife jealous and she left her husband and returned to the village.

On the 3rd of February, it was a fine day and the blacksmiths, John Shields and Alexander Willard, resumed work at the forge. Clark set out with a hunting party and was gone for ten days. Many natives came bringing corn in payment for iron work. They wanted iron blades for their battle axes, and pieces of sheet iron to be cut into arrow points and hide scrapers. For each 4 inch square piece of sheet iron, they traded 7–8 gallons of corn. The natives thought they had the better of the bargain.

Lewis took over Clark's journal entries, indicating that Clark's journal was the daily record of fort life. Lewis found out the wives of the interpreters were unlocking the fort gates and admitting visitors during the night, and he ordered the practice stopped. Then Thomas Howard, an old soldier who ought to have known better, was caught climbing over the stockade fence to enter the fort at night. He was court martialed, but mercy was recommended and his sentence of 50 lashes was forgiven. That was the last official disciplinary measure of expedition.

On February 11, 1805, Sacagawea gave birth to her baby, a "fine boy." Lewis wrote that it was her first birth and "her labour was tedious and her pain violent." René Jessaume advised Lewis that a small portion of rattlesnake rattles would hasten the birth, and ten minutes after she was given a portion in a drink of water, she gave birth. The baby was named Jean Baptiste Charbonneau, but he was called "Pompey" or "Pomp."

Clark and his men built two closed-in shelters to store the meat from the animals they killed. Wolves, magpies and ravens would devour any meat left in the open. After reaching the fort, he sent four men back with horses and sleighs on the 14th. They were attacked by 100 Sioux, who stole two horses. The next day, Lewis set out at sunrise with 24 men to apprehend the Sioux.

Most of the Indian men in the villages were away on hunting trips, but several accompanied Lewis. It was 16 degrees below zero. The robbers were so far ahead they couldn't be overtaken, but discarded moccasins indicated they were Sioux. Lewis returned on the 21st with 2,400 pounds of meat. The Sioux had burned one of Clark's "meat houses," but they hadn't discovered the other one. Clark learned they had seen him, but were afraid to attack his party.

Jessaume and his wife returned from the village, and several Hidatsa chiefs came to visit. Gravelines, the trader at the Arikara villages, arrived and said the Sioux were determined to kill "every white man they see" and were assembling several bands to attack the Mandans. Gravelines reported the Sioux chief Black Buffalo was on the side of the Americans and the Arikara had refused to give the Sioux anything to eat when they came to their villages after attacking the Americans. On March 1st, Clark wrote:

> a fine Day I am ingaged in Copying a map, men building perogues, makeing Ropes, Burning Coal, Hanging up meat & making battle axes for Corn [447]

On their departure, Lewis wrote to Jefferson, saying that although he had promised to send his vouchers and reports back with the soldiers from Kaskaskia in September, it would have risked the safety of the expedition to proceed without their additional manpower. He was now sending the papers with the keelboat, manned by ten able-bodied men. He had little doubt they would be fired upon by the Sioux, but they had pledged "they will not yeald while there is a man of them living."

He was sending 18 boxes and 8 large trunks containing skins, skeletons, painted Indian robes and other Indian goods; 61 specimens of salts, minerals, earths; and 60 plants. They were labeled with descriptions and the date and place where they had been acquired. Clark's "Estimate of Eastern Indians" and maps showing the expedition's campsite locations were going to the Secretary of War. There were three cages of living animals—four magpies in one cage; a "live burrowing squirrel" in another;

and a prairie hen in the third. Lewis also enclosed a personal letter to Jefferson, but its contents are unknown. The materials either went to the American Philosophical Society, Charles Willson Peale's Museum, or were retained by the president.

Only one magpie and the prairie dog survived the journey. When they arrived in Washington, Jefferson was in residence at Monticello. His French steward wrote to him that the staff was airing out the skins in the sun and—

> The magpie and the kind of squirrel are very well; they are in the room where Monsieur receives his callers.[449]

The keelboat arrived in St. Louis on May 20th, with a flotilla of boats carrying 45 chiefs of the Arikara, Yankton Sioux, Ponca, Omaha, Otoe and Missouria tribes, who were bound for Washington to meet the president. Pierre Chouteau advised that they should remain in the St. Louis area until the fall or winter, when it was a better season for them to travel.

Wilkinson was now in St. Louis, serving as the first Governor of Louisiana Territory. He was worried about the expense of sending Indian delegations to Washington. Pierre Chouteau was escorting a delegation of Osage and Republican Pawnee chiefs to Washington and Captain Amos Stoddard was escorting 26 chiefs from 11 nations, including the Indians of the Missouri.

On October 8th, 1805 Wilkinson wrote to Secretary of War Henry Dearborn reporting the Otoe and Arikara chiefs were dangerously ill and wanted to return to their homes, saying—

> Their safe return is deemed an Object of great Magnitude, not only to the safe return of Capt. Lewis, but to our future negotiations among the distant nations of the Missouri.

He was sending them back with a escort of 25 soldiers led by his son Lieutenant James Biddle Wilkinson. If the Otoe would permit it, they were going to build a fort at the mouth of the Platte. However, after one soldier was killed by Kansas Indians they returned to St. Louis. By late December, the Arikara chief had recovered his health sufficiently to go to Washington.

Wilkinson wrote to Jefferson that the chief was a *learned* savage because he spoke 11 different languages and was a "Master of the *Language* of Arms, Hands & Fingers." He was a great traveler, warrior and geographer, who attended the annual grand council of 10 to 15 Indian nations. The chief was to be returned to his own people, without regard for the other Indian delegates, with a military escort loaded with presents. He had gone on the trip "contrary to the will of his whole people."

Three months later, Henry Dearborn informed Wilkinson that on April 7, 1806 the "very respectable and amiable Ricara Chief" had died in Washington. Joseph Gravelines, who had accompanied the chief as his interpreter, returned to the villages with the news that Ankedoucharo had died. No amount of presents or expressions of sympathy could appease his people, and the chief's death was the cause of the attack on the expedition led by Pryor in 1807, and the subsequent hostilities which continued for many years to come.[450]

The expedition got underway on April 7, 1805. They traveled in the two large red and white pirogues and six small canoes carved out of cottonwood tree trunks. Lewis wrote:

> This little fleet altho' not quite so rispectable as those of Columbus or Capt. Cook were still viewed by us with as much pleasure as those deservedly famed adventurers ever beheld theirs . . . the picture which now presented itself to me was a most pleasing one. entertaining (now) as I do, the most confident hope of succeading in a voyage which has formed a da[r]ling project of mine for the last ten years (of my life), I could but esteem this moment of my (our) departure as among the most happy of my life. The party are in excellent health and sperits, zealously attached to the enterprise, and anxious to proceed; not a whisper of murmur or discontent to be heard among them, but all act in unison and with the most perfect harmony.

This was the start of what is known as the "permanent party" of the U. S. Army Corps of Volunteers for Northwestern Discovery—two captains, 26 enlisted men, the hunter and interpreter George Drouillard, the slave York, the interpreter and interpretress, Toussaint Charbonneau and Sacagawea, and their two month old baby, Pompey. Thirty three people and one dog would cross the Rocky Mountains, reach the Pacific Ocean, and return to the villages by August 14th, 1806, with no loss of life and their mission accomplished.

But from this time forward, they didn't know what to expect, and as Lewis wrote—

> we were now about to penetrate a country of at least two thousand miles in width, on which the foot of civilized man has never trodden; the good or evil it has in store for us was for experiment yet to determine, and these little vessells contained every article by which we were to expect to subsist or defend ourselves.[451]

Without being aware of it, they had the best defense of all—a woman and a baby traveling with them. They were not considered a war party because of them. They saw no Indians until meeting with Sacagawea's Shoshone tribe near the headwaters of the Missouri. Assiniboin bands watched them and made no contact. The oral history of the Assiniboin on You Tube relates they watched the corps travel up river with strong winds against them. The men were tired, and the Assiniboin could have easily attacked them but they didn't because there was a woman and baby with them.[452] The captains moved from army tents into larger quarters to share with the Charbonneau family and George Drouillard. They lived in a tent of dressed skins, made in the Indian style—in other words, a tipi.

They were beginning to see the the immense tracks of the "white bear" as they called the grizzly bear. They were anxious to meet one. The Indians, who hunted them in parties of six to ten hunters, had only bows and arrows and "indifferent" trade guns.

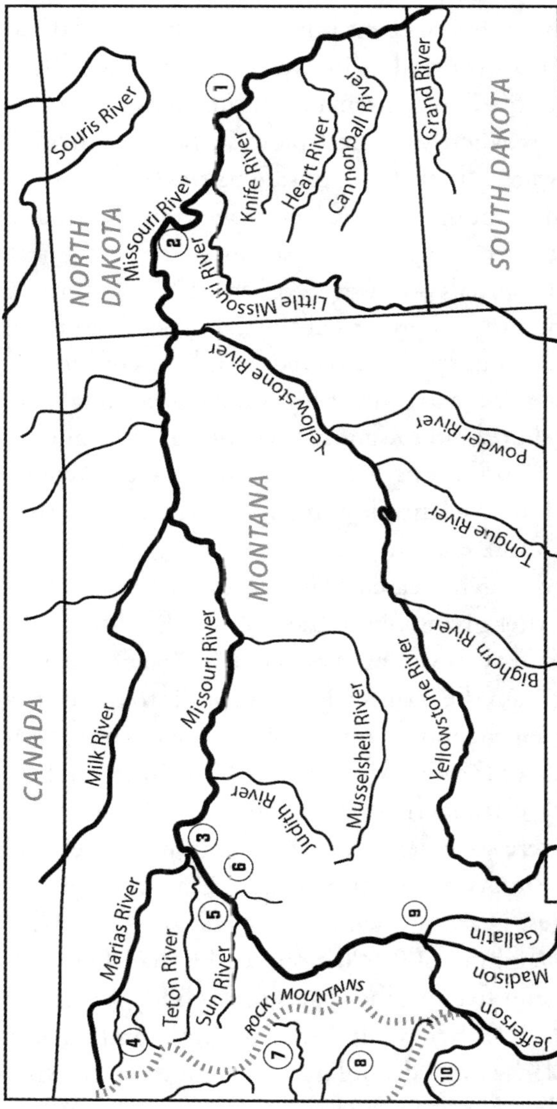

1. Fort Mandan, 2. Reunion Bay, 3. Decision Point, 4. Blackfeet fight, 5. Great Falls, 6. Portage Route, 7. Blackfoot River, 8. Clark's Fork River, 9. Three Forks of the Missouri, 10. Big Hole River.

On April 14th, they named a creek for Charbonneau, who had camped on it for several weeks on a previous trip. As far as they knew, it was the highest point on the Missouri ever reached by a white man. They were seeing Assiniboin camps all along the river. The Assiniboin were excellent hunters, and provided dried meat and grease to British traders in exchange for liquor. Their camps could be identified by the remains of rum kegs.

There were days when the wind blew so hard, with high waves and rough waters, that it was unsafe to travel. The swans and geese were flying south, and the beaver were abundant and with the best fur they had ever seen. Seaman doesn't often rate a mention in their journals—however, on the 24th of April he stayed out all night and was presumed "lost," but rejoined them in the morning. The country was covered with herds of buffalo, elk and antelope, who were so gentle they could pass among them as they walked on shore, and the animals would follow them.

On the 26th of April they reached the mouth of the Yellowstone, and stopped to take latitude and longitude readings. They thought it was a "most eligible site" for a trading post. Lewis wrote they all were "much pleased at having arrived at this long wished for spot." After giving out a dram of whiskey, the fiddle was brought out and the evening was spent in "much hilarity, singing & dancing." On the return trip, they split into five different groups as they left the mountains, with everyone reuniting at "Reunion Bay," east of the Yellowstone near New Town, N. D.

It was a "verry extroadernaley Climate" wrote Clark. The winds and waves were so violent the canoes couldn't travel, and then it snowed, one inch of snow on May 2nd. Clark and Drouillard killed the largest bear they had seen on May 5th, "a turrible looking animal," weighing 500-600 pounds. He had five balls through his lungs and five in other parts of his body, before he died. He was 8 feet 7 ½ inches tall. Seaman outran and killed a young antelope. Clark climbed a hill near the mouth of the Milk River and saw buffalo herds on the plains for sixty miles. When they walked among them they had to throw sticks and stones to clear a path. There were a great number of beaver on the river, and they passed four acres of trees cut down by beavers.

The country was covered with prairie turnip, or white apple, a staple of Indian food. Sacagawea showed them how to harvest the underground bulbs. Their main season was in July and August. Lewis loved a treat prepared by Charbonneau, called *boudin blanc*, or white sausage made from buffalo gut. He called it "white pudding," which "all esteem one of the greatest delacies." The men were suffering from boils and abscesses, caused by malnution because they lacked vitamin C. [453]

An Assiniboin dog was seen on May 10th and Lewis wrote:

> . . . we still believe ourselves in the country usually hunted by the Assiniboins, and as they are a vicious illy disposed nation we think it best to be on our guard, accordingly we inspected the arms and accoutrements of the party and found them all in good order.[454]

They were unaware they were often being watched by Assiniboin hunters. [455]

William Bratton, walking on shore because of a sore hand, shot a grizzly who pursued him for half a mile. Lewis wrote:

> these bear being so hard to die reather intimidates us all; I must confess I do not like the gentlemen and had reather fight two Indians than one bear.[456]

May 14th was their unlucky day, but it could have been much worse. It was the day of the boat overturn when Lewis's journal was lost. Sacagawea and Cruzatte saved the day when she rescued the articles that floated by and Cruzatte managed to bring Charbonneau to his senses and make him steer the boat, while they bailed out water and rowed despite the boat being nearly filled with water. Both captains were on shore, contrary to their usual strict policy of one captain remaining in the boat.

On that same day, six hunters shot a grizzly who pursued them vigorously. Two of the hunters put their guns down and jumped off a 20 foot cliff into the river, and the bear came into the water after them. Finally a shot through his head killed him.

A bullet in the brain was the only certain way to kill a grizzly. They counted eight balls when they butchered him.

On the night of May 28th they had another close call. A large buffalo bull swam across the river, and became alarmed when he hit the white pirogue anchored at the water's edge. He climbed over it and bolted through the camp at top speed, coming to within 18 inches of the heads of the men sleeping near the fires. He was headed for the captains' tipi when Seaman's barking scared him off and he changed course. They discovered the next morning that he had bent York's rifle, which had been left in the pirogue.

On May 29th, Clark named the Judith River for the girl he would marry, Julia Hancock of Fincastle, Virginia. Julia at that time was 13 years old. They would get married a year after his return in January, 1808, when she was 17.

They were using the tow ropes. The river was too deep to use poles, and the current too strong to row against. The river banks were continuing to fall in. The men were in the water up to their armpits pulling the boats by ropes. About one quarter of their day was spent in the water. The banks were slippery and muddy, and it was rainy and cold. The men couldn't wear their mocassins because of the heavy mud and sometimes had to walk on sharp fragments of rock. Lewis wrote:

> in short their labour is incredibly painfull and great, yet these faithful fellows bear it without a murmur.[457]

They were traveling through the White Cliffs, 200–300 feet high, soft sand stone cliffs worn into a "thousand grotesque figures" which they imagined to be the ancient ruins of elegant buildings. It was at the end of the month of May.

On June 3rd they came to a place they named "Decision Point." A look at the map will explain their uncertainty. The Marias ("Mah-rye-a's") River continued northwest and looked like the waters of the Missouri. The Missouri itself turned south and its waters were clearer. If they were mistaken, choosing the wrong river would cost them a season of travel. Two canoes and land

"Seens of Visionary enchantment"— the White Cliffs of Montana. (photo by Rick Blond)

parties were sent out to investigate. The captains strolled to the top of a hill and saw a most "inchanting view." There were snow covered mountains to the south and innumerable buffalo herds on the vast plains. The plains were covered with prickly pear, chokecherry bushes, and wild roses in full bloom.

The captains were alone in believing that the river heading south was the Missouri. The men were convinced the Marias River was the real Missouri because it looked the same. The captains believed the clearer waters of the south bend meant it originated in the mountains. The problem was their Indian informants hadn't told them they would see another river at this location, or that the Missouri would bend. Lewis wrote:

> . . . thus have our cogitating faculties been busily employed today. Those who have remained at camp today have been busily engaged in dressing skins for cloathing, notwithstanding that many of them have their feet so mangled and bruised with stones and rough ground over which they passed barefoot, that they can scarcely walk or stand; at least it is with great pain they do either. . . . they still remain perfectly cheerful.[458]

"Perfectly cheerful" indicates they were being medicated with some of the half pound of Turkish opium Lewis carried in his medicine chest. He also had four ounces of laudanum, which is a tincture of opium, or 1% morphine.[459] The journals mention using laudanum, but are almost silent on the opium. Opium is still the basis of pain medications today, and laudanum was made from 10% opium powder mixed with alcohol.

The exploring parties returned with inconclusive reports and on the next day, June 4th, the captains decided to see for themselves. Lewis chose the north fork, and Clark the south. Lewis took six men and Clark five. Lewis wrote it was the first time he ever carried a backpack, but he doubted it would be the last. The prickly pear thorns were "extreemly troublesome" and could pierce a mocassin with ease. They saw a barking squirrel (prairie dog) town that covered seven miles. By June 6th, Lewis was convinced the river was going too far north and they turned back. It rained day and night, it was cold, and they had no shelter.

On June 7th, Lewis nearly lost his life. He was walking on slippery clay on top of a bluff and nearly fell off the 90 foot cliff, but he used his espontoon to dig into the wet ground and save himself. An espontoon is a walking stick with a sharp spear point, a weapon. Scarcely had he recovered when the man following him cried out, "god god Capt. what shall I do." Windsor had slipped and was hanging over the cliff's edge. Disguising his feelings, Lewis spoke very calmly and advised him to use his knife to dig a hole in the cliff and place his foot in the hole to climb up to safety, and then crawl on his hands and knees over the rest of the ground. The men following them took to the river, wading breast deep.

Everyone in Lewis's party still believed the river was the Missouri, except for Lewis. He named it Maria's River, in honor of his cousin, Maria Wood. The river's edge was "one continued garden of roses." They left the red pirogue at Decision Point. Its crew of seven men were better employed in helping the others. The pirogue was tied to trees on a small island near the Marias.

Lewis's branding iron is 5.5 x 4 x 1.5 inches. Ir is on display at the Oregon Historical Society. It was found near the Hood River in the 1890's. (www.ohs.org and www.lewis-clark.org)

Lewis had a branding iron and branded trees near the boat. Blacksmith tools, extra ammunition, and other heavy goods were concealed in cache pits. A cache pit is a hole in the gound, wider at the bottom, with its opening carefully cut out and put back in place to look natural. Pierre Cruzatte knew how to build them because traders used caches to hide goods from robbers.

Cruzatte had accompanied Lewis on his tour of the Marias, and the experienced river navigator remained convinced it was the Missouri. The men trusted him, and not the captains. Lewis and Clark agreed between themselves that a new exploring party would continue until it reached either the Great Falls of the Missouri or the Snowy Mountains, and the boats would follow.

On June 11th, Lewis, who had been sick with dysentery for two days, set off with a party of four men. After suffering from violent stomach pains and fever, he improvised a "very bitter and astringent" remedy of chokecherry twigs boiled in water which cured him. He may have had a parasitic roundworm infection, *Trichinella,* caused by eating infected grizzly meat.[460]

The next day they saw the Rocky Mountains and realized there were several ranges of snow-covered mountains. Lewis called it an "august spectacle" and "formidable." He climbed a hill on June 13th and saw fifty to sixty miles of buffalo, "infinitely more" than he had ever seen before in one view, and on the same day he heard the tremendous roar of the Great Falls of the Missouri. When they reached it, the falls were the "grandest sight he had ever beheld" and wished he had the skill of a great artist to record it. He made a sketch, but it hasn't been found.

Clark was following behind Lewis with the boats. On the first day after leaving camp, three of the six canoes were in great danger. Lewis wrote he was going to take the land party because Clark was the "best waterman." Sacagawea was very ill, and Clark bled her the day before they left camp. She and the baby rested under the awning of the white pirogue as they traveled.

On June 14th, Lewis very nearly lost his life again. He was alone, exploring the area around the Medicine (Sun) River, when he encountered a herd of at least a thousand buffalo. He shot one and before he had a chance to reload his gun, a bear had crept up to about 20 steps from him. The bear chased him open mouthed and at full speed for about 80 yards until Lewis thought to get into the water—forcing the bear to swim—and use his espontoon to defend himself. When he turned around with his espontoon raised, the bear retreated.

Clark on the river was encountering a great number of dangerous places. The current was rapid and—

> . . . the fatigue which we have to encounter is incretiatable the men in the water from morning until night hauling the Cord & boats walking on Sharp rocks and on round Slippery Stones which alternately cut their feet & throw them down, not with Standing all this difficuelty they go with great chearfulness, added to those dificuelties the rattle Snakes innumerable & require great caution to prevent being bitten.[461]

Meanwhile, Sacagawea, was "sick and low spirited." He gave her a poultice of Peruvian bark and laudanum to apply to "her region." She was suffering from pelvic inflammatory disease, and had no doubt acquired a venereal disease from her husband.[462]

When Lewis rejoined the boats on June 16th, he found the Indian woman "extreemely ill." He was concerned for her, with her child in her arms, and for the help they needed from her in obtaining horses from the Shoshone. He decided to camp near a sulphur springs on the river, hoping its mineral waters would cure her. This camp became known as the Lower Portage Camp. With Lewis's prescription of opium and barks, and drinking sulphur water, Sacagawea recovered and was walking about and fishing by June 20th.

The Great Falls consist of a series of five waterfalls, ranging from the Great Falls height of 87 feet to 6 feet, extending for almost 20 miles. The boats would have to be portaged—that is brought over land—until they reached the place where the waterfalls ended. Clark mapped and staked out an 18 ¹/₄ mile route that crossed two creeks and went over hills covered with prickly pear cactus to reach the new camp.

The white pirogue was left at the Lower Portage Camp with another cache and the swivel gun hidden behind rocks. A large cottonwood tree supplied the wood to make eight truck wheels. The canoes on wheels were loaded with goods and dragged to the Upper Portage Camp at White Bear Islands. Sometimes they were able to use the sails because the winds were so strong—"sailing on dry land" Lewis called it. It was exhausting labor; the men would collapse and be "asleep in an instant" from fatigue. The ground was covered with sharp points of hard mud, from the hooves of the buffalo as well as the prickly pear cactus. The portage took 12 days, from June 21st to July 2nd.

Clark, York, and the Charbonneaus left the camp on June 29th to walk to the Upper Camp. Clark needed to revisit the falls while en route, because some of his notes had been lost. They were traveling along the river's edge towards the Great Falls when a tremendous storm came up and Clark and the Charbonneaus took shelter under a rock ledge in a deep ravine. York was out hunting. There was a huge torrent of rain and large hail and suddenly the ravine was filling up with water, threatening to drown everyone. Charbonneau pulled Sacagawea, carrying the baby, from above, and with Clark pushing from below, they managed to reach safety. They lost a few items, including Pompey's baby carrier and his clothing. Clark lost his umbrella and his large compass. The next day, the men found the compass in the mud and discovered that the ravine was now filled with huge rocks. Lewis wrote about their escape—

> one moment longer & it would have swept them into the river just above the great cataract of 87 feet where they must have inevitably perished.[463]

They returned to camp to get some dry clothing, and found the men had returned also, leaving the baggage on the plains. Lewis wrote:

> the men who were nearly all naked and [no] covering on the head were sorely mawled with the hail which so large and driven with such force by the wind that it nocked many of them down and one particularly as many as three times most of them were bleeding freely and complained of being much bruised. . . . Capt. C. gave the party a dram to console them in some measure for their general defeat.[464]

Grizzly bears very nearly caught Willard and Colter at the Upper Camp. Seaman barked continually, patrolling the grounds and sounding the alarm during the night. The men traveled in pairs around the island and slept with their rifles. Fortunately, the bears did not climb trees.

Lewis was still very hopeful about his iron boat, *The Experiment*, whose iron parts had been crafted at Harpers Ferry and carried across the country. He was proud of the concept—a lightweight, portable, metal boat frame that could be assembled on the spot and covered with animal skins. They began putting the boat together at Upper Portage Camp. It was 36 feet long, 4½ feet wide, and 26 inches deep.[465] *The Experiment* failed because he hadn't brought along a supply of pitch, or pine tar, to seal the skins, and they couldn't find any pine trees to supply the sticky resin. The boat's frame was reinforced with wood and bark, and covered with elk skins. When they tried it out in the water, it sank. They blamed it on making holes that were too large in sewing the skins together, and having makeshift glue.

It was a blessing in disguise. When Lewis and Jefferson dreamed up the idea of an iron boat, they expected the Rocky Mountains to be like the Allegheny Mountains. It was called

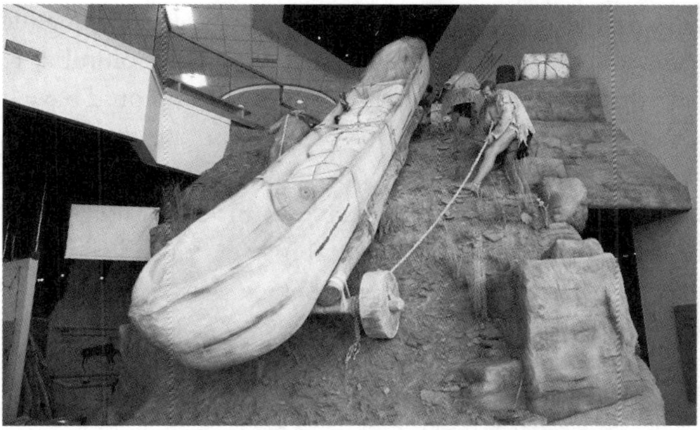

Great Falls, Montana: Exhibit at Lewis and Clark Interpretive Center, Prickly Pear Cactus, Iron Boat replica. (Photos by Rick Blond)

"symmetrical geography." They imagined the Rockies would be one range of not very tall mountains with the Columbia River on the other side flowing down to the ocean. It would have been impossible to portage a 36 foot boat across the Rockies to the Clearwater River in Idaho where, three months later, they began traveling on water again. The boat was intended for use by the 10-12 men authorized by the initial instructions. It didn't begin to answer the needs of the 33 member expedition.

Lewis and Clark decided they wouldn't send back a return party from the headwaters, as they could use the manpower, and it would have demoralized the men to have any of them sent back. They were a team. Lewis wrote on July 4th, 1805, while they were still dealing with the iron boat:

> . . . we all beleive that we are now about to enter on the most perilous and difficult part of our voyage, yet I see no one re-pining; all appear ready to met those difficulties which wait us with resolution and becoming fortitude.[466]

He described the mountains they were about to cross as the *shining mountains* for their "glittering appearance" with the sun shining on the snow. July 4th was also the end of their alcohol. They gave the men the last of the "sperits" and the men danced to the fiddle until heavy rain ended the dancing and they continued with songs and jokes.

By July 8th the buffalo were on the move. Lewis observed that wherever they saw buffalo, if they were undisturbed, they were moving in the same direction. The great migration south was beginning. They buried the iron frame boat and the truck wheels in a cache at Upper Portage Camp on July 10th. Clark had taken a crew of men ahead to find suitable trees to make two more canoes, replacing the white pirogue. The new canoes were 25 feet and 33 feet in length and about 3 feet wide. On the morning of July 15th, they set out for the headwaters region of the Missouri with eight canoes, all heavily laden. Lewis wrote:

> we find it extreemly difficult to keep the baggage of many of our men within reasonable bounds; they will be adding bulky articles of but little use or value to them.[467]

The canoes weren't big enough to carry all the men and their baggage—the captains and the men who weren't navigating the boats walked along the shore. The prickly pear and sunflowers were in full bloom, and the mosquitoes and gnats were as troublesome as always. Lewis went ahead with several men to discover where the river entered the mountains. Unfortunately, he forgot to bring along his mosquito netting, and wrote that he—

> . . . of course suffered considerably, and promised in my wrath that I never will be guilty of a similar peice of negligence while on this voyage.

Lewis described their "musquetoe biers" as being made of "*duck or gauze like a trunk—to get under.*" He wrote that without their "natural rest," it would have been impossible for the men to endure the fatigue of their daily life.[468]

They were now in Piegan Blackfeet territory. The Blackfeet were a confederacy of three Blackfeet nations, extending across the border of the United States and Canada east of the mountains. The Blackfeet obtained guns from British traders on the Saskatchewan River in Canada and were a nomadic, warlike people like the Sioux. They were enemies of the Shoshone and Flathead (or Salish) Indians, who had been forced to retreat into the mountains because they lacked guns and ammunition.

Lewis and his men had rejoined the boats, when on the morning of July 18th, Clark, York, J. Fields, and Joseph Potts set off to find the Shoshone, taking an Indian road over the mountains. Meanwhile the boats were going through a "dark and gloomy" place which Lewis named the "gates of the rocky mountains"—dark cliffs, about 1200 feet high, enclosing both sides of the river for 5 ¾ miles. Clark and his party walked 30 miles over two mountains on an old Indian road. Clark wrote:

> my feet is verry much brused & cut walking over the flint, & constantly Stuck full Prickley pear thorns, I puled out 17 by the light of the fire to night... Musquotrs verry troublesom.[469]

On July 20th, the plains were set on fire by the Shoshone to warn others of their approach. They later learned that the Indians, hearing gun shots from Clark's party, had fled into the mountains thinking they were a Blackfeet war party. Clark left signs along the road to indicate they were *white men.*

By July 22nd, Sacagawea was recognizing the country and assuring them they would soon reach the Three Forks of the Missouri headwaters. It was 80 degrees, the warmest day they had experienced that summer, except for one. They met up with Clark's land party, and even though his feet were terribly bruised and sore, Clark insisted on continuing despite Lewis's offer to switch places. Clark, the two Field brothers, Frazier, and Charbonneau set off to find the Shoshone, or Snake Indians, as they called them. Lewis ordered the canoes to hoist their small flags to announce to the Indians they were not their enemies. He did name his own enemies, however—"Musquitoes, eye knats and prickley pears." Lewis was also assisting the men in their labor, and said he had "learned to *push a tolerable good pole.*"[470]

Clark arrived first at the Three Forks of the Missouri River and left a note on a pole telling Lewis they were taking the north fork (the southwest river or Jefferson). They went 25 miles up the river. Clark's feet gave out, and Charbonneau hurt his ankle. When they limped back to Three Forks, Charbonneau—who could not swim—was swept away by the strong current, and had to be rescued by Clark as they were crossing the river to reach the middle fork (the Madison). They explored the middle fork for eight miles, and then returned to the junction, where they found Lewis and the others waiting for them. Clark was sick with a high fever, body aches, and infected feet. They decided to rest for a couple of days. The next day, on July 28th, they named the three forks the Jefferson, Madison and Gallatin, in honor of President Thomas Jefferson, Secretary of State James Madison, and Secretary of Treasury Albert Gallatin. Lewis called them all "noble

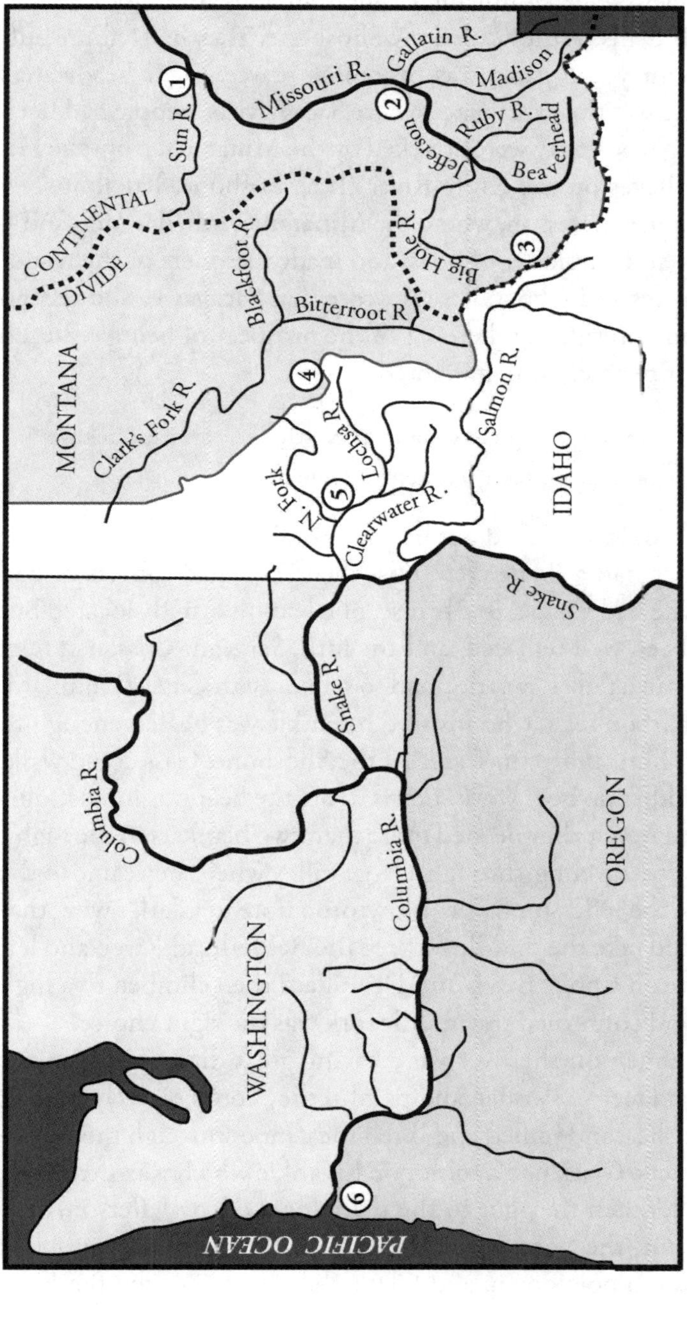

Finding a navigable water route to the Pacific Ocean: 1. Great Falls *Missouri River* 2. Three Forks *Jefferson & Beaverhead Rivers* 3. Lemhi Pass 4. Lolo Pass *Lochsa River* 5. Canoe Camp *Clearwater, Snake & Columbia Rivers* 6. Fort Clatsop.

streams" and recommended a fort be established there.

They were becoming increasingly anxious about locating the Snake Indians, as they called the Shoshone. They had left the buffalo country and game was becoming scarce. Their headwaters campsite was the exact place where Sacagawea's people had been camping when they were attacked by the Minataree from the Hidatsa villages on the Knife River. The Shoshone fled about 4½ miles up the Jefferson, where the Minatarees killed 4 men and 4 women and a number of boys, and made prisoners of the rest of the females and 4 boys. Lewis wrote that Sacagawea showed no emotion at returning there, or of the prospect of being reunited with her people. He commented:

> . . . if she had enough to eat and a few trinkets to wear I beleive she would be perfectly content anywhere.[471]

But he would soon find out he was mistaken.

They had a "lame crew" just then. Two had tumors or bad boils, one had a bad stone bruise, one had his arm dislocated but it had been well replaced, and the fifth, Sergeant Gass, had hurt his back in a fall. Lewis decided to set out with Gass, Drouillard, and Charbonneau, who insisted his ankle was healed enough to go with him, and go in search of the Shoshone. Gass could walk, but couldn't do boat work. Lewis wrote the heat was intense during the day, but they needed more than two blankets in the night. They were traveling through a large valley when they came to the forks of the Jefferson River. He wrote a note to Clark saying that he should take the middle fork, or the Beaverhead River, and left the note on a pole. Lewis and Drouillard then climbed to a high point and confirmed the middle fork was the right choice.

The men on the river were having great difficulty. Many of them had feet so swollen and painful they could scarcely walk as they pushed and pulled and lifted the canoes through the turbulent waters. Clark had a tumor on his ankle which was extremely painful. When they got to the three forks of the Jefferson, they didn't find the note Lewis had left on the pole. A beaver had chewed the pole down, and Clark chose to go up the Big Hole

River instead of the middle fork, the Beaverhead.

They had a boat accident on the Big Hole and the baggage got wet. Whitehouse almost lost his life when he was run over by a canoe. On August 6th they returned to the forks, where they reunited with Lewis and dried out their goods. Lewis had taken the precaution to secure their powder in small lead kegs sealed with cork and wax. The lead containers were melted down to make bullets and the powder was used to discharge the bullets from their guns. It was an ingenious way to keep their powder dry.

They called three forks the *Wisdom* River (Big Hole River), the *Philanthropy* River (Ruby River), and the middle fork, they continued to call the *Jefferson* River. They were right that it was leading to the ultimate source of the Missouri River. Lewis wrote these were the "cardinal virtues, which have so eiminently marked that deservedly selibrated character through life." [472] The middle fork, the Jefferson, is now called the Beaverhead River.

It happened their two objectives—to obtain horses from the Shoshone and to find the source of the Missouri River—both were fulfilled by continuing to ascend the middle fork. It was not the best route over the mountains. The Blackfoot River, which Lewis explored on his return journey, was the most direct connection to Lolo Pass and the Clearwater and Snake Rivers leading to the Pacific Ocean.

A mountain pass is a low route through the mountains. They would use Lemhi Pass (7,373 ft elevation) after obtaining horses from the Shoshone, but they had to then go back north through the Beaverhead Mountains (which are part of the Bitterroot Mountain Range) and follow the Bitterroot River to reach Lolo Pass (5,233 ft elevation) in the Bitterroot Mountains.

At Lolo Pass, they would cross over the Continental Divide into Idaho. A Continental Divide is a high point on a continent from which waters flow into the ocean on one side and flow into another ocean (if possible) on the other side. They had been going against the current of the Missouri River since leaving St. Louis in May, 1804. Contrary to popular belief, it is the Missouri River which enters the gulf waters at New Orleans, not the Mississippi. The Mississippi is a tributary of the Missouri River.

Once they crossed over the divide at the Bitterroots, they would be going with river currents all the way to the ocean. A look at the map will clarify their route decisions. Naturally, they hoped to use the Salmon River after going over Lemhi Pass, but the Indians told them it was impossible to navigate. They had to travel back north to reach Lolo Pass and find the rivers leading to the Columbia River and the Pacific.

Beaverhead Rock, a National Historic Landmark, resembles a beaver swimming through water. (Photo by Rick Blond)

But this was all in the future. On August 6th, at the forks of the Jefferson (Beaverhead) River, George Shannon got separated while hunting on the Big Hole, but turned up three days later. Since goods had been lost in the boat accident on the Big Hole, and others used up, they decided to leave one small canoe concealed in the cottonwood trees for their return trip. They needed hunters on the ground, not men on the river.

On August 8th, Sacagawea recognized Beaverhead Rock as a landmark, and said the summer camp of the Shoshone was on the other side of the mountains through Lemhi Pass. The next day, Lewis slung his pack on his shoulders and set out with

Drouillard, Shields and McNeal to find the Shoshones, vowing he would find them—or some other source for horses—if it took a month. It was "now all important with us to meet with those people as soon as possible." Clark wrote regretfully that he should have taken this trip, except for the "rageing fury of the tumer on my anckle musle."[473]

Lewis followed an Indian road and when he reached the forks of the Beaverhead River he left notes for Clark to camp there. Lewis's men explored both forks of the Beaverhead, the Red Rock River and Horsetail Creek. The 70 mile Red Rock River, bending towards the east, is the ultimate source of the Missouri River, at Hell Roaring Creek and Brower's Spring (9,100 ft elevation). Lewis believed Horsetail Creek, going west towards Lemhi Pass, was the source of the Missouri, instead of the Red Rock. The Red Rock River and Horsetail Creek are now part of the Clark Canyon Reservoir. Their confluence became the site of their later "Camp Fortunate."

On August 11th, as they were following Horsetail Creek towards Lemhi Pass, Lewis saw an Indian on horseback about two miles away coming towards them. He used his spyglass to discover he was dressed in a way unlike he had seen before, riding an "eligant" horse, and concluded he was Shoshone. Lewis continued to walk towards him, and at one mile tried to make contact.

The Indian, however, became alarmed at the sight of Drouillard and Shields approaching Lewis on his path. Lewis tried to signal them to halt. Drouillard saw it and stopped, but Shields didn't see the signal. Lewis was about 150 paces away, saying in a loud voice, "*tab-ba-bone,*" which he thought meant "white man." Instead it meant "stranger." He was holding trinkets, and rolling up the sleeve of his shirt to show he was a white man, when the Indian suddenly turned around, whipping his horse, and left.

Lewis decided it would be best not to follow him, as they were probably being observed from the hills. So they made a fire and had breakfast. He tied a bundle of assorted gifts to a pole and left it there. When they set out, McNeal carried a pole with a small American flag attached to it.

On August 12th, Lewis wrote they had reached the end of

Horsetail Creek, which he called—

> the most distant fountains of the waters of the mighty Missouri in surch of which we have spent so many toilsom days and wristless nights. thus far I had accomplished one of these great objects on which my mind had been unalterably fixed for many years, judge then of the pleasure I felt in allying my thirst with this pure and ice cold water which issues from the base of a low mountain. . . . two miles below McNeal had exultingly stood with a foot on each side of this little rivelut and thanked his god that he had lived to bestride the mighty & heretofore deemed endless Missouri.

They were soon at Lemhi Pass, from which he "discovered immence ranges of high mountains still to the West of us with their tops partially covered with snow." They began their descent on the western side of the mountain, and when they reached the Lemhi River, Lewis tasted it and declared it was the water of the great Columbia River. [474]

The Shoshone summer camp was located on the Lemhi River near today's Tendoy, Idaho. On August 13th they were following the Indian road when they saw two women, a man and some dogs. Lewis walked alone towards them, holding the flag pole. The Indians disappeared around a hill. Lewis, joined again by his companions, proceeded along the dusty road with twists and turns for another mile, where they suddenly encountered three women. The young woman immediately fled, but the old woman and a girl of about 12 sat down on the ground awaiting their fate and expecting to be killed.

Lewis repeated "ta-ba-bone" and rolled up his sleeve to show his white skin. His face and hands were as dark as their own. He gave them some trinkets—mocassin awls (to punch holes in leather), pewter looking glasses (mirrors), and vermillion face paint. They called to the girl who had fled and she came back to receive her presents. He then painted their cheeks with vermillion, which he had learned from Sacagawea was a sign of peace.

By means of signs, he indicated they wanted to go with them to their camp. Two miles down the road, about 60 mounted warriors on "excellent" horses, came at them at nearly full speed. Lewis left his gun with his men, and holding the flag he went alone to speak with the chief. The chief spoke with the women first, and then gave Lewis his first "national hug." He put his left arm around Lewis, hugging his back, and his left cheek against Lewis's cheek, saying "I am much pleased. I am much rejoiced." in the Shoshone language. [475]

They held a pipe council, seated in a circle and passing the pipe around. The Shoshone had a custom of pulling off their moccassins when doing this. Lewis commented that smoking a pipe with a stranger and going "bearfoot" was to indicate they were sincere, because going barefoot in this country was a pretty heavy penalty to pay if they were not sincere. He gave the principal chief, Ca-me-âh-wait, a flag as an emblem of peace and friendship. He later learned the warriors had expected to meet their enemies, the Minatarees of Canada. They were armed with bows, arrows and shields, and three small trade guns.

When they arrived at the camp, they were given an old tipi for their use, and a more elaborate pipe ceremony was held. They had eaten nothing since the evening before, and it wasn't until late on this evening they were offered dried cakes of serviceberries and chokecherries—the only food the Shoshones had. After dinner Lewis walked to the river and learned from the chief that the Lemhi River did not lead to "the great lake where the white men lived." He still hoped it wasn't true. He learned the Shoshone had been attacked by the Minataree of Canada in the spring, and about 20 had been killed or taken prisoner. They had lost a great part of their horses and their leather tipis. They were now living in willow brush huts. He saw enough horses that he thought some could be spared for them, or at least enough to carry their goods across the mountains. Then an Indian called Lewis to his lodge and gave him a small piece of antelope roasted and a piece of fresh salmon roasted. The salmon convinced Lewis they were on their way to the Pacific.

Lewis and his men had to wait four days for Clark and the

men to reach the forks of the Beaverhead. Drouillard and Shields were loaned horses and went out on a hunt. Forty to fifty young men would chase the antelopes for half a day, and finally succeed in killing two or three with bows and arrows. Lewis was able to watch most of the chase from his tent. On this day, they didn't kill any.

Cameahwait gave Lewis a geography lesson, using a stick to draw on the earth and making mountains out of dirt. He explained the Salmon River was impossible to navigate. Lewis didn't want to believe it. The chief introduced Lewis to an old man who knew the route to the "persed nosed Indians" who lived west of the mountains. The Nez Perce lived on a river which joined other rivers which led to the "stinking lake," the Pacific Ocean. The old man said the route through the mountains was very bad—there was no game, and they would almost starve to death while crossing the Continental Divide.

Lewis learned the Shoshone traded with the Spanish, whom they could reach by way of the Yellowstone River within ten days. The Spanish, however, would not provide guns and ammunition. This left them at the mercy of their enemies who had guns and ammunition supplied by British traders in Canada. Their enemies would murder all ages and sexes and take their horses. It was for this reason they had retreated into the mountains.

Lewis said they had received promises from the Indians of the Upper Missouri—the Minataree, Mandan and Arikara—they would stop making war on the Shoshone, and the United States would supply the Shoshone with arms and ammunition in abundance, along with articles for their comfort, on reasonable terms in exchange for beaver skins. The chief replied they had long wished to meet white men who traded guns.

The Indians were very reluctant to accompany Lewis to the forks of the Beaverhead. They feared it was a plot by the Minatarees of Canada to kill them. Lewis said he realized they were not acquainted with white men, and hoped there were some men brave enough to accompany him to meet the white men who were arriving at the forks. Cameahwait said he would go and he was not afraid to die. For a third time he "haranged his village" urging them to accompany Lewis. Six or eight men joined him and Lewis smoked a pipe with them and set out before they could change their minds. As they began to travel, others joined them and before long all the men of the village and some of the women were accompanying them.

> they were now very cheerfull and gay, and two hours ago they looked as sirly as so many imps of satturn.[476]

They stopped at Shoshone Cove to camp, and Lewis had the remaining pound of flour cooked in some boiling water for their dinner. Drouillard had been out hunting, but was unsuccessful. The next morning, on August 16th, Drouillard and Shields went out early to find game. Neither the Americans nor Indians had anything to eat. They asked the young Indian hunters to remain in camp to avoid their "hooping and noise" which would alarm the game, but this alarmed the Indians. So two parties of Indians set out to watch the hunters to ensure they weren't in contact with the Minatarees.

As the main group set out on their march, one Indian arrived whipping his horse to announce that one of the white men had killed a deer. In an instant they all gave their horses the whip and set out to feast on the deer. When Lewis arrived at the site, he found them devouring the raw intestines which Drouillard had discarded. He described them as a "parcel of famished dogs." He wrote, "I viewed these poor starving divils with pity and compassion" and told McNeal to skin the deer and give 3/4 of it to the Shoshone, who proceeded to eat it "nearly without cooking." After that, Drouillard killed two more deer which they distributed to the Indians, reserving only a small portion for themselves.

After the deer feed, they were deserted by most of the Indians, and only 28 men and 3 women remained as they approached the Beaverhead, where they would meet Clark. Before arriving, the chief halted their march and gave Lewis and his three companions tippets of fur, such as they themselves wore. A tippet was a fur collar consisting of 100–250 skins of white ermine. Lewis realized it was intended to disguise the white men as Shoshone.

Lewis said he was a brown as an Indian himself and wearing an Indian overshirt, so he looked like an Indian with the addition of the tippet. His men followed his example, and they were soon "completely metamorphosed." Lewis gave his cocked hat with a feather to the chief, and assured him that Clark and his party would be at the forks, or soon arriving.

He had an Indian carry the flag, but when they arrived at the forks, he discovered to his "mortification" that Clark wasn't there. He feared at any moment the Indians would leave and spread the alarm among other Indian bands, ruining all their chances to obtain horses that season. He now—

> determined to restore their confidence cost what it might and gave the Chief my gun and told him that if his enimies wre in those bushes before him he could defend himself with that gun, that for my own part I was not affraid to die, and if I deceived him he might make what uce of the gun he thought proper or in other words he might shoot me. the men also gave their guns to the other indians which seemed to inspire them with more confidence. . . . I thought of the notes which I had left and directed Drewyer to go with an Indian man and bring them to me which he did.[477]

Lewis read the notes (which he had earlier written to Clark), and told the chief Clark had sent a man ahead to leave a message saying they should wait for them at the forks. He said he would send one of his men to locate them and the chief should send one of his, which was done. Lewis and the others slept around a campfire that evening. He wrote —

> my mind was in reallity quite as gloomy all this evening as the most affrighted indian but I affected cheerfullness to keep the Indians so who were about me. . . I slept but little as might be well expected, my mind dwelling on the state of the expedition which I have ever held in equal estimation with my own existence, and the fait of which appeared at this moment to

depend in a great measure upon the caprice of a few savages who are ever as fickle as the wind.[478]

On the morning of August 17th, Lewis wrote that Captain Clark arrived with Charbonneau and Sacagawea, who—

> proved to be a sister of the chief Camahwait. the meeting of those people was really quite affecting, particularly between Sah-cah-gar-we-ah and an Indian woman who had been taken prisoner at the same time with her, and who had afterwards escaped from the Minatarees and rejoined her nation.[479]

Clark wrote Sacagawea danced for joy upon seeing her people.

The canoes arrived at noon, and the party was reunited at the place they named "Camp Fortunate." They held a council and gave out presents and medals, and promised to supply the Shoshone with firearms and other merchandise after they completed their mission to the Pacific. In return they needed horses and a pilot to guide them through the mountains. Clark wrote that Chief Cameahwait was a man of—

> Influence, Sence & easey & reserved manners, appears to possess a great deel of Cincerity. . . everything appeared to astonish those people. the appearance of the men, their arms, the Canoes, the Clothing my black Servent & the Segassity of Capt Lewis's dog.

The next morning, Clark set out in advance with the Charbonneau family and eleven men to make canoes if the Columbia River (in other words, the Lemhi and Salmon Rivers) proved feasible. They were accompanied by Indians who sang all the way back escorting Sacagawea to the village.

It was August 18th, Lewis's birthday. He wrote:

> This day I completed my thirty first year, and conceived that I had in all human probability now existed about half the period which I am to remain in this Sublunary world. I reflected

> that I had as yet done but little, very little indeed, to further the happiness of the human race, or to advance the information of the succeeding generation. I viewed with regret the many hours I have spent in indolence, and now soarly feel the want of that information which those hours would have given me had they been judiciously expended. but since they are past and cannot be recalled, I dash from me the gloomy thought and resolved in the future, to redouble my exertions and at least indeavour to promote those two primary objects of human existence, by giving them the aid of that portion of talents which nature and fortune have bestoed on me; or in the future, to live for *mankind*, as I have heretofore lived *for myself*.[480]

Lewis had just experienced the most crucial time of the whole expedition. They absolutely needed to obtain horses from the Shoshone. He had found the Indians to be starving and frightened, and had shown great skill and courage in dealing with them. They were now faced with the daunting prospect of traveling over the Continental Divide to the Pacific Ocean. On his 31st birthday, he fervently wished he was better prepared—intellectually and spiritually—for what lay ahead.

Lewis described the character of the Shoshone—that even in extreme poverty they were gay, fond of gaudy dress, and games of risk. The were frank, communicative, fair and generous, extremely honest and "by no means beggarly." Each individual acted from the dictates of his own mind. They would only follow the chief if he had their confidence. There were about 100 warriors and about 300 women and children living in the village. They seldom whipped their boys, as it broke their spirit and destroyed their independence of mind. The women did all the routine work. The men hunted and fished and rode the best horses.

Sacagawea was promised in marriage to a man when she was a child. The man was more than twice her age and had two other wives. He said because she had a child with Charbonneau, he did not want her. She was free to leave with her husband.

During the summer months, the Shoshone lived on the waters of the Columbia west of the mountains and fished for salmon. It was a safe place, where their enemies did not come. When the salmon left, they returned to the mountain valleys of the Upper Missouri. In the fall, Shoshone and Flathead bands would join together and enter the plains east of the mountains to hunt for buffalo. They alternated between risking their lives and retreating back into the mountains. The Blackfeet and Minataree would kill them if they encountered them.

The tribe was about to start on their journey to the buffalo country. Cameahwait was deeply concerned about delaying his starving people's travel to help the expedition. Lewis challenged him to keep his word. The Shoshone women hauled the expedition's baggage on their backs and on pack horses over Lemhi Pass. One woman gave birth, and was seen carrying her newborn baby along the road. If the women hadn't carried their baggage, the expedition would not have been able to cross the mountains in the fall of 1805.

Woman power played another role—Sacagawea stayed with the expedition. Pompey was now six months old. Their presence safeguarded their encounters with Indians, and it must have helped greatly with the morale of the men. The presence of a woman, baby and dog were a welcome reminder of home and a distraction from their own troubles.

The Shoshone lacked horses and mules because of the Minataree raid. Lewis said their mules were the finest he had ever seen, and when Clark set out to explore the river route, he traded a waistcoat for a mule. Tobe ("Toby"), an elderly Shoshone chief, was Clark's guide as they explored the area of the Salmon River for four days (August 21-24, 1805). Clark discovered the Salmon River was indeed impassible, and saw in the distance the loftiest mountains he had ever seen, entirely covered with snow.

Tobe would guide the expedition north through the mountains to Lolo Pass, and then west through the pass to the Nez Perce villages on the Clearwater River. On the 31st of August, they begin their journey over the mountains with 29 horses and one mule. It would take over three weeks to reach the Nez Perce.

"Lewis and Clark Meeting the Flatheads at Ross' Hole, September 5, 1805," a 12 x 26 foot mural by Charlie Russell at the Montana State Capitol in Helena. (Wikimedia Commons)

On September 4th they encountered the Flathead, or Salish Indians. The Flatheads of the Bitterroot Valley did not have flattened heads. Their name came from the Coastal Salish, who lived near the ocean and who flattened the heads of their infants by binding their foreheads in cradleboards. A sloping, pointed head was considered attractive.

Patrick Gass wrote "They are the whitest Indians I ever saw." They wondered if the Flatheads might be the fabled Welsh Indians. They spoke a guttural, harsh-sounding language unlike any Indian language they had heard. Whitehouse said they spoke with a "burr" or speech impediment, and wrote—

> we take these Savages to be the Welch Indians, if their be any Such from the Language. So Capt. Lewis took down the Names of everrything in their language . . .

The translation chain involved five people—Salish to Shoshone by Tobe; Shoshone to Hidatsa by Sacagawea; Hidatsa to French by Charbonneau; and French to English by Labiche. There were at least seven people engaged in making the vocabulary, counting Lewis and the Salish language speakers.[481]

The Flatheads had lived in the Bitterroot Valley for thousands of years. Their oral history dates back to the end of the last Ice Age. Their history of Lewis and Clark relates they felt pity for them because they were white and had beards. They were the first white people they had ever seen. Because York had black skin they thought they were a war party, as warriors painted themselves black. Their hair was cut short, a sign of mourning among Indians. They had never seen men wear pants, and they gave them buffalo robes because they were men without blankets.[482]

Clark wrote they possessed "ellegant horses." They traded for eleven horses and exchanged seven more. The tribe consisted of about 400 people, with 80 warriors. They had about 500 horses and were on their way to hunt buffalo when the expedition left on September 6th. The expedition now had 40 pack horses, and three colts which they intended to eat when they ran short of provisions.

They named the Salmon River for Lewis and the Bitterroot River for Clark. Neither name has endured, except for the river they named the Clark Fork River, thinking it was a fork of the Clark (Bitterroot) River. In fact, the Clark Fork River was a tributary of the Columbia River, with The Clark Fork's own tributaries being the Bitterroot, Blackfoot and Flathead Rivers. Tobe told them the Clark Fork would take them east over the route of today's Interstate 90 through Hell's Gate Canyon to the Three Forks Headwaters. Both the Clark Fork and Blackfoot Rivers were much easier routes through the mountains, but they were also prime territory for Blackfoot ambushes.

They arrived at a creek they named "Traveler's Rest" in the Bitterroot River Valley near Lolo Pass on September 9th, where they rested for a day and made lat-long readings. They would travel 120 miles west through the Pass, over steep and rugged terrain, during the next 16 days.

On the morning of September 13th, Lewis's horse ran away with her colt. The colt was being used by Tobe's son. The horse and colt were never located. They visited the famous Lolo Hot Springs on the 13th, but didn't have time for a hot bath.

Tobe chose the wrong trail among several trails at the springs, and they descended into the valley of the Lochsa River near today's Powell Ranger Station, where they killed a colt and ate it. The main trail followed the mountain ridges. On September 15th, they climbed back over 3,000 feet to camp on Wendover Ridge (6,692 ft. elevation).[483] Several horses slipped and rolled down the mountains. Clark's writing desk was broken, but the horse survived unhurt. They melted snow to drink and cooked the remaining horseflesh to eat.

It began to snow on September 16th, and 6–8 inches of snow fell during the day. Clark wrote:

> . . . we are continually covered with Snow, I have been wet and as cold in every part as I ever was in my life, indeed I was at one time fearfull my feet would freeze in the thin mockersons which I wore.[484]

On the 18th they decided that Clark should take six hunters and go on ahead to hunt on level ground.

Lewis and the rest of the party dined on a "skant proportion of portable soupe" of which they still had a few cannisters. They also had a little bear oil and about 20 pounds of candles. They had been eating portable soup twice a day since the 14th. Lewis had purchased 193 pounds of the soup in Philadelphia, packed in 32 tin cannisters. The "portable soup" was a chunk of concentrated beef stock to be heated in water. Its nutritional content was about 360 calories per serving. On September 19th, the remaining soup would run out in two days. Tallow candles were a high calorie fat, as was bear oil. Research has revealed they did not eat their candles, instead they used about one candle a day to write by candlelight. Lewis mentioned them only as a last resort. Their average daily caloric intake on a working day would have been about six pounds of meat (about 5,100 calories.)[485]

Clark and his men killed a horse which they found on September 19th. After breakfasting on their portion, they left the rest for Lewis and his party, who on the next day, "ate a hearty meal . . . much to the comfort of our hungry stomachs." Clark, on September 20th, met up with three Nez Perce boys after arriving at a level open plain, the Weippe ("WEE-eyep") Prairie of the Nez Perce village.

The arrival of Clark and his hunters caused "great confusion" at the Indian camp. About 100 men, women and children accompanied them back to their village, located about two miles away. They passed fields of camas roots being harvested and gathered in heaps. Camas ("Quamash") roots are small onion-like bulbs, tasting something like baked sweet potato. After being roasted in deep earth pits they could be pounded into flour, or raw bulbs could be boiled. It was a main food of the Nez Perce. Their other main food was dried salmon.

The camp consisted mostly of women and children because three days earlier, the Nez Perce warriors had ridden off to make war against the Lemhi Shoshone. The Shoshone had recently killed three peace emissaries sent by the Nez Perce. Most of the able bodied men were away. The Nez Perce, contrary to their name, "pierced noses," did not routinely wear ornaments in their noses, but Clark does record one young woman doing this.

The boys who met Clark remembered their first impressions. He had red hair, a red beard covering his face, and a red braid hanging down the back of his neck. His hands were covered with red hairs. He wore a strange head dress on top of his head. All of the men had different colored hair and hair on their faces like buffalo hair. They wore clothes made of cloth and skin and their horses were skinny. On their saddles they carried "sticks-that-shoot-black-sand-with-thunder." Others described some men as "upside down men" because they had hair on their chins and none on their heads. Five teenagers tackled York when he was alone and scrubbed his skin until he bled to determined whether his skin was really black. The men with buffalo hair on their faces, when they met the men of the tribe, put their right hand in the right hand of the man and moved it up and down. The captain's dog followed behind him and was as big as a black bear. These are some of the Nez Perce stories handed down and found in two fine books, *Do Them No Harm!* and *Lewis and Clark Among the Nez Perce.*[486]

Upon their arrival, Clark and his men ate heartily of dried

salmon and camas roots and then suffered from diarrhea and abdominal cramping. When Lewis and his men arrived on the 22nd, Lewis became seriously ill. He was "Scercely able to ride" on the 24th. Several men were so unwell they just lay on the side of the road. On the 29th Clark wrote that Lewis was still very sick and most of the men were "complaining very much of ther bowels & Stomach." On October 4th, Lewis was able to "walk about a little." The Corps of Discovery finally set out for the coast on October 7th, while Lewis was still recuperating.

Clark was pretty much in charge throughout this stay because Lewis was so ill. Clark was giving Dr. Rush's Bilious Pills—laxatives called "Thunderbolts"—to the men and also cathartic salts to induce vomiting. Dr. David Peck in his book, *Or Perish in the Attempt*, says it is "unclear whether the men were more sick from the diet or from the medical treatment." He writes the sun-dried salmon could have harbored Staphyloccucus bacteria causing food poisoning, and the fish could have been carrying tapeworms and other worms inducing peritonitis.[487]

They were with the Nez Perce for 17 days. At first, the Indians considered killing them. As sick as they were, it would have been easy to do. An old Nez Perce woman, Wat-ku-ese. who had known white men in her youth, saved their lives. As a girl, she had been captured by the Blackfeet and was eventually sold to a white trader living in the Great Lakes area. He treated her well, and they had a son together. He wanted to take her and the baby across the "Big Water" back to his home, but she fled with the baby back to her homelands. The child died along the way. Now, when she heard them plotting to kill the men of the expedition, she told them:

> These men are So-yap-pos! [white men] Good men! Men like these were good to me. Do not kill them! Do them no harm! Do them no harm! [488]

These words lived on in the stories and councils of the Nez Perce, who entered into a formal alliance with the United States on May 12th, 1806 when the corps returned to the villages.

The Nez Perce had acquired six guns in trade with the Minataree of the Missouri (Hidatsa), but they no longer had any ammunition. They may have been trading at the Mandan-Hidatsa villages while Lewis and Clark were there. A hatchet made by John Shields preceded the corps's arrival at the Nez Perce villages. The tribe may have sent peace emissaries to the Shoshone in response to the United States's message of peace between tribes. It would be to the advantage of the Shoshone, Flatheads and Nez Perce if they stopped making war against each other, and instead concentrated on their mutual defense against the Blackfeet and Minataree of Fort de Prairie in Canada. [489]

From September 26th until their departure on October 7th, they were busy making five dugout canoes and oars at "Canoe Camp" on the Clearwater River west of Weippe Prairie. The Nez Perce showed them how to use fire to hollow out the center of the giant Ponderosa Pines they cut down, which speeded up the work. They made four large canoes, as long as 50 feet in length, and a small canoe which would serve as lead canoe. Several men were unable to work at all, and the rest were so "weak and feeble" that the work proceeded slowly. The Indians helped them.

The 38 horses and one mule were branded using a "stirrup iron" and a piece of hair was cut off of their forelocks. The herd was left in the safekeeping of Chief Twisted Hair's family. Twisted Hair and his younger brother, Tetoharsky, would act as the expedition's guides to Celilo Falls on the Columbia River.[490]

The war party against the Shoshone returned, but only Patrick Gass mentioned it in his journal, noting the captains gave out medals to 3 or 4 chiefs. Any victory celebrations would have been held at another village. It was obviously not politic to celebrate in front of Americans, with their message of peace, and the presence of their Shoshone guides, old Tobe and his son.

On the second day after their departure from Canoe Camp, they were joined by the two Nez Perce chiefs who had gone ahead to another village without telling them. After they were safely in the care of their new guides, Tobe and his son left without saying goodby or receiving his pay. When the captains asked Twisted Hair to send a messenger after them with Tobe's pay, the chief advised against it, saying "his nation would take his things from him before he passed their camps."[491]

In order to understand the corps's different relationships with the tribes they encountered, geopolitics has to be considered. A primary goal of the United States was to enter into the fur trade of the Upper Missouri. The Upper Missouri trade had already been established by British traders in Canada who supplied the Indians with guns and other articles in exchange for furs. The Teton Sioux dominated the region between the Yellowstone and Platte tributaries of the Missouri. They simply were more numerous, and used force to obtain what they wanted. But all the tribes had guns supplied by the British. The Mandan-Hidatsa villages were the major trade center for the northern plains.

The Blackfeet and the Minataree of Fort de Prairie in Canada had been supplied with guns by traders on the Saskatchewan River since the 1750's, and they terrorized the tribes who had no guns. This was the balance of power that Lewis was proposing to correct by supplying the Shoshone, Flatheads and Nez Perce with guns and ammunition. The mountain tribes traded for horses with the Spanish in the Southwest, but the Spanish would not trade for guns. The situation was particularly awful because there were literally millions of buffalo on the Great Plains. The Blackfeet and Minataree (Gros Ventre) were representing British interests and protecting their special trade relationship. But the consequences were that the Shoshone, Flatheads and Nez Perce hunted with bows and arrows and lived in near starvation for many months of the year out of fear of enemy attacks.

Entering into the fur trade of the Upper Missouri had a larger, global, geopolitical significance for the United States. It meant the establishment of a world wide trading empire. Jefferson's instructions to Lewis said:

> Should you reach the Pacific Ocean inform yourself of the circumstances which may decide whether the furs of those parts may not be collected as advantageously at the head of

> the Missouri (convenient as is supposed to the waters of the Colorado & Oregan or Columbia) as at Nootka sound, or any other point on that coast; and that trade be consequently conducted through the Missouri & U.S. more beneficially than the circumnavigation now practised.[492]

In other words, Jefferson proposed collecting furs on the Upper Missouri and bringing them through the mountains to the mouth of the Columbia River to be shipped to China, and then bringing articles from China back to the United States by transporting them on the Missouri River.

Captain Robert Gray, who discovered the mouth of the Columbia in 1792, was the captain of the first American ship to circumnavigate the globe in 1790. His ship, the *Columbia,* traveled almost 42,000 miles after leaving Boston harbor in 1787. The *Columbia* delivered furs from the Pacific Northwest to China and brought back almost 22,000 pounds of tea to Boston. It was the beginning of the Boston triangular trade with the Pacific Northwest. Articles from Boston were traded to coastal Indians for sea otter pelts and seal fur skins, which were then traded in China for silks, porcelain and tea, and the Chinese goods were brought back to Boston to be sold.

In 1801, of the 23 ships which sailed to the Northwest Coast, 21 were American, 2 were British, and 1 was Russian. The coastal Indians called all Americans "Boston Men." Profits ranged between 300 to 500%, and went as as high as 2,200%. Nearly18,000 otter and seal pelts were carried to Canton annually until the animals were almost extinct by the 1830's.[493]

Immediately upon his return to St. Louis, Lewis wrote to Jefferson on September 23, 1806 about the possibilities of the fur trade:

> The Missouri and all it's branches from the Cheyenne upwards abound more in beaver and Common Otter, than any other streams on earth . . . The furs of all this immence tract of country . . . may be conveyed to the mouth of the Columbia by the 1st of August in each year and from thence be shiped

> to, and arrive in Canton earlier than the furs at present shiped from Montreal annually arrive in London.[494]

He proposed that a fort be built at the mouth of the Columbia for the purpose of collecting furs. An American trading fort was subsequently built at Astoria near the mouth of the Columbia by John Jacob Astor's Pacific Fur Company in 1811, but the enterprise was halted by the War of 1812.

The contest between the United States and Great Britain for control of the Pacific Northwest would go on for many years, and would not be resolved until the Oregon Treaty was signed in 1846. This treaty set the boundary line between the United States and Canada at the 49th parallel, with the exception of Vancouver Island being retained by the British.

There were three territorial missions for Captain Lewis and his army Corps of Volunteers for Northwest Discovery—

> (1) to find a navigable water route from the Missouri to the Columbia and Pacific Ocean. Regardless of whether an all-water route was discovered, it was necessary to establish a fort at the mouth of the Columbia in order to reinforce the United States' claim to the Pacific Northwest under the Doctrine of Discovery.

> (2) to determine the source of the headwaters of the Missouri, which would be used to establish America's claim to the northern boundary of the Louisiana Purchase.

> (3) to see how far a tributary river of the Missouri flowed down from the north, as the latitude of that river could be used to determine the boundary line between the United States and Canada. If Lewis had followed the Milk River instead of the Marias, the United States might have gained 17 additional miles.[495]

The expansion of the fur trade into the Upper Missouri and

Another picture of a Newfoundland, from Thomas Bewick's *A General History of Quadrupeds* (1800). Discovering Lewis and Clark website, www.lewis-clark.org

the building of trading forts were necessary to secure America's continental empire. For this reason, Lewis was determined to break the hold that Britain's Indian allies had over the region. There were enormous commercial profits to be made, but the underlying issue was America's Manifest Destiny. The support of the Shoshone, Flatheads and Nez Perce was needed against the Indians supported by the British. The Montana buffalo plains and rivers teeming with beaver and otter were the scene of conflicts between Americans and Indian allies of British traders for many years to come.

The expedition was now traveling with river currents instead of against them. They would arrive at the Pacific Ocean on November 18th, after a journey of 500 miles from Canoe Camp. Traveling on the Clearwater and Snake Rivers, they reached the Columbia River on October 18th.

They were buying dried salmon and dogs to eat from the natives, whose fishing camps lined the river shores. Clark and the Nez Perce chiefs were the only ones who didn't like to eat dogs. He said all the rest of the party had a great advantage as they "relished the flesh of dogs." According to journal entries, the expedition ate 193 dogs. The salmon weren't counted. Journal entries total 1,048 deer, 382 elk, 259 buffalo, 28 bear, and 111 beaver among the animals killed and eaten.[496] The slaughter of animals to eat was a major occupation of the corps. Seaman must have taken it for granted.

They were being observed by Indians all along the Columbia River Plateau, who lined the shores watching the passing of their canoes. Clark wrote

> "the woman of Shabono our interpetr we find reconsiles all the Indians as to our friendly intentions a woman with a party of men is a token of peace." [497]

They were taking chances on the river because the season was so far advanced and they were unwilling to take the time to portage around the falls and rapids. On October 14th, they had a canoe accident at a rapids on the Snake in which they lost all the roots acquired from the Nez Perce and half of their goods. They had been following a strict rule of taking nothing from the Indians without permission—stores of dried fish and firewood were left untouched—but now they took firewood found on an island to dry out their remaining goods.

On October 16th, at the confluence of the Snake and Columbia, 200 men of the Wanapum tribe came to their camp drumming, singing, and dancing to greet them. The Nez Perce chiefs had gone ahead to prepare the way. The Wanapam were one of the Sahaptin language speaking tribes who lived along the Snake and Columbia Rivers—now known as the Confederated Tribes and Bands of the Yakama Nation. They met the Yakama tribe at the confluence of the Yakima and Columbia Rivers. Lewis took a vocabulary, and held a council on the 18th. On the 19th, they met with the great chief of all their tribes, the Walla Walla Chief Yel-lep-pit, who urged them to stay so his tribe could meet them. They promised on their return trip to stay a day or two, but now they were in a hurry to reach the coast.

They were catching occasional glimpses of Mount Hood in the distance, the highest volcano in the Cascade Mountain Range at 11,000 feet. They were leaving the treeless region of the Columbia Plateau, a lava plain of basaltic rock created by the eruption of ancient volcanoes, to enter the mountain ranges of the Pacific Northwest.

Clark wrote that John Collins made "Some verry good beer"

Indians fishing at Celilo Falls in the 1950's, before The Dalles dam was built. (Wikimedia Commons)

out of the last of the camas root bread that had become "wet molded & Sowered." It must not have been a very great quantity, because none of the other journal keepers mention beer.[498]

On October 22nd they reached Celilo Falls and hired Indians to take their heavy articles around the falls by horseback. The Celilo Falls, inhabited for at least 11,000 years, was the oldest continuously inhabited settlement in North America until The Dalles Dam destroyed it in 1957. The waterfalls stretched for about 12 miles. Historically, an estimated 15-20 million salmon passed through the falls every year, making it one of the great fishing sites on the continent.

The villages at the falls were the great trading mart of the West Coast. The expedition arrived late in the season after the trading fair was almost over. At its peak, 3,000 Indians would be in attendance. The Nez Perce brought horses, buffalo meat, and skin clothing from the mountains. The Chinooks brought European goods from coastal traders and wapato roots. The locals supplied dried salmon, roots and berries.[499]

Every man of the expedition stripped naked to portage the canoes around Celilo Falls, in order to brush fleas off of their legs and bodies as they carried the canoes. While on the water they were not bothered by fleas, but on land the fleas, which were part of the salmon industry, were as bad as the dreaded mosquitoes and gnats of the plains.

At the end of the portage day, Twisted Hair, the Nez Perce chief, informed them they had learned the Chinook Indians below the falls were plotting to kill the Americans. They put their arms in order and prepared a 100 rounds of ammunition. On the same day, Lewis traded the small lead canoe for a beautiful Chinook canoe, wide in the middle, tapering at both ends, and decorated with carved figures. The next day, the Nez Perce chiefs said they were at war with the Chinook, who would kill them if they continued. The captains persuaded them to stay two more days, in the hopes of brokering a peace for them.

They had to take the boats through the Short and the Long Narrows. The series of falls were among the largest in North America, dropping 82 feet at high water over the course of 12 miles. The Columbia River, a mile wide on the plateau, was confined to narrow gorges, creating turbulent, boiling whirlpools. Clark wrote that the Short Narrows squeezed down to 45 yards wide and that he—

> thought (as also our principal watermen Peter Crusat) by good Stearing we could pass down Safe, accordingly I deturmined to pass through this place not withstanding the horrid appearance of this agitated gut Swelling, boiling & whorling in every direction (which from the top of the rock did not appear as bad as when I was in it) however we passed Safe to the astonishment of all the Inds of the last Lodges who viewed us from the top of the rock.[500]

The men portaged the guns and papers and other valuables around the narrows, and the best swimmers and watermen handled the boats. It wasn't possible to portage the large canoes.

The plank houses of the Chinooks were the first wood houses they had seen since leaving the St. Louis area. Scaffolds outside the houses contained dried, pounded, salmon wrapped in reed packaging. Clark counted 107 packages and estimated their weight at 10,000 pounds. They held a council between the chiefs of the Wishram and Wasco villages and the Nez Perce chiefs and arranged a peace. The natives were delighted when Cruzatte played the violin and the men danced in the evening.

On October 25-27, they camped at Fort Rock, an ancient fortification overlooking the Dalles. Archeological investigations have found evidence of tools and campfires going back 10,000–13,000 years; and seventy five sandals were discovered here dating back 9,000 years. Clark wrote about Fort Rock—

> our Situation well Calculated to defend (us) our Selves from any designs of the natives, Should They be enclined to attack us.[501]

Fort Rock is one of the few places we know for certain that the expedition camped. A riverfront trail and signage mark the site, the most historic Lewis and Clark campsite in the country.

The two chiefs of the Nez Perce left the party. They had served the Corps of Discovery for five weeks, since the making of the canoes at Canoe Camp, and had done what they promised, guiding the expedition through the lands of the Sahaptin peoples to The Dalles. The Nez Perce spoke a Sahaptin language, and after the falls, the Chinooks had their own language.

They were now meeting the Coastal Salish, the people who really did flatten their heads. They spoke with the same "clicking tone" as the Flatheads of the Bitterroot Valley. Lewis took a Chinookan vocabulary at the Wishram village. They may have also been encountering the "Chinook Jargon," of coastal Indians. Clark wrote "maney words of those people are the Same."

People who speak different languages often develop a simplifed version of one of the languages, mixed in with words from different languages, which is called "pidgin" or "jargon." Members of the corps most likely used some variations of English and French pidgins themselves. Different languages played a part in the daily life of the corps. They were a traveling group of linguists engaging in trade with all the people they encountered.

Pacific Northwest Indian canoe (Wikimedia Commons). The canoes they saw had carved figurines in the bow and stern, as tall as five feet in height.

They were finding brass tea kettles in the Chinook homes. The Indians were friendly. They passed "Sepulchar Island," where their dead were buried in vaults. They were now in the Columbia Gorge, a land of mountains and timber, of clouds, mist and rain, and of cascading waterfalls.

On October 31st they observed their first tidal waters at Beacon Rock, an 848 foot landmark named by Lewis and Clark. The massive rock made out of basalt was purchased by Henry Biddle, a descendant of the Philadelphia Biddles, who donated it to the state of Oregon. A 3/4 mile hiking trail lead to its top today.

They passed the "Great Shute" or rapids of the Cascades, where the river dropped forty feet in two miles. (The Bonneville Dam was built to control the Cascades in 1938.) They portaged their baggage around the Cascades, and moved the four large canoes through the rapids by sliding them on poles placed across rocks in the water. Three of the canoes were damaged and had to be repaired.

They were encountering Indians who had heads flattened to a point and pierced noses with white shells pushed through their noses—genuine flathead and pierced nose people. They were sharp traders who said they could get better prices from white people on the coast. They prized blue and white beads most of all. The Indian men were dressed in sailor's jackets, trousers, shirts and hats. They were having trouble with Indians stealing things.

The weather was cool, wet, rainy, and foggy. These conditions would continue for months to come. On November 7th they were traveling through the Columbia estuary, when Clark wrote "Great joy in camp we are in View of the Ocian." An estuary, however, is the transition area between fresh water rivers and the saltwater of an ocean. They wouldn't reach the ocean until November 18th, but they were feeling the effects of the ocean.

Several of the men got sea sick from the high waves that came in with the tide. The winds blew hard, and driftwood trees were being tossed about in the water. Many of the trees were 200 feet long and 4–7 feet through. They managed to get the canoes out of the water and unloaded. It rained day and night, but all were cheerful, anticipating reaching the coast.

The waves were so high, however, that on November 10th they were forced to take refuge in a small cove that Clark called "this dismal nitch" on the north side of the river. They formed a camp on top of the giant drift logs and stored their goods in the rocks above tide water. He wrote on November 11th:

> . . . we are all wet as usial and our Situation is truly a disagreeable one; the great quantites of rain which has loosened the Stones on the hill Sides, and the Small Stones fall down upon us, our canoes at one place at the mercy of the waves, our baggage in another and our Selves and party Scattered on floating logs and Such dry Spots as can be found on the hill Sides, and Crivices of the rocks.

They stayed at Dismal Nitch for six days (November 10-15). They were miserable. It was wet and cold, their bedding was wet, and their skin clothes were rotting. It rained continually. While they were there, the natives came by in their canoes and sold them salmon. The Indians traveled across the five mile wide river in the highest waves Clark had ever seen. He wrote, "Certain it is they are the best canoe navigators I ever Saw." [502]

Three of the men used their Indian canoe and entered a bay of water with a sandy beach near the ocean. A Chinook Indian village was located there, and the Indians were "thievous." Shannon and Willard had their guns stolen while they were sleeping, but when Captain Lewis and a second exploring party arrived they got their guns back. Finally the rain stopped and the expedition's camp was moved to a point of land at the east entrance to the bay (Baker Bay, Ilwaco, Washington). Cape Disappointment was on the west side of the bay, on the ocean.

On November 17th Lewis and his men went around Cape Disappointment to enter the ocean waters. Cape Disappointment was named by an Englishman in 1788 when he was disappointed in not finding the Columbia River. He didn't realize that the 700 foot tall promontory of land was actually the north side of the entrance to the river. Four years later, the American ship captain Robert Gray managed to get through the Columbia River Bar and enter the river, claiming it for the United States. The bar is a treacherous place, six miles long and three miles wide, where wind, waves and current have destroyed over 2,000 vessels since 1792, giving it the name "the Graveyard of the Pacific."

On November 18th, the captains and several of the men marked their names and the date with the words "by land 1804 & 1805" on the soft basalt rock of Cape Disappointment. On November 20th, they met with two Chinook Chiefs and gave them medals and a flag. The saw the most beautiful fur they had ever seen, a robe made out of sea otter skins. After much bartering, they persuaded Sacagawea to relinquish her belt of blue beads so they could obtain the robe. In return, they gave her a coat of blue cloth.

They were visited by natives in the evening, with women eager to "gratify the passions of the men." The women were tatooed, and one had "J. Bowmon" on her arm. Ribbons were distributed to the men, rather than have knives and more valuable

Fort Clatsop replica, Lewis and Clark National Historial Park, Astoria, Oregon (Photo by Rick Blond).

articles given to their "favorite lasses." There was evidence of venereal disease, scabs and ulcers, among the group that accompanied the girls. A few days before they started on their return journey in March, 1806, Lewis forbade the men to have sex with women who came to the fort, because it was the same group—

> that had communicated the venerial to so many of our party in November last. and of which they have finally recovered.[503]

On November 24th, the captains asked the members of the party to vote as to where they should spend the winter. Everyone except the captains and the baby cast a vote. "Janey"—Clark's name for Sacagawea—voted in favor of a place in which there were plenty of wapatos (Indian potatoes). York voted in favor of examining the other side of the river. The majority opinion was to go take a look at the south side of the river where the Clatsop Indians had told them there was plenty of game. Other alternatives were to go back to Celilo Falls, or to the confluence of the Sandy River near Mount Hood. If the south side of the coast wasn't suitable, they would go back to the Sandy.

It was "truly disagreeable" weather after they crossed over to the south side. Their tents, tipi, sails, bedding and clothing were rotting. They were continually wet and there were violent winds and rain. They were exploring the coastal area, looking for game. On December 2nd they killed an elk, the first since the Rocky Mountains. They discovered there were plenty of elk and deer and Lewis found a location to built a fort. On December 10th they started cutting down trees to built Fort Clatsop. By Christmas Eve, on December 24th, they were moving into their huts. On Christmas morning, they were awakened by a "Shout and a Song" and presents were given out after breakfast—tobacco to the men who smoked, and handkerchiefs to the others. Dinner consisted of some poor elk boiled, spoiled pounded salmon and a few roots. Their blankets were infested with fleas and the huts were filled with smoke, but they were living under a roof.

On the 27th, they set up a salt making camp at the ocean, 15 miles to the southwest. The Salt Works consisted of five kettles of ocean water boiling 24 hours a day to extract the salt from the water. Salt was necessary to preserve the meat from spoiling and for use as flavoring. For the next two months, three men tended to the salt works, producing about 20 gallons of salt. 12 gallons were put in small kegs to be used on the return journey.

Lewis resumed writing in his journal on January 1st, 1806. He abandoned writing at the Nez Perce village after he became ill, and made only a few notes in the weather records after that. On January 1st, the captains issued orders for the fort. They were concerned about safety—a sentinel was to keep watch day and night for the approach of any Indians, and a sergeant and three privates were on guard duty day and night. The natives were to be treated in a friendly manner, but the sergeant could expel any native who became troublesome. No Indians could remain in the fort after sunset and the main gate was to be locked. All tools were to be kept in the room of the commanding officers, and brought back the moment they were not being used. Only John Shields, the gunsmith, was exempt from tool regulations.

Lewis described the bargaining methods of the locals, calling them "great higlers in trade." They always refused the first offer. He tested his theory by offering a lot of merchandise for an inferior otter skin. The seller refused, but the next day accepted a much smaller amount of merchandise for the same skin. He found them different from all other Indians he had ever dealt with, as they would trade for anything that caught their fancy.

It would appear to be a cultural conflict. The Indians were long accustomed to bargaining with white traders in the otter skin trade. They didn't understand white men who came on an exploring expedition—white men who were hungry and who lacked merchandise. Bargaining was a sport, and theft was a game. They would spend a whole day bargaining over a handful of roots, and steal small things if they had the opportunity.[504]

Two men who had helped carry kettles and supplies to the salt works returned with the news that a dead whale had washed up on the beach, and brought back some of its blubber meat. It was white and spongey, resembling the flavor of beaver and dog meat. They also brought back the first salt that had been made, which was a "great treat" for Lewis and most of the party, except for Clark who didn't care about salt. Lewis wrote about eating all kinds of food on the expedition—

> I have learned to think that if the chord be sufficiently strong, which binds the soul and boddy together, it dose not so much matter about the materials which compose it.[505]

Lewis had dined at Jefferson's table for two and a half years, and now he was eating roots and spoiled salmon.

The next morning Clark set out with a party of 13 men and Sacagawea and 11 month old Jean-Baptiste in two canoes to see the whale. Lewis wrote that Sacagawea had complained—

> she had traveled a long way with us to see the great waters, and that the monstrous fish was also to be seen, she thought it very hard she could not be permitted to see either (she had never yet been to the ocean).[506]

When the party finally reached the whale, they found it was entirely stripped of its "valuable parts." It measured 105 feet.

They went to the Tillamook village where the blubber was being boiled down for its oil. Trading all the merchandise they brought with them, they were able to obtain 300 pounds of blubber and a few gallons of oil.

On their way back to Fort Clatsop, they stopped at a small Chinook village to spend the night. At 10 o'clock in the evening, Clark heard screams. His local guide said someone's throat had been cut. Hugh McNeal had gone by himself to visit a Chinook woman, "an old friend," without telling anyone. Other Indians, who pretended friendship, planned to kill him for his blanket and clothes. His woman friend learned of the plot and started screaming for help, scaring off his attackers. Clark sent out four men to investigate, who met McNeal coming back in haste. McNeal was simply puzzled about what had happened until he learned of the plot.

Lewis wrote that the traders visiting the coast must be either English or American. The Indians knew many words of English:

> musquit, powder, shot, nife, file, damned rascal, sun of a bitch, etc.

He was grateful the traders had not introduced any "sperituous liquors," as the locals never asked for it, calling it a "very fortunate occurrence" for the natives as well as the "quiet and safety of thos whites who visit them." [507]

They lost their prized Chinook canoe. Some of the men were negligent in tying it up after using it. It was so light that four men could carry it a mile without resting. It could hold four men and 1,000–1 200 pounds of goods. They lost a large pirogue, when a tie rope broke, but fortunately it was recovered. As a precaution, they drew three of the large boats out of the water, and secured the fourth with a strong rope of elk skin.

Drouillard killed seven elk in one day. Lewis said he didn't know how they could survive without his skill as a hunter. They dried (jerked) the meat so as to ensure the supplies would last. Otherwise, the men would eat too much when they had the chance. They used the last of their candles on January 13th, but they had candle molds and wicks, so they made new candles out of rendered elk fat.

There was "nothing worthy of notice" on many days in the journal entries. The men filled their time with hunting and making skin clothes and mocassins. Lewis wrote many pages on the Indian way of life, scientific observations, and suggestions for trade. Clark worked on maps and drawings and copied Lewis's journal entries almost verbatim. The captains wrote two lengthy reports—an *Estimate of Distances from Fort Mandan to the Pacific Coast* and an *Estimate of Western Indians*, which was less detailed than the *Estimate of Eastern Indians*.

They were anxious to start the return journey, but it would be "madness" to start out before the first of April, because the snow was deep on the Columbian Plains until then and timber for firewood would be hard to find. The Nez Perce had warned them that snow was 20 foot deep in the Rocky Mountains until the first of June.

As they prepared to leave in March, Lewis praised the work of John Shields, saying that except for his skill in repairing guns, most of their guns would be entirely unfit for use. Lewis had ordered extra gun parts at Harpers Ferry, and his idea to put the gun powder in lead kegs had managed to keep their powder dry. Each lead keg contained 4 pounds of gun powder, and 8 pounds of lead when it was melted down to make bullets. They had an "abundant stock" of ammunition for the return journey, and they always took the precaution to divide the lead cannisters between the canoes. Guns and ammunition were all they had for subsistence and defense for their journey of over 4,000 miles back to St. Louis. [509]

On March 17th, Lewis wrote that—despite the "winning graces" of the girls brought to the fort by their old "baud" who were camping nearby and laying "sege" to the fort—the men were living up to their promise not to engage in sex. They did not want to deal with venereal disease on their return journey.

He wrote the weather was so uncertain they were going to start on their return journey whenever the weather broke. It had rained almost every day of the 141 days they spent at Fort Clatsop. Only 12 days were without rain.[511]

Drouillard had traded Lewis's uniformed laced coat and nearly half a carrot of tobacco (a carrot was 3 pounds of tobacco) for a canoe at the Cathlamet village. Lewis wrote he thought his uniform coat should be replaced by the U' States as it was but "little woarn."

The Clatsops refused to part with any canoe, for the small amount of merchandise they had left. Lewis decided to take a canoe from the Clatsops, "in lue of the six elk which they stole from us in the winter." Lewis wrote about their merchandise—

> . . . two handkercheifs would now contain all the small articles of merchandize which we possess; the ballance of the stock conists of 6 blue robes one scarlet do.[ditto meaning robe] one uniform artillerist's coat and hat, five robes made of our large flag, and a few old cloaths trimmed with ribbon. on this stock we have wholy to depend for the purchase of horses and such portion of our subsistence from the Indians as it will be in our power to obtain. a scant dependence indeed, for a tour of the distance of that before us.[512]

Lewis left a list of all of their names with the Chinook chief who accompanied his wife and their "female band." The statement was a proclamation announcing they had been sent out by the government of the U' States in May, 1804 to explore the interior of the continent. They had reached the mouth of the Columbia River on November 14, 1805, and they were departing on ___ day of March, 1806 to return to the United States. They made multiple copies of the proclamation to distribute and added sketches of their route. He wrote:

> . . . our party are also too small to think of leaving any of them to return to the U" States by sea, particularly as we shall be necessarily divided into three of four parties on our return in order to accomplish the objectives we have in view. [513]

They gave Fort Clatsop and their furniture to the Clatsop chief

Coboway, who had been "much more kind and hospitable to us than any other indian in this neighborhood," and left the fort on March 23rd, homeward bound.[514]

They had two canoes and three pirogues to paddle against the Columbia River's strong current, which was made more intense by the spring snow melt. After they arrived at the village of a Chinookan tribe at the Great Cascades, Lewis characterized them as the "greatest theives and scoundrels." On April 11th, three Indians stole Seaman. They also stole an axe, but Thompson wrestled it back from them. Three men set out to recover Seaman, with orders to fire on the Indians if they made any trouble. When they discovered they were being pursued, the thieves abandoned Seaman. The chief apologized, and the captains made him a present of a small medal. Lewis said he hoped it would keep the peace, as the "men seem well disposed to kill a few of them."

In bringing the boats up over the rapids, they were portaging the baggage around the cascades and carrying their "short rifles" to prevent theft. The boats were being hauled by the means of cords over the raging waters, when they lost one pirogue when it hit a rock. They were able to purchase two small canoes to replace it. The fleet now consisted of four small canoes and two large pirogues. They were beginning to encounter Indians with horses as they approached The Dalles, but were unable to purchase any of them. They halted at Rock Fort (Fort Rock) and Clark went out with a party to purchase horses at neighboring villages.

Lewis and his party prepared saddles and baggage cords for 12 horses and continued with the boats up to the Long Narrows. There was no possibility of river travel through the Narrows, and so the pirogues were cut up for firewood, much to the chagrin of the natives—who, however, were unwilling to pay for them. Clark returned with four horses, for which he had been forced to pay high prices. They portaged around the Narrows with the four pack horses on April 19th. On that day, the natives reported the first salmon had arrived in the river. According to custom, this first salmon was cut in pieces and given to every child. The salmon would arrive in great quantities in about five days.

They were forced to trade two kettles to acquire two more horses. They now had one kettle left for each mess of eight men. Clark was extremely provoked when one of the horses couldn't be found. The horses were stallions, who had not been gelded, and were at their "most vicious" during this season. It turned out the horse had been taken by an Indian who had won it gambling prior to its trade to Clark. Clark took back his goods from the seller. The Indians stole six tomahawks and a knife from them during the night. After that, he wrote they slept with their merchandise under the heads and arms ready to use, "as we always have in Similar Situations." He called the natives, "great jokies and deciptfull in trade." [515]

Meanwhile Lewis was having trouble at camp. Tomahawks were stolen and though they searched the natives they couldn't find them. Lewis said they would kill anyone who stole any more property and ordered all the spare poles, paddles and balance of the canoe burned rather than let the Indians have it.

They now had ten horses. William Bratton rode one, as he was suffering from back trouble and was too weak to walk. He had injured his back at the Salt Works and it hadn't yet healed. One horse broke away, losing his baggage, but retaining his saddle and robe. They found the horse and its saddle, but not the robe. The robe had been stolen by Indians who denied taking it. Lewis said he would "birn their houses." He wrote:

> they have vexed me in such a manner by such repeated acts of villany that I am quite disposed to treat them with every severyty, their defenseless state pleads forgiveness so far as rispicts their lives.

Fortunately, at this point, one of their men discovered the robe hidden in a village hut. [516]

By April 22nd they had acquired 13 horses. Clark wrote the horses, 10 of which were studs (male horses) were "very troublesom." It was mating season. Over the next days they hired horses, bought horses, traded horses, lost horses, found horses and were given a horse. On April 27th, they arrived at the Walla Walla village of Chief Yelleppit, the head chief of the region, whom they had promised to visit on their return.

Yelleppitt gave a very elegant white horse to Captain Clark, hoping to get a kettle in return. Instead, Clark gave him his sword, which the chief had dearly wanted, a hundred balls (bullets) and gun powder, and some small articles. The chief insisted they stay around for a dance and the arrival of their Yakama neighbors. Everyone wanted to see the Corps of Discovery dance. The fiddle was played and the men danced for about an hour. Then all the Indians, about 350 of them—men, women and children—danced and sang until about 10 in the evening.

While they waited, Clark doctored sick people. Like all the Indians of the region, they suffered from eye disease, which he relieved with eye wash, made from sugar of lead (lead acetate) and white vitriol (zinc sulfate) dissolved in water. Clark wrote he was giving them—

> Some eye water which I believe will render them more essential Sirvece than any other article in the Medical way which we had it in our power to bestow on them Sore eyes Seam to be a <serious> universial Complaint among those people.[517]

On April 30th they "took leave of these friendly, honest people," and accompanied by 23 horses, proceeded on to the west. While they were en route, three Indians came with a steel trap they had forgotten to take. While at camp the Indians had returned knives which had been carelessly left about. Lewis wrote:

> I think we can justly affirm to the honor of these people that they are the most hospitable, honest, and sincere people that we have met with in our voyage.[518]

On May 3rd they met the Nez Perce Chief they called "Big Horn" (he wore the horn of a bighorn sheep tied to his left arm by a cord), or "We-ark-koomt." He had been very instrumental in helping them on their outbound journey, and now had arrived with ten young men to meet them. The Walla Walla guide, who had showed them the short cut route from the Walla Walla River to the confluence of the Snake and Clearwater Rivers, and the men who returned the trap, now left them.

When they arrived at the Clearwater River they found the Nez Perce Chief Tetoharsky, the younger brother of Chief Twisted Hair, and a Nez Perce man who had guided their lead canoe, waiting for them. The Nez Perce were on their way to fish for salmon, but until the salmon arrived, they were harvesting the cous root growing on the plateau. Cous root was used to make bread, and to flavor soup.[519]

Clark was receiving many requests to resume his doctoring with the Nez Perce, and they decided to use his services to obtain food to eat (dog and horse meat). They had almost nothing left to barter. Clark was given a "very eligant Gray mare" for a tube of eye wash and other small items. Lewis received a elegant horse, in an exchange of horses with Chief Big Horn. It was a sorrel (red) horse, "strong, active and well broke." He gave the chief a small flag, and the Nez Perice chief left them in the care of Tetoharsky.

Tetoharsky acted as their guide along the Clearwater River. They learned they wouldn't be able to cross over the mountains because of the snow before the first of June or even at a "more distant period." When they met up with his brother Twisted Hair, they found he had quarreled with another chief, Cut Nose, over taking care of the expedition's horses and saddles during the winter. Twisted Hair abandoned the care of their horses because of this quarrel. He would now begin to collect the horses which had scattered, and locate the saddles. They said they would pay him the two guns and ammunition he had been promised if he delivered them. They met with both chiefs together Cut Nose and Twisted Hair. Cut Nose said Twisted Hair was a "bad old man" who had allowed his young man to ride the horses to go hunting and the horses were much injured. They were collecting their horses on May 10th when the snow began to fall. It was 8 inches deep.

William Clark's Nez Perce son was sketched by Gustavus Sohon at the 1855 Walla Walla Treaty Council. He labeled the drawing, Pinākua-ut-at, or Capon Rouge, Red Head. (Washington State Historical Society)

On May 11-12, they held a Grand Council with the chiefs and entered into a formal alliance with the tribe. A war chief, Red Grizzly, participated in the council. William Clark had a romance with the chief's sister—or perhaps she was a half-sister, or daughter. All stories agree that a baby with reddish hair and blue eyes was born after they left and was called "Clark," as well as "Daytime Smoker" or *Capon Rouge* (French for Red Head). Corps members were called Daytime Smokers by the Nez Perce, who did not smoke during the day. Clark's son was both photographed and sketched in later life. He was with Chief Joseph's band when they were captured in 1877 in the Nez Perce War. Tribal tradition relates that York had children with two Nez Perce women, and one child died, but a son lived and his descendants live on the reservation today.

Twenty five years later, in 1831, when William Clark was Superintendent of Indian Affairs, a delegation of Nez Perce visited him in St. Louis. The delegation consisted of four men, three Nez Perce and one Flathead. They came requesting a missionary be sent to teach them the "white man's book." It marks the beginning of the American missionary movement in the Far West.

Two of the men, the Nez Perce chief Black Eagle, son of Chief Red Grizzly, and the Flathead man, Man of the Morning, died in October and November, 1831 and were given Catholic burials. A Warriors Memorial honoring them was erected at Calvary Cemetery in St. Louis in 2003.

William Clark's second wife, Harriet Kennerly Clark, died in December, 1831. She had nursed Black Eagle at the Council House, while she was in weakened health herself. William Clark had erected a large building behind his two story home where his Indian artifacts were displayed. It was both a museum and a council chamber for visiting Indian delegations. Clark left no record of the Nez Perce visit. His wife was ill, and their 13 year old son Julius had died in September. It was a very sad time.[520]

But in 1806, it was springtime on the Clearwater. The men enjoyed themselves as they waited for the snow to clear out of the mountains. They set off for home on June 10th, although they were warned the snow was still too deep. After discovering it was indeed too deep, they had to backtrack. On June 24th, they set off again, accompanied by Nez Perce guides. It took them six days traveling through the dense forests of Lolo Pass to reach Traveler's Rest, a journey of 156 miles.

At Traveler's Rest, in the Bitterroot River Valley, they followed Lewis's plan to split into several groups with different missions. Initially there were two groups—Lewis headed north with nine men, and Clark went south with the rest of the party. Lewis crossed the Continental Divide through the Blackfoot River corridor, which the Nez Perce called "the Road to the Buffalo." Clark crossed at Lemhi Pass, arriving at Camp Fortunate at the forks of the Beaverhead (Jefferson) River on July 8th.

On July 5th, Lewis named Seaman's Creek (now Monture Creek) along the Blackfoot River for his dog. Historian Donald Jackson realized Lewis's dog must have been named "Seaman" rather than "Scannon" when he was identifying place names in Montana. The error had occurred years earlier through a misreading of journal handwriting. ("Seaman" referred to the Newfoundland dog's primary occupation as a fisherman's dog on the Atlantic Coast.) Lewis and his group were much "on their guard" as they expected to meet the Minatarees of Canada (Gros Ventre of Fort de Prairie) on the Blackfoot River.

When they reached the plains, the continual roar of bellowing male buffaloes could be heard for miles. It was the mating season. Lewis estimated there were at least 10,000 buffalo surrounding the location of their old Upper Portage Camp. They used buffalo skins and willow branches to make two "bull boats." One boat was "Mandan style," and the other was their own invention. When they opened their cache site from 1804, they found many contents destroyed or damaged by flooding. Fortunately, the Missouri River Chart was intact. A phial of laudanum had spilled open and destroyed most of the medicine. All of his plant specimens were lost. His papers were damp, and they were spread out to dry.

The iron boat frame and the wooden wheels were all right. He sent McNeal down to the Lower Portage Camp to see if the white pirogue and cache site were safe. McNeal had another exciting adventure, being chased up a tree by a grizzly bear, after he hit the bear on the head with his musket. Meanwhile, Indians had stolen seven of their best horses, leaving them only ten. Drouillard tracked the horse thieves, but they had two days head start and he abandoned the chase.

Lewis left Sergeant Gass in charge of four men at the Upper and Lower Portage Camps. It was their job to transport the canoes and baggage over the portage route. The "mosquetoes" were so bad that Lewis spent 3/4's of his time under his mosquito netting. Seaman howled with the torture of them. They were so thick that they frequently got caught in their throats.

On July 16th, Lewis set out with Drouillard and the two Field brothers, Joseph and Reuben, bound for the Marias River. They took one bull boat and six horses. When they arrived at the Great Falls, Lewis again made a sketch of them, but no drawings have been found. On July 17th, he wrote after leaving the falls—

> . . . the Minnetares of Fort de prarie and the blackfoot indians

> rove through this quarter of the country and as they are a vicious lawless and reather an abandoned set of wretches I wish to avoid an interview with them if possible. I have no doubt but they would steel our horses if they have it in their power and finding us weak should they happen to be numerous will most probably attempt to rob us of our arms and baggage; at all events I am determined to avoid them if possible.[522]

It was a prophetic warning of events to come.

As they traveled towards the Marias River they saw an immense herd of buffalo covering 12 miles. On July 19th they began traveling on horseback along the Marias. By the 22nd, Lewis realized the Marias headed west into the mountains instead of continuing north towards Canada. They were following the north fork of the Marias called "Cut Bank Creek." This was one of his three territorial missions—to see if he could establish a latitude reading for a tributary of the Missouri that might reach 50 degrees north. He named their camp, "Camp Disappointment," where he took latitude-longitude readings and they explored the country. Lewis speculated the White Earth and Milk Rivers probably extended further north than the Marias. He was right about the Milk, but not about the White Earth.[523]

After Drouillard discovered an Indian camp of eleven lodges had been occupied about ten days earlier, Lewis decided to avoid the Marias and travel through open plains on their way back to the Missouri. On July 26th, they headed south towards the other fork of the Marias, Two Medicine River, and crossed over it. On the south side of Two Medicine River, Lewis climbed a hill to take a look with his spy glass. He saw about 30 horses and some Indians on top of a hill about a mile away who had spotted Drouillard, who was traveling alone along the river.

Lewis directed Joseph Field to display the flag, and they rode towards the Indians. When the Indians became aware of their presence they were alarmed. Lewis realized that their own horses were too "indifferent" to escape capture by fleeing, and because

Drouillard was separted from them, he would most probably be killed. He told the Field brothers—

> I expected that we were to have some difficulty with them; that if they thought themselves sufficiently strong I was convinced they would attempt to rob us in which case be their numbers what they would I should resist to the last extremity prefering death to that of being deprived of my papers instruments and gun and desired they should form the same resolution and be on their guard.[524]

At the first encounter, Lewis shook their hands and they asked to smoke a pipe together. He said Drouillard had the pipe and they should send one man and he would send one man to go in search of Drouillard and the pipe. This was done. There were eight Indians, and he asked by sign language if there were any chiefs among them. They designated three of their party as chiefs. Lewis didn't believe them, but gave a medal to one man, a flag to the second, and a handkerchief to the third. He believed they were the "Minatarees of the North" (Gros Ventre of Fort de Prairie), but they were Piegan Blackfeet. Since there were only eight Indians, Lewis decided they "could manage that number should they attempt any hostile measures."

He suggested they camp together, as he had a "great deel to say to them." They went to the Indians camp near the river, where they had temporary lodges set up, skins over domed shaped lodges. Drouillard returned with the men sent to find him, and with Drouillard's sign language, Lewis had "much conversation with these people in the course of the evening." The Indians were part of a large band, 1½ days march from there, who had a white man with them. There was a larger band near the mouth of the Marias hunting buffalo. They got arms and ammunition from traders on the Saskatchewan River.

Lewis told them he had come a great way from the east and had been to the great waters where the sun sets. He had arranged peace between many tribes, and had invited all of them to come trade with him on the eastern side of the Rocky Mountains. He was going to set up a trading establishment at the mouth of the Marias River on the Missouri. He wanted them to be at peace with their neighbors on the west side of the mountains. They replied these Indians had recently killed several of their relations, and pointed to those present who had cut their hair (a sign of mourning), but readily agreed to a trading post and making peace with their neighbors. They smoked the pipe until late at night. Lewis asked them to bring their leaders and the white man to a council at the mouth of the Marias, and requested some of them accompany him to the mouth of the Marias where he would meet his own men, who were anxious to see him. If they went with him, he would give them ten horses and some tobacco. They made no reply to this proposition.

Lewis and Drouillard slept in a shelter and the Field brothers lay near the fire in front of their shelter. Lewis wrote:

> [July 26, 1806] I took the first watch tonight and set up untill half after eleven; the indians by this time were all asleep, I roused up R. Fields and laid down myself; I directed Fields to watch the movements of the indians and if any of them left the camp to awake us all as I apprehended they would attempt to seal [steal] our horses. this being done I fell into a profound sleep and did not wake untill the noise of the men and indians awoke me a little after light in the morning.

> [July 27, 1806] This morning at day light the indians got up and crouded around the fire, J. Fields who was on post had carelessly laid his gun down behid him near where his brother was sleeping, one of the indians the fellow to whom I had given a medal last evening sliped behind him and took his gun and that of his brothers unperceived by him, at the same instant two others advanced and seized the guns of Drewyer and myself. J. Fields seeing this turned about to look for his gun and saw the fellow just running off with her and his brothers he called to his brother who instantly jumped up and pursued

> the indian with whom they overtook at the distance of 50 or 60 paces from the camp sized their guns and rested them from him and R Fields as he seized his gun stabed to the indian to the heart with his knife the fellow ran about 15 steps, and fell dead . . .

After that, in the ensuing commotion, they retrieved their guns and ammunition. Lewis had an extra gun, a pistol. Lewis forbade the men to shoot any Indians, "as they did not appear to wish to kill us." However, the Indians then stole Lewis's horse and another horse and Lewis set off in pursuit. At 300 paces he was out of breath, and called to them saying—as he had several times—he would shoot them if they didn't return his horse.

The Indians who had his horse stopped, and one of them had a gun, a trade musket. At a distance of 30 steps, Lewis shot the Indian "through the belly." The Indian fell to his knees, and returned fire. His bullet overshot Lewis, who wrote—"being bearheaded I felt the wind of his bullet very distinctly."

Lewis was unable to reload his pistol, being without ammunition. He returned to camp, where they began to catch the Indians' horses. They still had four of their own. They left one of theirs and took four of the best Indian horses, leaving nine. They found the other Indian gun and kept it, and burned all the rest of their possessions. The Indians had two trade muskets. Lewis took back the American flag, but left the medal around the neck of the dead man, so "they might be informed as to who we were."

They began their escape to the confluence of the Marias and Missouri River, where they planned to meet up with Sergeant Gass and his men. They rode as fast and as hard as they could until about two o'clock the next morning, through immense herds of buffalo. The moonlit night sky had heavy thunderclouds as they rode through the buffalo. They covered 80 miles on that single day, and 20 miles the next day.

Lewis wrote that his Indian horse was better than his own horse, the one which had been stolen. When they awoke the next morning they could "scarcely stand— they were so stiff and sore from the long ride. He encouraged them to keep going, saying their lives and the lives of their friends depended on their exertion. They had to get to the mouth of the Marias in time to warn the men coming down the Missouri of the danger.

When they approached the Missouri, they heard the sound of rifle shots, and had "the unspeakable satisfaction to see our canoes coming down." The party coming down river consisted of both Gass's party of six men, and ten men led by Sergeant Ordway, who had separated from Clark at Three Forks. Seaman would have remained with Gass during Lewis's exploration of the Marias.

British traders left two accounts regarding the incident. On December 23, 1806, James Bird, manager of the Hudson's Bay Company's trading post on the Saskatchewan River, wrote:

> A party of Americans was seen last Summer where the Missoury enters the rocky Mountains & tis reported by the Muddy or Missoury River Indians that four of them set off with the intention to come here but that they kiled one and the rest returned.

Historian John C. Jackson found this reference and writes, "this suggests the gut-shot man actually survived." [525]

There are two other accounts, one by the fur trader and map maker David Thompson, and the other by Wolf Calf, a Piegan Blackfeet. Both men were old when they recounted their stories. Thompson, born in 1770, was writing his memoirs after his retirement in 1845. Wolf Calf, born in 1793, was interviewed by anthropologist George Bird Grinnell in 1895 at the age of 102.

Wolf Calf was one of the horses thieves who encountered Lewis's party. He would have been 13 years old. Grinnell wrote:

> He told me that he was with a war party to the south when they met the first white men that had ever come into the lower country. They met these people in a friendly fashion, but the chief directed his young men to try and steal some of their things. They did so, and the white men killed the first man with their "big knives." This was the man killed, I suppose, by

> R. Fields. Afterwards the Indians ran off some of the horses of the white men. The name of the first man killed was Side Hill Calf, or Calf Standing on a Side Hill.

Wolf Calf does not say that *two* men were killed in this account. He names one man, the dead man, whom Reuben Field killed with a knife—the one who was designated their chief and who was wearing the medal Lewis gave him. It is Grinnell who calls him the *first* man killed, "I suppose by R. Fields." There is no name supplied for the second man. Grinnell is adding his own interpretation.[526] The journals do not say there were two men killed. They say that one man was killed, and one man was wounded, who fired back at Lewis. The contemporary account by James Bird refers to one man killed.

History has not been kind to Meriwether Lewis, because of the death by suicide story. Lewis did not say he killed a man, but it has been "assumed" he killed a man and *two* men were killed at the Blackfeet Fight. These are the kind of assumptions that have plagued the story of his life for over 200 years, ever since the suicide story was accepted. His accomplishments have been downplayed and negative interpretations of his life are commonplace, seeking to justify his supposed suicide. His character has been assassinated, without knowing the facts.

After the fight on Two Medicine River, David Thompson, an explorer, mapmaker, and fur trader working for the North West Company, wrote:

> . . . the murder of two Peagan Indians by Captain Lewis of the United States, drew the Peagans to the Mississouri to revenge their deaths; and thus gave me an opportunity to cross the Mountains by the defiles of the Saskatachewan River, which led to the head waters of the Columbia River, and we there builded Log Houses . . .[527]

Thompson wrote this account some 40 years later, when it was already an accepted fact that two Piegan Blackfeet had died. Thompson built Fort Kootenai at the headwaters of the Columbia in July, 1807, securing the British claim to British Columbia. At the same time, Manuel Lisa's Missour Fur Company was building a trading post on the Upper Missouri in Montana.

Two of Lewis and Clark's men were killed by the Blackfeet after they returned to the Upper Missouri on Missouri Fur Company expeditions. John Potts and John Colter were on the Jefferson River in 1808, trapping for beaver, when they encountered a war party of about 500–600 Blackfeet. Potts shot and killed an Indian, and then Potts was killed by the Indians. John Colter was taken captive, stripped naked, and offered the chance to run for his life. Colter, who could speak some of the Blackfeet language, told them he was a very bad runner.

He was given an advantage of about 3–400 yards and was told *to save himself if he could*, before young warriors with spears set out after him. He was in fact a very fast runner, but ran at a speed that surprised even him, over prickly pear cactus. He was racing towards the Madison River at Three Forks, six miles away, when an Indian caught up with him, one mile from the forks. Colter, with blood gushing from his nostrils from his exertion, took the spear away and killed his pursuer. Taking the Indian's blanket, he set off again. At the Madison River he dove underwater to take shelter under a "raft of drift timber" lodged against an island. The Indians stood on the raft throughout the day looking for him, but didn't discover him. Another account calls it a two story beaver house, ten feet high, with the beavers building a lodge on top of the driftwood. Colter then traveled through Bozeman Pass and across Montana, eating roots and the barks of trees for eleven days, before reaching the safety of Manuel Lisa's fort at the junction of the Yellowstone and Bighorn Rivers, 300 miles away.[528]

In 1810, George Drouillard was the second member of the expedition killed by Blackfeet. The fort at Three Forks was established by Lisa in the spring of 1810, despite unrelenting Indian harrassment and attacks. Drouillard was part of a group of 18 men who went up the Jefferson (Beaverhead) River in April to trap for beaver. Two of the men were killed by the Gros Ventre of Fort de Prairie.

It had proven impossible to meet with the Blackfeet, who continued to roam around the Three Forks Fort. They wanted to tell the Blackfeet they would build them a trading fort at the Marias, as promised by Lewis. They considered taking a prisoner, and giving him the message to take back to his leaders. However, this plan was not put into effect.

An overconfident Drouillard set out with two Delaware Indians to trap for beaver on the Jefferson. He was killed sometime in May, and his body hacked to pieces. His companions were also killed. It was a sad end for a man, whom Lewis called a "man of much merit . . . who uniformly acquitted himself with honor." After he returned from Three Forks, Pierre Menard, a partner in the Missouri Fur Company, gave an interview to the St. Louis *Louisiana Gazette*, published on July 26, 1810. Menard blamed Blackfeet hostility on the company's decision to built the first American fort on the Yellowstone and Big Horn Rivers, rather than on the Marias and Missouri. John Colter had persuaded Lisa that the Yellowstone River was the best beaver country. It was also the home of the Crow Indians, bitter rivals of the Blackfeet. The Blackfeet wanted better access to guns and ammunition than they had with British traders on the Saskatchewan, and deeply resented that the trading post Lewis had promised to them was given to the Crows.[529]

Other reasons have been proposed for the Blackfeet attacks on Americans over the next years—blaming it on the fight at the Two Medicine River, or because Colter fought on the Flathead side in a battle between the Flatheads and Blackfeet—but Menard's explanation makes the most sense.

Lewis, after reuniting with Gass and Ordway's parties, went to Decision Point near the mouth of the Marias River. He discovered the red pirogue was beyond repair, and had the "nails and other ironwork" taken from the boat. Gass's party would have been hauling the iron frame of *The Experiment* left at Upper Portage Camp. Their forge, ironworking tools, and salvaged iron was given to the Mandans when they reached their villages.

On July 28th, they set out from the Marias in the white pirogue and five small canoes. They killed some bighorn sheep at the White Cliffs. It was rainy, cold and disagreeable, but the men were anxious to get home and "plyed their oars faithfully," traveling at a rate of 7 miles per hour. The river was rising and very muddy. It continued to rain and they took shelter in Indian lodges and built fires to dry and air their skins.

On August 7th, they arrived at the mouth of the Yellowstone River where they found a torn note left by Clark, who informed Lewis that because the musquetoes were troublesome and game was scarce, they were going to move ahead a few miles. They had been camping there 7 or 8 days earlier. Soon they found a campfire and part of a Chinook hat belonging to Gibson.

After not finding Clark the next day, Lewis halted so as to allow the men time to make some clothes for themselves out of skins, as most of them were "extreemly bare." The pirogue and a canoe also needed repair.

On August 11th, they began traveling again. Lewis and Cruzatte went out to kill some elk. Lewis killed an elk, and Cruzatte wounded one. They reloaded their guns and started through the willow thickets to find the wounded elk. Lewis wrote:

> I was in the act of firing on the elk a second time when a ball struck my left thye about an inch below my hip joint, missing the bone it passed through the left thye and cut the hinder part of the right thye; the stroke was very severe; I instantly supposed that Cruzatte had shot me in mistake for an elk as I

> was dressed in brown leather and he cannot see very well; under this impressions I called out to him, damn you, you have shot me.[530]

Cruzatte was afraid to answer him, and Lewis then assumed he had been shot by Indians. Lewis ran back to the pirogue and called the men to arms. He told them he was wounded, but he "hoped not mortally." Lewis set out with the men to rescue Cruzatte but the pain became so unbearable he returned to the pirogue, where he prepared his pistol, rifle, and air gun, determined "to sell my life as deerly as possible."

When the men and Cruzatte returned, Cruzatte denied hearing Lewis call to him, but Lewis knew better. His wounds were dressed, and he was happy to find that the bullet had "touched neither bone nor artery." He slept on board the pirogue with a very high fever.

The next day they set out in haste to find Clark and his party. Instead they found two hunters from Illinois, Joseph Dickson and Forest Hancock, who reported that Clark had passed them about noon the day before. Lewis described the river ahead, and told them where the beaver "most abounded." He gave them a file and some powder and lead. He then proceeded on to find Clark and his party, reporting on August 12th that—

> as wrighting in my present situation is extreemly painfull to me I shall desist until I recover and leave to my frind Capt. C. the continuation of our journal.[531]

Clark, writing about the incident, said that Lewis was dressed in leather very nearly the color of elk, and reported—

> This Crusat is near Sighted and has the use of but one eye, he an attentive industerous man and one whome we both have placed the greatest Confidence in dureing the whole rout.—[532]

It's interesting that—despite his eyesight—Cruzatte was the navigator and bowman who brought them safely through many dangerous waters on the rivers.

William Clark and his party separated from Lewis at Travelers Rest on July 3rd to travel south through the Bitterroot River Valley to begin their exploration of the Yellowstone River. They had fifty horses. Charbonneau and Sacagawea would serve as interpreters with the Crow, because the Hidatsa language was similar, and Sacagawea would interpret with the Shoshone. Crossing at Lemhi Pass, they arrived at Camp Fortunate on July 8th. Clark wrote the men who used tobacco could scarcely wait to get back to the cache deposits where they had left tobacco. He called the route from Traveler's Rest to Camp Fortunate, an "excellent road."

On July 10th, they set off using the canoes that had been stored at the Jefferson forks. Clark chose the group which would go with him on the *Roche Jaune* (French for Yellowstone), from among applicants. Sergeant Pryor was put in charge of bringing the horses along the Jefferson to Three Forks and then along the Yellowstone. When they reached Three Forks, Segeant Ordway took six canoes and ten men to join Lewis at the Marias river junction. Five different groups were active at this time, including Sergeant Gass's party at Upper Portage Camp. It was a complicated and risky maneuver to split into small groups, but it demonstrates the captains' faith in the abilities of their men.

At Three Forks, Clark's party set out by horseback, guided by Sacagawea, to follow an old buffalo road. After going through Bozeman Pass, they arrived at the Yellowstone on July 15th. They proceeded along the river hunting for timber, and on July 19th Clark set up camp to make canoes. They saw Indian smoke signals, but only one Indian in the distance. The trees near their camp were small, and Clark decided to have two small canoes lashed together, side by side, to make a sturdier vessel.

On July 21st, they discovered half their horses were gone. By the next day, Clark realized the Indians had stolen 24 horses and taken them over hard ground to conceal their tracks. On the 24th, Clark sent out Sergeant Pryor with two men to take the

Pompeys Pillar National Monument and Clark's inscription, "Wm Clark July 25 1806" (Wikimedia Commons)

remaining horses to the Mandan Villages. Hugh Hall, who couldn't swim, wanted to help with the horses but "observed he was necked." Clark gave him one of his own two remaining shirts, leather leggings and three pairs of mocassins.

On July 25th, William Clark named a very large rock, "Pompey's Tower," for a very small boy, 17 month old Jean-Baptise Charbonneau, nicknamed "Pompey," and signed his name, "Wm Clark July 25 1806." The 200 feet tall rock was covered with petroglyphs (Indian drawings) and in later years, pioneers added their names. Clark's name can still be seen.

By August 3rd, they had reached the junction of the Yellowstone with the Missouri River. On August 8th, Sergeant Pryor and three men came down the Yellowstone in two bull boats. All of their remaining horses had been stolen by Indians. They had planned to use their horse herd to trade for corn and beans with the Mandans. Now they began to collect skins to use for trade.

On the 12th of August, Lewis and Clark were reunited at Reunion Bay. Lewis told Clark his wound was slight, and he would be recovered in 20 to 30 days. Clark took over the job of dressing Lewis's wound. Clark then proceeded to write a detailed report of Lewis's travels, Lewis's recommendations of the best routes to follow, and his account of their fight with the Blackfeet.

By August 14th, they had arrived at the upper villages of the Mandan and Hidatsa. When they met with the Hidatsa chiefs, they invited them to return to the U. States with them. The chiefs said they were afraid to go because they would be killed by the Sioux. The captains gave their swivel gun to "One Eye," the great chief of the Hidatsa, because it couldn't be fired from the pirogue. Since they last saw them, the Hidatsa had attacked the Shoshone; been attacked by the Sioux and Arikara; and had quarreled with the Mandan.

The expedition had been joined by the two trappers from Illinois who returned to the villages with them. Dickson and Hancock offered Colter shares in their company if he would go back to trap on the Yellowstone with them. The captains agreed, on the condition that no other members of the expedition would receive an early discharge from military service.

The chief of the Mandans offered them six loads of corn, more than their canoes could carry. They thanked him for his kindness. The Hidatsa chiefs still refused to go, so they asked René Jessaume to use his influence to persuade one of the Mandan chiefs to go to Washington, and they would employ him as an interpreter. Jessaume persuaded Big White, or Sheheke (White Coyote), a Mandan Chief to go, on the condition that Big White's wife, Yellow Corn, and their toddler son, and Jessaume's wife and their two children would accompany them.[533]

Charbonneau really wanted to go, but couldn't persuade any Hidatsa chiefs, and he didn't speak the Mandan language. They settled up with him and paid him $500 dollars and 33 ⅓ cents. Clark offered to take their little son, Pompey, "a butifull promising Child," and raise him in St. Louis. Toussaint and Sacagawea agreed to bring him to St. Louis a year later, after he was weaned. He was 19 months old.

They set out from the Mandan villages with Big White and his entourage on August 17th. The chiefs of all the villages held one more council and said they would strictly attend to the captains' advice and not engage in warfare unless they were attacked. As they traveled down the Missouri, Big White pointed out places where his nation had formerly lived, and told "Some extraordinary Stories of their tredition"—which unfortunately were not recorded.

On August 21st, they met some French traders on the river, who informed them that the Arikara Chief who had gone to Washington had died on his journey, and that 700 Sioux were on their way to attack the Hidatsas and Mandans. On the same day, Clark held a council with the Arikara chiefs at their villages. The Arikara were not aware their chief had died, and he did not tell them. After they learned of his death, it caused the Arikara to attack Americans for many years afterwards.

On the 22nd Clark reported:

> I am happy to have it in my power to Say that my worthy friend Capt Lewis is recovering fast, he walked a little today for the first time.[534]

They were traveling rapidly down the Missouri, making over 40–60 miles a day. Their fleet consisted of several canoes, the white pirogue, and two canoes lashed together in which Big White, Jessaume, and the two wives and three children were riding, under the care of Sergeant Pryor.

On the 30th of August, they saw 80–90 Indian warriors, armed with Indian trade muskets and bows and arrows. Clark took three French men who could speak Indian languages (Cruzatte, Labiche, and Jessaume) to find out who they were. When they learned they were Teton Sioux, Clark rebuked them for their treatment of the expedition in 1804, and said they had given a cannon to the Mandans and Menetarres (Hidatsas) to defend themselves. He warned them to keep away from the river, or they would kill every one of them. Black Buffalo, the Teton chief who had accompanied them for two days on the river, approached them. Clark observed that he was "Said to be a friend to the white people," but paid him no attention.

On September 1st, they heard gunfire from Indians, and formed groups to defend themselves. They discovered they were Yankton Sioux who had been shooting at an old keg the expedition had discarded on the river. When they smoked pipes together, Clark said if they had been Teton, he planned to kill them all. Their great chief and many brave men had gone to Washington with Pierre Dorion, and had not yet returned.

On the 3rd, they met the British trader Jame Aird, who told them their first news of the United States. They learned that General Wilkinson was the Governor of Louisiana at St. Louis, and there were 300 troops stationed at the junction of the Missouri and Mississippi, and that—

> . . . Some disturbance with the Spaniards in the Nackatosh Country is the Cause of their being Called down to that Country . . . and that Mr. Burr and Genl. Hambleton fought a Duel, the latter was killed &c. &c.[535]

Aird gave them tobacco, enough to last until they reached St. Louis for every man who smoked, and traded a barrel of flour for six bushels of corn.

On the 4th of September they stopped and visited Sergeant Floyd's grave, where they found the grave had been opened and left half filled. Clark later told Nicholas Biddle, editor of the journals, he had learned that a Sioux chief had lost one of his sons, and had opened Floyd's grave and buried his son with him—

> . . . for the purpose of accompanying him to the other world believing the white man's future state was happier than that of savages.[536]

On the 6th they met a trading boat of Auguste Chouteau's and obtained a gallon of whiskey from him, giving every man a dram to drink, the "first Spiritious licquor" they had tasted since the 4th of July, 1805 at Upper Portage Camp. On the 8th of September they made 78 miles, being anxious to reach the Platte River. They camped again at White Catfish Camp and the next day passed the mouth of the great river Platte. They were entirely unaware there were 300 Spanish troops being detained at the Pawnee village on the Republican River who had been sent to capture them. They were simply anxious to get home to see their families and friends.

On the 10th they met another trader on the river who told them General Wilkinson had descended the Mississippi with his troops and Zebulon Pike and young Mr. Wilkinson had set out to explore "the Arkansaw River or in that direction." The river was now filled with sand bars and moving sands (quick sands) which required great caution to navigate.

On the 12th they met Robert McClellan, an old army comrade of Lewis and Clark. He was escorting Joseph Gravelines, the Arikara trader, and Pierre Dorion, the Yankton trader, back to their homes. They were now regularly meeting boats on the river, and being given gifts of whiskey. On the 17th they met Captain John McClallan, an associate of General Wilkinson. McClallan told them he was going to New Spain to establish trade with Sante Fe, after first establishing a post at the mouth of the Platte. He was going to enlist the services of Pawnee chiefs, and trade his merchandise with the Spanish for the "Silver & gold of which those people abound." Wilkinson had provided him with a speech to make to the Pawnees and Otoes. They found him "a man of information" and learned they had long since been given up for dead. They talked until nearly midnight about the political state of their country.

When they arrived at the little French village of La Charette on September 20th, the men raised a shout of joy at seeing cows along the banks of the river. On September 21st, they saluted the town of St. Charles with three rounds of fire, and the next day proceeded on to the new army cantonment on the river. On the 23rd they provided Big White with clothes from the army store, and went on to St. Louis, where they received a "harty welcom" from the inhabitants of the village.

It had been 863 days since they left Wood River Camp near St. Louis on May 14, 1804. The United States Army Corps of Volunteers for Northwestern Discovery had successfully completed their mission to explore the Upper Missouri and reach the mouth of the Columbia River on the Pacific Coast and they had returned with maps, journals, Indian artifacts, plant, animal and mineral specimens, and a lifetime of stories to tell.

EXPEDITION ROUTES
1804–1807

1. Lewis and Clark
2. Zebulon Pike: Mississippi
3. Zebulon Pike: Southwest
4. James Biddle Wilkinson
5. Dunbar and Hunter
6. Freeman and Custis
7. Aaron Burr

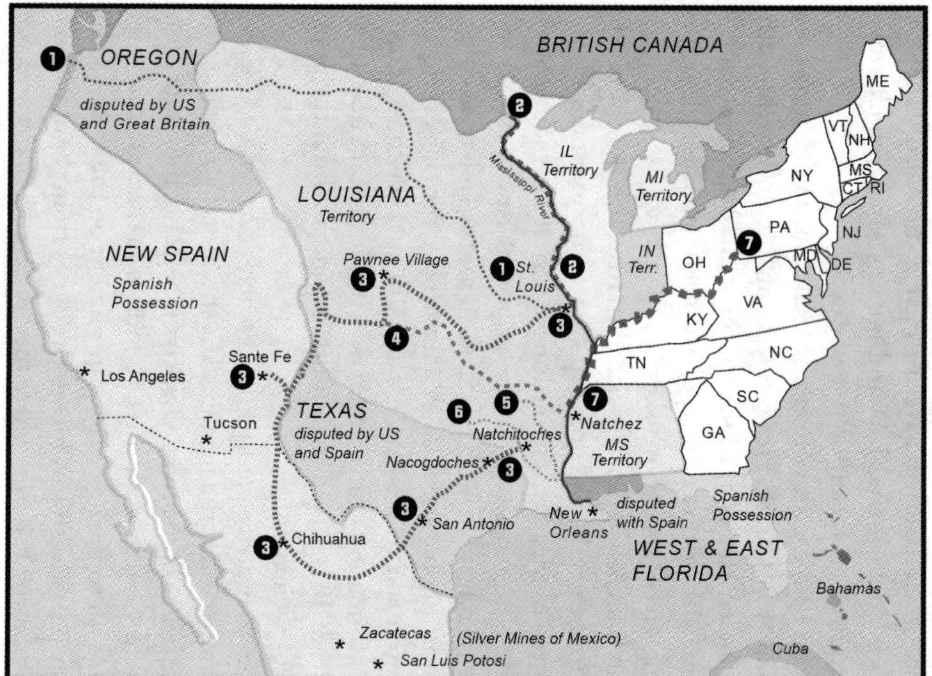

Disputed boundaries and expedition routes, 1804-1807

Chapter 11

The Burr-Wilkinson Conspiracy: 1804–07

General Wilkinson paid a secret visit to Aaron Burr in New York City on May 23rd, 1804. The general was on his way to Washington to report to the president, bringing the American version of his Louisiana report and 28 maps of the Southwest made by Philip Nolan. Wilkinson sent Burr a note:

> To save time of which I need much and have little, I propose to take a bed with you this night, if it may be done without observation or intrusion—Answer me and if the affirmative I will be with [you] at 30 after the 8th hour.[537]

This was the start of the renewed conspiracy of Wilkinson and Burr to invade Spanish territory. Historian Thomas Fleming in his book, *Duel: Alexander Hamilton, Aaron Burr and the Future of America,* proposes that Burr killed Hamilton over this issue in the famous duel, held on July 11th, 1804.

Aaron Burr, while still Vice President of the United States, had run for the office of Governor of New York in the preceding months. He ran as a Republican with the support of Federalists, but lost the election. Alexander Hamilton had worked strenuously to defeat him. Hamilton, the founder of the Federalist Party, supported Jefferson's Louisiana Purchase because he believed in American expansionism, but most Federalists opposed the purchase.

Federalists argued the purchase was unconstitutional, but really feared it would increase the number of slave states and end their chances of regaining national power. The 3/5's Compromise, brokered between slave and northern states at the Constitutional

Map of the Northern Confederacy

Convention in 1787, established that a slave counted for 3/5's of person, increasing the slave states' representation in the House of Representatives. All of the Louisiana Purchase states formed between 1803–1845 became slave-holdng states.

There was open talk of seceding from the union among the Federalist congressmen from northern states, who advocated forming a Northern Confederacy consisting of New Jersey, New York, the New England states of Massachusetts, Connecticut, New Hampshire, Vermont and Rhode Island, and the British provinces in Canada and Nova Scotia. Hamilton opposed the election of Burr for governor of New York because if Burr became governor, he would lead the break away from the union. After Jefferson's anticipated reelection in the fall, a secessionist meeting was going to be held in Boston.[538]

What were Wilkinson's true feelings regarding Republicans and Federalists? When Meriwether Lewis made out his coded review of army officers, five out of six of the general's staff were "opposed most violently to the administration and still active in its vilification," and the sixth was "opposed more decisively." Although Lewis left the general's evalutation blank, Wilkinson was a staunch Federalist in his sympathies. When the general approached Aaron Burr in May, he must have been willing to cooperate with the Federalists planning to secede from the union, and willing to support the establishment of a Northern Confederacy with close ties to Great Britain.

The general and Burr had been planning an invasion of Texas during the election of 1800. The inconvenient truth, however, was that—if the Northern Confederacy succeeded in taking over Spanish territories—they would allow slavery. It was simply that Federalists and the British would be in control, rather than the French-loving Republicans.

Wilkinson had written to Alexander Hamilton from New Orleans on March 26, 1804, in response to a letter from Hamilton. Hamilton had asked the general to use his influence in aiding two young men who were moving to New Orleans. Wilkinson replied to Hamilton—with whom he had planned the establishment of the cantonment on the Ohio in 1799 when Hamilton was commanding general of the U. S. Army—

> I would give a *Spanish Province* for an Interview with you. My Topographical information of the So. West is compleat. The infernal designs of France are obvious to me, & the destinies of Spain are in the Hands of the U. S. Name me if you please to G. Morris.

Gouvernour Morris, a firm supporter of George Washington, had cast the deciding vote during the Conway Cabal to keep Washington as Commander in Chief of the Revolutionary Army. He was also a supporter of taking Louisiana and the Floridas by force.[539]

Wilkinson believed the time was ripe to stage an invasion of Mexico. The geographical boundaries of the Louisiana Purchase were disputed by Spain, and a filibuster expedition to seize control of Sante Fe would settle the matter. It would also prevent any "infernal designs" of Napoleon to take back Louisiana and invade Mexico. Privately financed invasions of foreign territory by armed expeditions were popular during the first half of the 1800's. Filibusters were associated with Free Masonry, and thousands of young Americans were ready to invade the Spanish countries surrounding the Gulf of Mexico, in order to aid their fellow masons in fighting for independence from Spain.[540]

After Wilkinson's overnight interview with Burr, he left New York within the week. The general and his wife visited their son Joseph Biddle at Princeton, and arrived in Washington on June 25th. It was three weeks before the duel which ended Alexander Hamilton's life.

It seems likely that another visitor to Burr's home stimulated his plan to force a duel with his arch enemy Alexander Hamilton. The visitor was Charles Williamson, a Scottish-born American citizen who had acted as land agent for one million acres of land in western New York owned by British investors. Aaron Burr had done legal work for Williamson and was a speculator in New York's western lands himself. Williamson, after ending his employment as land agent, had gone to England where he associated with high ranking government officials, and was now back in the U. S. promoting a new project.

He was planning a filibuster, which he called "The Levy." On May 16th he wrote to Burr that he expected to see him in one month's time, when he would be in New York City before returning to England. His plan for "The Levy" involved using young Englishmen who had recently arrived in the U. S. to serve as a corps of volunteers to invade the French West Indies and Spanish possessions on behalf of Great Britain.[541]

Aaron Burr had strong connections to a prominent British family, the Prevosts. His wife Theodosia was first married to Jacques Marc Prevost, a British military officer. She was a widow with five children when she married the 26 year old Burr in 1782. Theodosia died in 1794, leaving him to bring up their only child, also named Theodosia, born in 1783. His wife's first husband was the uncle of George Prevost, who was serving as Governor of the British island of Dominica in the West Indies in 1802-05. Sir George became the Lieutenant Governor of Nova Scotia in 1808, and then served as Commander in Chief of British North America during the War of 1812.

Thomas Jefferson had a bust of Alexander Hamilton placed in the entrance hall of Monticello opposite his own portrait bust. (www.monticello.org)

Williamson may have brought Burr a copy of an upstate New York newspaper, which contained a letter in which the writer said that he heard Hamilton speak of Burr as a "dangerous man and [who]ought not to be trusted." After the letter was published, the writer defended his comments by saying, "I could detail you a still more despicable opinion which Gen. Hamilton has expressed of Mr. Burr." Since newspapers of New York State had been filled with outrageous remarks during Burr's run for governorship, by Burr taking offense at the word "despicable," it was Burr's means of challenging Hamilton to a duel.

Burr sent a letter to Hamilton on June 18th asking him to deny the allegations. Hamilton replied it was "too indefinite to discuss." The situation continued to escalate until it ended in a duel which took place across the Hudson River in New Jersey, because New Jersey was more lenient in prosecuting duelists.

It appears that Hamilton had decided to end his life in order to destroy his rival's chance of breaking up the union. On June 27th Hamilton wrote he had decided "to *reserve* and *throw away* my first fire," but he was not going to tell Burr. In his last letter to his wife he said he was unwilling to take the life of another, and would rather "die innocent than live guilty." On his last night, Hamilton wrote to Thomas Sedgewick of Massachusetts:

> I will here express but one sentiment that Dismembrement of our Empire will be a clear sacrifice of great positive advantages without any counterbalancing good.

After the duel on July 11th, it was disputed whether Burr or Hamilton fired the first shot. Burr's shot hit Hamilton in his liver and diaphragm. Hamilton's gun discharged, but it was a wild shot which hit a tree 12½ feet from the ground. His comment that he was going to reserve fire was reported to the doctor in attendance and others. Hamilton died the next day, leaving behind a widow and seven children. She was the daughter of General Philip Schuyler of upstate New York, and he wrote he trusted her father would save her from "indigence." Unfortunately, her father was also deeply in debt.[542]

After the death of Hamilton, Burr was indicted for murder by a coroner's jury. Before the jury handed down its verdict on August 2nd, Burr fled to Philadelphia where he met with Charles Williamson who promised to enlist British government support for an invasion of Texas and Mexico. The British Ambassador Anthony Merry was in Philadelphia for medical treatment. On August 6th, the ambassador wrote a "*Most Secret*" letter to the British foreign secretary, Lord Harrowby:

> My Lord,
>
> I have just received an offer from Mr. Burr the actual vice president of the United States (which Situation he is about to resign) to lend his assistance to His Majesty's Government in any Manner in which they see fit to employ him, particularly in endeavoring to effect a Separation of the Western Part of the United States from that which between the Atlantick and the Mountains, in it's whole Extent.—His proposition on this and other Subjects will be fully detailed to your Lordship by Col. Williamson who has been the Bearer of them to me, and who will embark for England in a few Days. . . . [543]

On August 11th, Burr left Philadelphia to visit Saint Simeon's Island in South Carolina, the home of politician Pierce Butler.

For ten days in September the two men explored Spanish Florida by horseback and canoe. When he returned north, Burr took his place in the Senate on November 5, 1804. His house in New York City had been auctioned off and sold to John Jacob Astor to satisfy his creditors. He was ruined politically in the East, but the rest of the country looked more favorably on dueling and was anti-Federalist in its sympathies.

It was during the winter of 1804–05 that General Wilkinson and Vice President Burr poured over maps of the Southwest together in Washington. Jefferson and Wilkinson must have thought Hamilton's death had ended Burr's role in heading up a Northern Confederacy, and that—in the event of war with Spain over the boundary disputes—Burr and his followers could be employed in an American filibuster to invade Texas and Mexico. Burr's term as vice president came to an end in March.

Wilkinson's life style was recorded by a neighbor, who lived next door to him in Washington—

> Genl. Wilkinson was an elegant Gentln in person & manner. He was of medium size, probably 5 feet 8 or 9 inch. . . . He was sumptuous & hospitable in his living; not very nicely balancing his means & ends—He appeared much abroad with his aide [his eldest son, James Biddle Wilkinson] both in full uniform, and generally on horseback. His array was splendid; he having gold stirrups & spurs, & gold leopards claws to his leopard saddle cloth. [544]

In March, 1805, Jefferson appointed Wilkinson to the post of Governor of Louisiana Territory, based in St. Louis, while retaining his military command, a combination which many considered a dictatorship. Jefferson appointed Aaron Burr's relatives to other government posts. Burr's brother-in-law, Dr. Joseph Browne, became Secretary of Louisiana Territory under Wilkinson, and his stepson, James Prevost, became a Superior Court Judge in New Orleans Territory. After receiving his appointment, Wilkinson wrote to Charles Biddle, his wife's cousin—

> I can only say the country is a healthy one, and I shall be on the high road to Mexico.[545]

Wilkinson and Burr were supposed to journey together down the Ohio River, but the general was delayed with family matters. Burr, who was setting out on a 3,000 mile tour to explore the western country, had a "floating house" built for him at Pittsburgh. He stopped at Marietta to inspect the boat-yards and tour the Indian mounds, and then visited Blennerhassett Island in early May. The island was the site of a magnificent mansion, built by an ecentric and radical Irish aristocrat and his wife, Harmon and Margaret Blennerhassett. The island location would become the focus of Aaron Burr's treason trial in 1807.

Burr abandoned his boat in Louisville and began a tour of Kentucky and Tennessee by horseback, making friends and being introduced to men of consequence everywhere. At Nashville he was a guest of Andrew Jackson, who provided a boat for him to return to the Ohio River. Burr arrived at Fort Massac near the junction of the Ohio and Mississippi Rivers on June 6th, where he met with General Wilkinson. They spent four days together, before the general went to St. Louis to assume his post as governor, and Burr went to New Orleans with a military escort provided by Wilkinson. At New Orleans, he met with the Mexican Association, a secret society of about 300 Americans who were planning to "liberate" Mexico.

The leaders of the association were Edward Livingston and Daniel Clark. Edward Livingston was Aaron Burr's long time friend. Livingston had served as both the Mayor and U. S. District Attorney for the City of New York in 1801–03. A confidential clerk stole or mismanaged money from the district attorney's office and Livingston was responsible for the loss. He resigned his positions and moved to New Orleans early in 1804. The choice of New Orleans was more than a coincidence— his brother, Robert R. Livingston, had negotiated the Louisiana Purchase in 1803, and there was said to be a Mexican Association in New York City on which the New Orleans association was modeled.[546]

Daniel Clark, the Irish-born nephew of a wealthy New Orleans merchant, was Wilkinson's business partner. Clark, after the Burr Conspiracy collapsed in 1807, wrote a book called *Proofs of the Corruption of General James Wilkinson and of His Connexion to Aaror Burr*, published in 1810. The next year Wilkinson published a rebuttal, *Burr's Conspiracy Exposed and General Wilkinson Vindicated Against the Slander of His Enemies on That Important Occasion.* Daniel Clark fought a duel with Governor William Claiborne in 1807, while serving as the congressman from Orleans Territory in 1806-09.[547]

Daniel Clark tried several times to expose Wilkinson as a Spanish agent, but the general successfully defended himself. He was found not guilty by a military court of inquiry in 1808. Congressional investigations in 1808–10 were inconclusive, and the general demanded a military court-martial in 1811, which again found him not guilty. Of course he was guilty of being a Spanish agent, but his activities in defending New Orleans against the Mexican Association, Burrites, and the French and British were the real reasons he escaped censure.

Burr stayed in New Orleans for 17 days as a guest of Edward Livingston. The Catholic Bishop designated three Jesuits to aid him in revolutionizing Mexico and the Ursuline Mother Superior gave her support. Daniel Clark pledged $50,000 ($2 million in today's money). The army of liberation would consist of 8,000 militia from Louisiana, 10,000 militia from Kentucky, 3,000 regular troops and 5,000 slaves who would be promised their freedom. There were already fifty five cannons left behind by the French from their short-lived occupation of Louisiana which would be used.[547]

Burr was no slacker. He chose to travel by horseback along the Natchez Trace for his 400 mile return journey to Nashville. He wrote he found the road almost non-existent, and where it did exist, it was overgrown with bushes. General Jackson gave Burr a public banquet and ball, and everywhere he went he renewed his friendships with westerners. From Nashville Burr returned to Louisville and took a boat to St. Louis, arriving for a reunion with Wilkinson on September 12th.

It is at this point that divergent paths occur. Members of the Mexican Association wanted to capture Baton Rouge and the American Fort Adams, and then take control of the banks and shipping of New Orleans. The British fleet in the Bahamas would be utilized in an invasion of Mexico. The plan was based on a proposal by Burr's rival, General Francisco de Miranda, a Venezulean Freemason who had plotted for many years with the British government to revolutionize the Spanish colonies in the Americas. New Orleans would serve as capitol of a new country.

When Burr entered into discussions with the association, he brought new energy and new plans to the table. Because of his ties to Wilkinson, he could claim he had the secret support of the United States government. War between the U. S. and Spain was widely expected in 1805, and General Wilkinson would lead the army in an invasion of Mexico in support of Mexican independence. The army would travel from St. Louis to Sante Fe.

At the same time, Burr's volunteers and the Mexican Association, could take over New Orleans, and—while the U. S. Army was too busy fighting in Mexico to oppose it—they would establish a new independent country headquartered in New Orleans. Burr's schemes had many moving parts which could be altered to suit any audience he was talking to, but they depended upon a war with Spain.[548]

Historian Thomas Crackel, author of *Mr. Jefferson's Army,* points to a letter that Secretary of War Henry Dearborn wrote to General Wilkinson on August 24th, 1805.

> There is a strong rumor that you, Burr, etc., are too intimate . . . You ought to keep every suspicious person at arm's length, and be as wise as a serpent and as harmless as a dove.[549]

Daniel Clark warned Wilkinson, writing on September 7th:

> Many absurd and wild reports are circulated here. . . . You are spoken of as [Burr's] right hand man. . . . The tale is a horrid one, if well told. Kentucky, Tennessee, the State of Ohio, with

> part of Georgia and part of Carolina, are to be bribed with plunder of the Spanish countries west of us to separate from the union. . . . Amuse Mr. Burr with an account of it.[550]

On September 8th, Wilkinson wrote to Secretary of War Dearborn describing his plans for an invasion of Sante Fe. Some excerpts from his detailed report are—

> Should we be involved in a War (which Heaven avert) and it should be judged expedient to take possession of New Mexico . . . . These dispositions with judicious order and rapid movements attended by the Cross for your Banner, and a band of Irish Priests who have been educated in Spain (of whom I have a dozen) and we should take possession of the Northern Provinces without opposition, after which we might increase our force according to our means and at our discretion—The uncertainty of human life and the instability of political affairs, induce me to lodge this information with you in its present crude State . . .[551]

The *Gazette of the United States*, published on August 2nd in Philadelphia, an article headlined "Queries," which accused Burr of plotting a revolution with a series of inflammatory questions, including—

> How soon will the forts and magazines in all the military ports of New Orleans and on the Mississippi be in the hands of Col. Burr's revolution party? How soon will Col. Burr engage in the reduction of Mexico, by granting *liberty* to its inhabitants, and seizing on its treasures, aided by British ships and forces?

Newspapers across the country quickly republished "Queries" and it was even published in the Lexington *Kentucky Gazette* while Burr was in town on his return journey. The most likely author of the anonymous article was Spanish ambassador, Marqués de Casa Yrujo, married to Sally McKean, daughter of the governor of Pennsylvania, Thomas McKean.[552]

From August, 1805 onward there were widespread rumors across the country of revolutionary intrigues against Mexico linked to Aaron Burr. Wilkinson reacted by keeping his distance, but Burr appeared not to notice it. Finally on April 6th, 1806 Burr wrote to Wilkinson saying, *Nothing has been heard from brigadier since October*.[553] It still didn't seem to raise any doubt in Burr's mind as to whether he had the general's support.

When Burr returned to Washington in November, he met with British Ambassador Anthony Merry and learned Merry's letter written in August, 1804 regarding Burr's proposals had never reached London due to a boat accident. Merry reported in a letter to Lord Mulgrave on November 25, 1805 that Burr planned to launch his operation four months later, in March, 1806. Burr wanted British ships of war at New Orleans by April 10th, and a loan of 110,000 pounds stirling. He warned Merry if his demands were not met, France would regain the country and annex the Floridas. Merry believed Burr's statements that once Louisiana and the western country became independent, the northern states would separate from the southern states and a new country would be formed.[554]

Burr's friend, Jonathan Dayton, met with the Marqués de Casa Yrujo, the Spanish Ambassador, the next month in Philadelphia. Dayton, a former congressman and senator from New Jersey, told Yrujo the British were supporting Burr and that, after New Orleans and the Floridas fell, Mexico would be invaded. After several meetings, Dayton revealed a new plan, saying he had lied about British support. Dayton told Irujo that Burr planned to take control of the government in Washington with armed adventurers, seize control of the banks, take the best ships and burn the rest, and sail to New Orleans, where they would leave the Floridas alone and make a satisfactory border settlement with Spain. Yrujo paid Dayton $2,500 towards this project.[555]

Burr was having trouble raising money in New York because General Francisco de Miranda had arrived there with his ship, the *Leander,* for a filibuster expedition to invade Venezuela and South America. Burr's supporter in London, Charles Williamson, had become involved in Miranda's plan. President Adams's son-in-law, William S. Smith, was a supporter of Miranda, along with Ogden family members who would later support Burr. The *Leander* became an embarassment to Jefferson when it sailed from New York in February and was captured by the Spanish. William S. Smith and Samuel G. Ogden stood trial in July, 1806 in a U. S. Circuit Court, charged with violating the Neutrality Act of 1794; both men were found not guilty.[556]

Meanwhile President Jefferson was trying to negotiate a deal with Spain to buy West Florida, with an agreement to recognize the Rio Grande River as the boundary of Louisiana. He received a secret appropriation from Congress of two million dollars in February, 1806 to be used for this purpose.[557]

The Mexican government was taking no chances. In October, 1805, they established two small outposts east of the Sabine River ("Say-bean") on ground claimed by both nations. When news of this was received in Washington, Wilkinson was ordered to begin reconnaissance missions. The president in his annual message to Congress in December, 1805 announced he had ordered troops to be in readiness to protect U. S. citizens in Orleans and Mississippi Territories, and there were upwards of "300,000 able-bodied men between the ages of 18 and 26 years," who could be used for "offense or defense" if they were needed.[558]

Previously, in 1804, the president had authorized William Dunbar to lead a expedition to explore the Ouachita River. Dunbar was a Scottish-born surveyor with whom he had carried on a scientific correspondence for years. The Dunbar-Hunter Expedition of 1804 was cut short by hostile Osage Indians and Spanish officials.

General Wilkinson, after arriving in St. Louis to serve as governor of Upper Louisiana, sent Lieutenant Zebulon Pike up the Mississippi River on July 30, 1805. Pike's mission was very similiar to the Lewis and Clark expedition, but his assignment was limited to one season of exploration. Wilkinson had other plans in mind for the young officer. When he returned on April 30th, 1806, he was assigned to exploring the Southwest.

In 1806, Dunbar organized the Freeman-Custis Expedition to explore the Red River. The mission of the expedition, led by surveyor Thomas Freeman and botanist Dr. Peter Custis, was to map the Red River and discover whether it was a suitable water route to Sante Fe. Freeman was the surveyor who had been fired by Andrew Ellicott and hired by Wilkinson to construct Fort Adams. The Freeman-Custis Expedition was halted by Spanish troops 615 miles up the Red River in northeastern Texas, and forced to turn back in late July, 1806. The expedition had a military escort of 45 soldiers.

On July 15th, 1806, Captain Zebulon Pike, Jr. accompanied by Wilkinson's son, Lieutenant James Biddle Wilkinson, set out from St. Louis on a multifaceted mission. They escorted Indians back to the Osage village on the Osage River in Kansas and to the Pawnee village on the Republican River in Nebraska, where the Spanish army expedition had been halted a few weeks earlier by the Pawnee chief. The Spanish expedition was hunting for Pike's party, which arrived a few weeks after the Spanish left.

Pike's party headed south to the Arkansas River, and split into two groups. Lt. Wilkinson followed the Arkansas back east to the Mississippi, and Pike followed the Arkansas west to the mountains; Pike's Peak near Colorado Springs is named for his exploits. After making a large circle, Pike returned to the Arkansas and dropped south to the headwaters region of Rio Grande, where he built a fort near Alamosa, Colorado. They stayed there from January 31–February 26, 1806. Pike claimed he mistakenly thought the Rio Grande was the Red River.

Pike had written to Wilkinson shortly after setting out from St Louis, on July 22, 1806 that he expected to be in the vicinity of "St. Afee," and if they should encounter the Spanish, he would —

> . . . give them to understand that we were bound to join our Troops near Natchitoches but have been uncertain about the Head Waters of the Rivers over which we passd—but now if the Commandt. desired it we would pay him a visit of politness —either by Deputation, or the whole party—but if he

> refused; signify our intention of pursuing our Direct route to the posts below—this if aceded would gratify our most sanguine expectations; but if not I flatter myself, secure us an unmolested retreat to Natchitoches. [559]

If they were made "prisoners of War—(in time of peace)" he trusted their release would be secured by the United States.

The 27 year old Pike was accompanied by Dr. John Hamilton Robinson, a 24 year old medical doctor who for many years was involved in filibusters for the liberation of Mexico.[560] Dr. Robinson set out for Sante Fe on February 7th, 1807, leaving Pike and five soldiers struggling for survival and suffering from frostbite in the small fort on the Rio Grande. When Robinson arrived at Sante Fe, a detachment of 100 soldiers was sent to bring Pike and his men back to Sante Fe. Altogether, Pike's party numbered 16 men, some of whom were scattered at other locations.

Governor Alencastar at Sante Fe hosted Zebulon Pike at a splendid dinner on March 3rd, and in parting asked Pike to—"remember Allencaster in peace or war." Pike and his men were escorted to Albuquerque where they met up with John Robinson and Pike learned that Captain Falcundo Melgares would escort them to Chihauhau. Melgares had been the leader of the 300 man expedition which had gone to the Pawnee Village to capture Pike. They were entertained with *fandangos* (dances) with beautiful women and parties along the way, reaching Chihauhau on April 3rd.

For over three weeks they remained at Chihauhau, headquarters of General Nemesio Salcedo, Commanding General of the Internal Provinces. His nephew Manuel Maria de Salcedo became Governor of Spanish Texas in 1808. Manuel de Salcedo's father, Juan Manuel de Salcedo, had been governor of Spanish Louisiana in 1801-03, based in New Orleans.

On leaving Chihauhau, Pike found that he was forbidden to take notes. He secretly wrote detailed notes of the people and places along their travel route and hid them in his men's gun barrels and under their shirts. Along the way Pike encountered many priests and community leaders who supported independence.

On June 7th they reached San Antonio, in Spanish Texas, where they were greeted by Governor Antonio Cordero of Texas and Governor Simon de Herrera of Nuevo Leon. Pike wrote "we were received like their children," and were offered any means of leaving Mexico they desired. Pike and Robinson were guests in Governor Cordero's home for five days before leaving for Natchitoches and the United States.

Pike wrote a most interesting description of the two governors, and the conversations he had with them:

> The two friends agreed perfectly in one point, their hatred to tyranny of every kind; and in a secret determination never to see that flourishing part of the New World, subject to any other European lord, except *him* [King Charles IV of Spain], whom they think their honor and loyalty bound to defend with their lives and fortunes. But should Bonaparte seize on European Spain, I risque nothing in asserting, those two gentlemen would be the first to throw off the yoke, draw their swords, and assert the independence of their country.[561]

In regards to the Neutral Ground Agreement that Governor Herrera had reached with General Wilkinson seven months earlier, on November 6, 1806, Pike recorded an anecdote told by Herrera, who was in command of 1,300 troops at the Sabine River when he met with Wilkinson. His superior, General Nemisio Salcedo, had ordered Herrera to attack the U.S. forces if they should cross the *Arroyo Hondo* ("Deep Stream," now known as the Calcasieu River). Herrera told Pike, that—

> . . . he called a council of war on the question to attack or not; when it was given as their opinion, that he should immediately commence a predatory warfare, but avoid a general engagement; yet notwithstanding the orders of the viceroy, the commandant general, governor Cordero's and the opinion of his officers, he had the firmness (or termerity) to enter into the agreement with general Wilkinson, which at present exist

> relative to our boundaries on that frontier. On his return he was received with coolness by Cordero, and they both made their communication to their superiors. Until an answer was received, said Herrera, "I experienced the most unhappy period of my life, conscious I had served my country faithfully, at the same time I had violated every principle of military duty." At length the answer arrived, and what was it, but thanks of the viceroy and commandant general, for having pointedly disobeyed their orders, with assurances that they would represent his services in exalted terms to the king. What could have produced this change of sentiment to me [Zebulon Pike] is unknown, but the letter was published to the army, and confidence again restored between the two chiefs and the troops.[562]

The change in sentiment was no doubt facilitated by the large amount of money—$16,883.12—that General Wilkinson received from the federal government while on the Sabine. The War Department's accountant, William Simmons, wanted to know more about this irregular transaction, and why it was not being charged to Wilkinson's account for secret service money. Captain Moses Hooke delivered amounts of $2,500 and $1,196 for secret service expenses in January, 1807 while Wilkinson was at New Orleans, but this expense money was different. Captain Hooke delivered this money in three installments to the general in October and November, 1806 and January, 1807. Simmons complained that Wilkinson had appointed Hooke as "quartermaster general to the expedition on the Sabine," but—

> . . . no such appointment as quarter-master general is authorized by law, neither is there any particular appropriation, or head of expenditure, to which such payments are chargeable. I have repeatedly called upon Captain Hooke to produce vouchers for the expenditures of the above sum, which he has never done. In conversation with him, not long since, he

Captain Moses Hooke (1777-1821) was Lewis's choice of a second in command if William Clark couldn't join him. Hooke was the commander of Fort Fayette in Pittsburgh. He resigned from the army on January 20, 1808, married a wealthy widow, and settled down in Mississippi. (franceshunter.wordpress.com)

> observed, that a considerable sum of money had been drawn out of his hands by General Wilkinson on account of secret services, and for which he only had the general's receipt. Captain Hooke is now out of service, and his account has been reported to the Treasury for a suit.

In short, the accountant wanted to know under which column of expenditures he should list the $16,883.12.[563] The logical explanation is the money was used to bribe Mexican officials, and given the timing, it must have been arranged between Jefferson and Wilkinson months earlier.

The Neutral Ground Agreement between the two generals, Wilkinson and Herrera, established a "no man's land" which was inhabited by renegades, outlaws, filibusterers, escaped slaves, and its original inhabitants, the Attakapas Indians. It became the center of the slave trading operations of the Lafitte pirate brothers—whose headquarters were on Galveston Island—after the War of 1812. The Neutral Ground (approximately 55 miles at its widest point, and 110 miles in length) endured until the Adams-Onis Treaty of 1821 fixed the boundary line between Lousiana and Texas at the Sabine River. This pragmatic solution to the boundary dispute was desired by both the United States and Spain, as a war would have allowed Aaron Burr and his followers

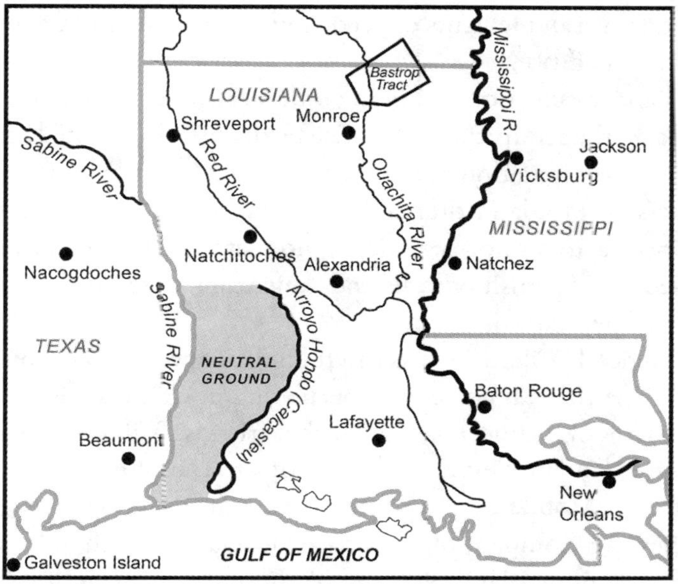

Neutral Ground and Bastrop Tract areas

the opportunity to seize New Orleans and invade Mexico by way of Vera Cruz, a seaport on the Gulf of Mexico.

In 1806 Burr entered into a contract to buy a piece of land known as the "Bastrop Tract" on the Ouachita River. The 350,000 acre tract had been acquired by American speculators from the "Baron De Bastrop," a Dutch con artist. The "Baron" had no title to the land owned by the Spanish government, but was acting as their agent in selling land to settlers. He was to earn his profits from the exclusive right to mill the settlers' wheat and sell the flour in Havana and Cuba.[564] The fraudulent real estate deal provided Burr with a reason to bring a large group of men to Louisiana, who were promised free land.

Historian Thomas Abernethy, in his classic book, *The Burr Conspiracy*, says the purchase was "sheer speculation" and "Burr did not expect to make his claim good under the government of the United States." In October, 1806 Burr wrote to a friend enclosing a map, showing where a 30 mile road was to be built from the Bastrop Tract to the Mississippi River and a longer one to Natchitoches. Burr's men were given tools for clearing roads, but no farm implements or seed, and women were not allowed on the expedition.[565]

Land speculation was going on everywhere, as investors anticipated the acquisition of Spanish territory by the United States, and/or the separation of the western states from the union. The possession of a land title and the occupation of the land were the two factors to be considered, no matter how circumstances played out. Spanish officials and Americans were both engaged in land speculation.

General Wilkinson was careful not to have his name on any Louisiana titles, but he was speculating in claims registered in the names of Daniel Clark and other friends. Wilkinson openly acquired land in West Florida—the title to Dauphin Island at the mouth of Mobile Bay—through the English trading firm, John Forbes and Company of Pensacola, Florida on March 6, 1806.[566]

Dauphin Island was one of the most strategic locations on the Gulf. It was a major trading port, the former home of a French fort established in 1702, and later, the home of the American Fort Gaines established in 1821. During the War of 1812, General Wilkinson seized Dauphin Island for the United States after the Spanish garrison at Mobile Bay surrendered peacefully.

Historian James Ripley Jacobs, author of the classic biography of Wilkinson, *Tarnished Warrior*, says that how he acquired Dauphin Island is not clear, but it was probably in payment for keeping the Spanish informed about Burr's movements. Ironically, in 1813, after Dauphin Island became the property of the United States, Wilkinson's claim to ownership was disallowed on the grounds that he was not a "Spanish citizen." [567]

The Aaron Burr Expedition of 1806–07 has been the subject of many full length books. Though the title of this chapter is "The Burr-Wilkinson Conspiracy, 1804–07," Wilkinson was actively working against Burr during the latter half of the conspiracy—mostly likely from the time of their meeting in St. Louis in September, 1805, after Burr had visited New Orleans and met with the Mexican Association. The general, however, continued to lead Burr on, in order to learn his plans, and because the president wanted to make certain that a charge of treason would stick when Burr was eventually brought to trial.

In the summer of 1805, Burr had met with the Mexican Association leaders—Edward Livingston, Daniel Clark, and Burr's stepson, J. B. Prevost—in New Orleans. The semi-secret society, whose goal was the liberation of Mexico, had about 300 American members who moved to New Orleans after the Louisiana Purchase. They were intending to work with Burr's rival, the Venezuelan revolutionary, Francisco de Miranda. Now they were also working with Aaron Burr.

Earlier that year, in March, 1805, President Jefferson had appointed Wilkinson, Governor of Louisiana Territory; Burr's brother-in-law, Joseph Browne, Secretary of Louisiana Territory; and his stepson, J. B. Prevost, Judge of Superior Court in Orleans Territory. Burr could create a new career for himself by being elected to congress from one of the western territories or states. Jefferson intended that Burr's filibusterers would join with Wilkinson's army if a war with Spain broke out, but when Burr met with the Mexican Association, it became clear that Jefferson's peace offering wasn't going to be accepted.

When Burr returned to the east coast in late 1805, he found everyone—including President Adam's son-in-law, William S. Smith—was supporting Miranda's filibuster to liberate Venezuela. Miranda had arrived in New York in November and hired a ship, the *Leander,* from Samuel Ogden. The *Leander* was captured by the Spanish on its way to Venezuela. Miranda later accused Burr of betraying him, and the Spanish Ambassador, the Marqués de Casa Yrujo, reported to his superior that he had paid Burr and his associates $2,500 for revealing Miranda's plans, among other things.[568]

At this time, Burr's associate, Jonathan Dayton, the former senator from New Jersey, was often seen meeting with Yrujo and another Burr associate, Dr. Erich Bollman, in a tavern where Burr was staying. Bollman was a German adventurer who had attempted to liberate the Marquis de Lafayette from an Austrian prison in 1794. With Bollman's help, Lafayette had managed to escape, but he was quickly recaptured. (Lafayette was released from prison in 1797.) Bollman was imprisoned for eight months, and came to the United States after this adventure.[569]

In 1805, Erich Bollman became involved in the Burr Conspiracy. General Victor Moreau, a French general banished by Napoleon, had arrived in United States in September, 1805 and was living at Morrisville, Pennsylvania near Philadelphia, and across the Delaware River from Trenton, New Jersey. In June, 1806, Burr vacationed near Morrisville with his daughter and her family, and met with Bollman and General Moreau frequently. By the end of July, Burr was ready to set his plans in motion.[570]

Dr. Bollman and other associates were sent by sea to New Orleans. Dayton wrote to Wilkinson on July 24th that it was well know Jefferson was going to replace Wilkinson as commanding general of the army in the next session of Congress— "Are you ready? Are your numerous associates ready? Wealth and Glory! Louisiana and Mexico!"

Burr wrote to Wilkinson in cipher on July 29th. The famous cipher letter told Wilkinson he had received funds and was starting operations. He promised there would be British and American naval support. He would bring 500–1,000 men in light boats to "seize on or pass by Baton Rouge," and would meet Wilkinson in mid-December at Natchez. He wrote:

> It will be a host of choice spirits. Wilkinson shall be second to Burr only; Wilkinson shall dictate the rank and promotion of his officers. Burr will proceed westward first August never to return. With him goes his daughter; her husband will follow in October, with a corps of worthies. . . . The people of the country to which we are going are prepared to receive us; their agents now with Burr, say that if we will protect their religion, and will not subject them to a foreign Power, that in three weeks all will be settled. The gods invite us to glory and fortune; it remains to be seen whether we deserve the boon.[571]

Burr set out in late August accompanied by his German secretary, Charles Willie, and his chief-of-staff, Colonel Julien De Pestre, a French military officer who had served in both the French and British armies. Pittsburgh was a main rendezvous place for the small groups of men who began traveling to join them. At Blennerhassett Island in the Ohio River near Marietta they began assembling boats and supplies. Burr went to Kentucky, joining Theodosia and her family at Lexington. There were large sums of money being spent on supplies. Burr had promised six months of provisions to Wilkinson. A contractor said Burr and his wealthy New York friends purchased $200,000 worth of provisions in Kentucky, enough for an army of 25,000.[572]

The United States District Attorney for Kentucky, Joseph Hamilton Daviess, a brother-in-law of Chief Justice John Marshall, had been alerting President Jefferson of the dangers posed by both Wilkinson and Burr, sending him information as he received it. Daviess's objective was to stop the expedition before it got started. On the 5th of November, Daviess made out a warrant for the arrest of Aaron Burr on the charge of preparing a military expedition against Mexico. Burr stood trial in Frankfort in early December. The grand jury exonerated Burr, who proceeded on to another meeting with Andrew Jackson in Nashville.

There were reports of 1,000 young men on the river in December, 1806 traveling between Pittsburgh and Marietta. Historian Thomas Abernethy writes:

> Judging by the quantity of supplies contracted for and the number of boats built, a total force of at least 1500 men must have been contemplated.[573]

Burr's eastern supporters were described as Federalists and Royalists who opposed the Louisiana Purchase—but many intended to settle there. It is no wonder there are countless, conflicting stories of what Burr was planning—separating the western states, starting a colony at Bastrop, invading the Floridas and Mexico, seizing New Orleans. He told different stories to everyone, but they all thought it was going to be a *Grand Adventure*.

The fortress guarding the entrance to Vera Cruz harbor, where the gold and silver of Mexico was stored to be shipped out on Spanish treasure ships. (Wikimedia Commons)

The most logical explanation is that Burr intended to arrive at New Orleans with as many men as he could recruit, and take over New Orleans in a bloodless coup organized by the Mexican Association, seizing the banks and ships. The rumors must have been absolutely electric regarding an invasion of Mexico. There would be no war between the United States and Spain because General Wilkinson and Colonel Herrera signed the Neutral Ground Agreement on November 5th. Now Burr's plan depended entirely upon reaching New Orleans.

This is where it gets interesting when you consider the international associates of Burr—the Marqués de Yrujo, General Victor Moreau, Erich Bollman, Charles Willie, Harmon Blennerhassett, and Colonel Julien De Pestre. What would happen after they arrived in New Orleans? It appears they were working with French agents in Mexico and their Mexican collaborators. In 1807, France and Spain were allies in the war against Britain. In 1808, Napoleon would turn against Spain and put his brother on the Spanish throne, but that was in the future.

Ambassador Yrujo had been recalled by the Spanish government in February, 1806; but he remained in the United States, living in Philadelphia, the home of his wife's family. His father-in-law, Thomas McKean, was the pro-French, Republican Governor of Pennsylvania from 1799–1808. Philadelphia was a center of French refugee revolutionaries, and there would have been ample opportunity to connect with the Spanish French party.

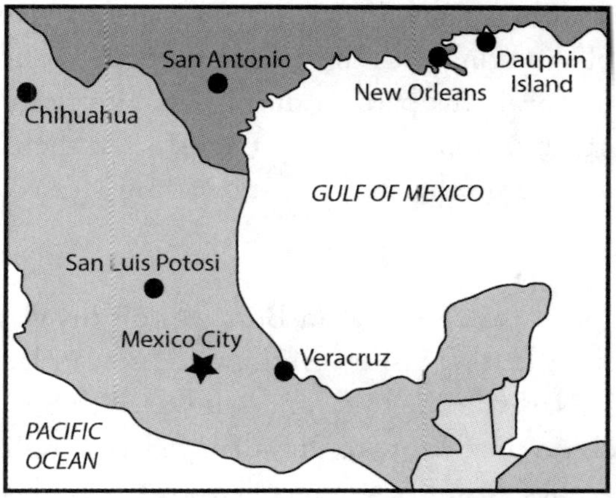

Vera Cruz and the silver mines of Mexico. San Luis Potosi is where some of the richest silver mines are found.

Secretary of State James Madison received an anonymous letter warning him Burr was working with Yrujo, and Madison thought there was a plan to establish Burr as the head of a sub-confederate state of France in the Spanish Americas. In 1808, when Burr was in exile he worked closely again with Yrujo in London. Senator William Plumer's noted in his diary on January 27, 1807:

> Dined with the President of the United States—[he said]—he did not believe either France, Great Britain, or Spain was connected with Burr in this project, but he tho't the marquiss de Yrujo was—That he had advanced large sums of money to Mr. Burr—& his associates. But he believed Yrujo was duped by Burr. That last winter there was scarse a single night but that, at a very late hour, those two men met & held private consultations. I have since then ascertained the fact.[574]

Burr spent a lot of time in private meetings with both Ambassador Yrujo and General Moreau, Napoleon's rival who had been sent into exile by the emperor. His chief of staff was an experienced French officer, Julien De Pestre. New Orleans was a French city. Burr spoke French. It is altogether plausible that Burr was collaborating with the French party in Mexico. In fact, Burr would repeat this pattern during his European travels in 1808–12—first he approached the English, and then Napoleon with his plans for another attempt at invading Mexico.

The trap was set, and Aaron Burr was on his way to being caught. When he visited Andrew Jackson in late September, 1806 at Nashville, he told him that war with Spain was imminent. Jackson then wrote to President Jefferson telling him he had three Tennessee militia regiments ready to march against the Spanish. Jefferson replied by asking Jackson to explain his offer.

Burr and his entourage were based in Lexington, Kentucky as plans for the invasion went ahead on multiple fronts. Abernethy says "much collateral evidence" supports the idea that Burr was planning to wait with his forces at Natchez until the Mexican Association made a "declaration of independence" in New Orleans.

The liberation of Mexico made strange bedfellows, as Burr's western supporters were Jeffersonian Republicans and his eastern supporters were Federalists and British loyalists. It would not have worked out. But, of course, he believed he was about to set himself as the leader of a new country and the past would be forgotten. Above all else, Aaron Burr was a very good salesman. The prospect of being in on the ground floor of an independent Mexico, or a "liberated" Spanish Empire, excited the greed of everyone involved and blinded them to their own obvious conflicts of interest.

On the 27th of September, Colonel Herrera pulled back his force of 600 soldiers to the west side of the Sabine, disobeying his government's orders. General Wilkinson had arrived at Natchitoches on September 22nd. On the 26th and 28th, Wilkinson wrote to Senators John Smith of Ohio and John Adair of Kentucky urging them to join him with militia troops for an army invasion of Mexico. Both men would later be accused of participating in the Burr Conspiracy.

The general was entrapping the two senators. He wrote to

his long time friend Adair—"unless you fear to join a Spanish intriguer [Wilkinson] come immediately—without your aid I can do nothing." Wilkinson would arrest Adair when he arrived in New Orleans on January 14, 1807, and ship the senator out to Fort McHenry in Baltimore. Two years later, Adair sued Wilkinson from "wrongful arrest," and won a judgement of $2,500. The judgement was paid by Congress on Wilkinson's behalf. [575]

Colonel Herrera, who was in charge of the Spanish forces on the Louisiana frontier, was a staunch supporter of the Spanish king. Herrera—who was married to an English wife—had met and admired George Washington when he traveled in America. He won the commendation of his superiors for disobeying orders because by withdrawing his forces to the west side of the Sabine, he defused the war situation and prevented the French faction of Mexican revolutionaries and the Burrites hastening to join them from overthrowing the royalist government.[576]

On October 8th, the general received Aaron Burr's cipher letter of July 29th and the letter from Jonathan Dayton saying he would be replaced as commanding general at the next session of Congress. Wilkinson had already sent home the militia who had reached Natchitoches, on the grounds of economy. He retained a force of 100 mounted dragoons.

Burr's messenger, Samuel Swarthout, remained at Wilkinson's camp for ten days. Swarthout said 7,000 men were expected to join Burr from New York and the western states and territories. Orleans Territory would then be "revolutionized," and money seized from its banks would be used to outfit ships which would sail to Vera Cruz on the 1st of February. They expected naval protection from Great Britain and officers of the American navy were ready to join them. The general believed that British naval support would be forthcoming.[577]

Wilkinson continued to send out misleading information in support of Burr in October, while making preparations to defend Mississippi and Orleans Territory. He sent artillery to Fort Stoddert in Alabama and ordered a blockhouse built in New Orleans. Because there are contradictory documents, historians have said Wilkinson was wavering in his support of Burr, and only began to make up his mind at this time.

On October 20th, Wilkinson wrote to the president, enclosing an anonymous document, obviously written by the general:

> A numerous & powerful association, extending from New York through the Western States to Territories bordering on the Mississippi, has been formed with the design to levy & rendezvous eight or ten thousand men in New Orleans at a very near period, and from thence, with the cooperation of a Naval Arrangement, to carry an Expedition against Vera Cruz—Agents from Mexico, who were in Philadelphia at the beginning of August, are engaged in this Enterprize; those Persons have given assurances that the landing of the proposed expedition will be seconded by so general an insurrection, as to insure the subversion of the present government. . . .[578]

The general was perplexed as to who was really behind this enterprise. The combination of French revolutionaries and British naval support was unlikely, but believable. War ships in the invasion force—with the implied threat of attacking the fortress at Vera Cruz if it wasn't peacefully surrendered—would be of enormous advantage. The revolutionaries were not connected to Napoleon. General Victor Moreau, because of his fame as a military leader had escaped execution, but he had been banished to the United States by Napoleon. The Federalists and British Loyalists of the Northern Confederacy and the Mexican Association had the connections for a British "Naval Arrangement"—as Burr himself had, through the Prevost family.

The simple answer is that everyone would want to be in on the invasion from the beginning because, literally, that's where the money was—stored in the fortress at Vera Cruz. The hard currency of international trade was entirely based on the Mexican silver peso (dollar), mined and minted in Mexico.

After the Neutral Ground Agreement was signed on November 5th, Wilkinson sent his aide, Captain Walter Burling, on a mission to Mexico City. Burling brought a letter asking the

Spanish government to pay Wilkinson 121,000 pesos for his efforts in defeating Burr's expedition. The Viceroy of New Spain, José de Iturrigaray, returned the letter, thanking him and saying he was not authorized to pay it. Wilkinson's real motive was reconnaissance. Burling traveled overland to Mexico City and returned by ship. The general wanted to know the routes and the state of military preparedness for Mexico's capital city. Captain Burling was paid $1,500 for his travel expenses by the United States government.[579]

Wilkinson reached New Orleans on November 25th. Though others advised stopping Burr up river before he reached Natchez, or before he reached Baton Rouge where Spanish military supplies were stored, Wilkinson believed the story about British naval support and made his stand at New Orleans. He knew Burr was planning to take over the territorial government in a bloodless coup with the help of the Mexican Association. The leaders of the association were hostile to Governor William Claiborne and actively opposed him.

However, one of Burr's major supporters, Daniel Clark, had deserted him. Clark was now the Territorial Delegate to Congress. In the middle of October, before leaving for Congress, Clark called together French leaders of the old establishment and told them to take no part in Burr's plans and support the United States. Governor Claiborne assured Secretary of State James Madison of the "fidelity of the Ancient Louisianians to the U. States."

After learning the true nature of Burr's plans, Andrew Jackson wrote to Governor Claiborne on November 12th—

> . . .—put your Town in a State of Defence, organize your Militia, and defend your City as well against internal enemies as external. . . .—beware the month of December—I love my Country and Government, I hate the Dons [Spanish nobility] I would delight to see Mexico reduced, but I will die in the last ditch before I would yield a part to the Dons or see the Union disunited.[580]

Jefferson finally acted, issuing his *Proclamation on Spanish Territory* on November 27, 1806. The proclamation stated that citizens of the U. S. were unlawfully conspiring to begin a military expedition against the dominions of Spain, and warned them to withdraw from the criminal enterprise without delay. Civilian and military authorities were ordered to seize their supplies and detain all persons engaged in the enterprise.

Aaron Burr and Senator John Adair arrived in Nashville around December 12th. Although Andrew Jackson had learned enough of their plans from an unsuspecting confederate to alert Governor Claiborne in New Orleans, he still provided some help to Burr. Burr had paid for five boats to be constructed at Nashville. He got two. Jackson eventually refunded the difference. He was given a few supplies, but the 75 promised recruits failed to appear. At this time, the president's proclamation was still unknown in the western country.

Burr and a few men, some horses, and the two flatboats arrived at the junction of the Cumberland and Ohio Rivers on the 27th of December. As Burr waited for Blennerhassett to join him, he sent messengers to nearby Fort Massac. (One of his messengers was a man named Mr. Hopkins—Mr. Hopkins would turn up again as an advance agent in Burr's 1810 invasion plot.)[581] Burr's boats passed by the fort on December 30th, but he declined an invitation to visit, saying he was a "suspicious character" and wished to avoid causing further suspicion. He had already learned of the peace arranged by Wilkinson at the Sabine. Burr chose to believe that the general was still part of the conspiracy. Upon learning Wilkinson was in New Orleans, he thought the general was planning to start the invasion from there.

Harmon Blennerhassett and his men, and another group of men who joined them at the Falls of the Ohio, had arrived at the Cumberland with more bad news. The Ohio state militia had been called out against the conspirators on Blennerhassett Island. On December 6th, they seized ten of their eleven boats. Burr's Kentucky treason trial and alarming stories in the newspapers were causing many to back out of the expedition. On December 10th, in the middle of the night, Blennerhassett and his men made their escape from the island on the remaining boat. The Ohio militia on the west shore and Virginia vigilantes on the east shore were preparing to stop them in the morning. The activities on Blennerhassett Island would form the basis of the federal government's charge of treason again Burr. The island was chosen because it fell within the jurisdiction of the state of Virginia. (West Virginia didn't become a state until 1863.)

Their combined flotilla had ten boats, with about a hundred men or more in the party. The boats were capable of carrying fifty men each. Fifteen boats had been left at Marietta and three at Nashville. Burr tried to recruit additional men as he traveled down river, and gained a few. On January 10th, they arrived at Bayou Pierre on the Mississippi River, about forty miles above Natchez (near today's Port Gibson, Mississippi).

Burr visited his old friend Peter Bruin there. At Bruin's home he read the latest newspapers, which carried Jefferson's proclamation; Wilkinson's version of Burr's cipher letter to him; the Governor of Mississippi Territory's statement that all conspirators should be seized; and the news that Wilkinson had arrested his co-conspirators in New Orleans. Burr was enraged. Finally he understood that Wilkinson had betrayed him.

Wilkinson had called out the militias when he arrived at New Orleans, repaired the fortifications, and impressed seamen. He arrested Erich Bollman and Samuel Swarthout for "misprision of treason," or concealing knowledge of treason, and shipped them out on a war vessel to Baltimore. He did the same to John Adair and Peter Ogden. The city was in an uproar over this imposition of martial law.

At the same time, the general suffered a great personal tragedy. His wife Ann died of tuberculosis on February 23, 1807 and was buried in New Orleans. She had been seriously ill since the summer of 1806 in St. Louis. They had been married for 28 years. Wilkinson had been facing grave national emergencies and his wife's terminal illness for months.

On February 3rd, the president wrote to Wilkinson, praising his activities. He wrote that they now knew Burr had ten boats; and that there were between 80 and 100 men in his party and there were 60 oarsmen, "not all of his party." Jefferson wrote:

> Although we at no time believed he could carry any formidable force out of the Ohio, yet we thought it safest that you should be prepared to receive him with all the force which could be assembled, and with that view our orders were given; *and we were pleased to see that without waiting for them, you adopted the same plan yourself, and acted on it with promptitude. . . .*[582]

Jefferson was waiting to spring his trap, which added to the confusion as Burr traveled around the country. The $16,883.12 brought to the Sabine by Moses Hooke is the clearest indication the president wanted to avoid a war with Spain and Wilkinson was acting on advance instructions in negotiating the Neutral Ground Agreement. Jefferson originally supported Burr and Wilkinson's filibuster activities because it was cheaper than war. Filbuster expeditions eventually did establish independent republics in West Florida in 1810, and in Texas in 1836–46, which were then annexed by the United States.

Historians have blamed the general and discounted Burr's international intrigue in attempting to unravel the story of the conspiracy. Wilkinson was both an instigator and a collaborator, but it was an American-based conspiracy when he was involved. He was a die hard Federalist, who liked the Spanish people. He wanted to be in on the profits to be made in the revolution that would establish Mexico's independence from Spain. He was ready to start a new country as part of the Northern Confederacy, and to lead the United States Army in an invasion of Mexico in the event of a war with Spain. He preferred seeing Burr as the leader of the country, and hoped it would happen.

What Wilkinson couldn't abide, however, was the "infernal French." It must have come as a shock to learn that Burr was working with the French. Jefferson was pro-French, but he was as firmly opposed to the French "revolutionizing" Mexico as his general was. It was all a question of neighborhood, and both men believed it should be an American neighborhood.

Burr surrendered to the Mississippi authorities on January 18th, 1807 and was taken to the town of Washington, just north of Natchez. Burr agreed to surrender on the condition it was to a civilian authority, rather than to the military authority of Wilkinson. Burr's men hid their rifles under water until the boats passed inspection. They were entertained with balls and dinners in Washington, a wealthy planter's town. The two judges hearing the case and a grand jury were Federalists and in full accord with Burr's plan to attack Spanish territory. Their verdict, absolving him of all charges, was delivered on February 4th.

Wilkinson had offered a $5,000 reward for Burr's capture in December. After Burr surrendered at Washington, Moses Hooke with five other men were sent to seize Burr and bring him back to New Orleans. They were dressed in civilian clothes, armed with pistols and dirks, and had no warrants. Burr escaped custody in the middle of the night on February 5th, forfeiting his bail bond of $10,000. A reward of $2,000 was offered for his capture by the Mississippi authorities.

Burr went into hiding, and sent a message to his men to keep together, put their arms in order, and he would join them the next night. The message was discovered hidden in the lining of a coat owned by Burr and worn by a slave who was sent to deliver the note. His men were arrested temporarily, forcing Burr to abandon his plans to reunite with them.

On February 18th, Burr and a companion, Robert Ashley—who had been with Philip Nolan's group of adventurers in Texas—were spotted by a local man near Mobile, Alabama and captured. Burr was riding a "tackey" horse, and wearing a white floppy hat and homespun clothes in an attempt to disguise himself as a river boatman, but his piercing eyes betrayed him. Ashley escaped from confinement after his arrest.

Burr remained under arrest at Fort Stoddert for two weeks, playing chess with the commander's wife. An eight man escort brought Burr to Richmond, Virginia on March 25th, to await his trial for treason. The trial will be discussed in the next chapter, as Meriwether Lewis attended it as President Jefferson's observer. [583]

Aaron Burr
(1756–1836)

James Wilkinson
(1757–1825)

Thomas Jefferson
(1743–1826

What was President Jefferson's role in the Burr-Wilkinson Conspiracy of 1804–07? The following is a review of the new interpretations offered in this book—

Jefferson was aware that Burr and Wilkinson were meeting secretly and studying maps of the gulf coast and southwest in the winter months of 1804–05 in Washington D. C. Burr was his vice-president, and Wilkinson was the commanding general of the army. Burr's term as vice president would end in March, 1805.

For over a year, secession from the union was being openly talked about by the northern states. In 1804, Burr had been defeated in his run for Governor of New York, and then had killed the founder of the Federalist Party, Alexander Hamilton, in a duel. Hamilton had opposed the secessionists and Burr's election to the governorship. If Burr had become Governor of New York, he would have served as the leader of the breakaway states in a union with the British provinces of Canada.

When General Wilkinson met in secret with Burr at his home in New York City in May of 1804, he must have been willing to support the Northern Confederacy. For many years he had been pursuing the goal of revolutionizing Mexico. Wilkinson was a Federalist, with loyalties to Federalists officers in the army—the same officers Jefferson was replacing with his own Republican officers. The general was ready to switch sides.

Jefferson made a strategic move in the winter of 1804–05. He offered General Wilkinson the post of Governor of Louisiana

Territory, based in St. Louis. The Louisiana Purchase—whose extent was unknown—had been divided into Orleans and Louisiana Territories. He did not appoint Burr to any federal posts, but gave his relatives and friends jobs in both territories. Many people were outraged by Jefferson's giving the top civilian post to the general, saying it was a military dictatorship.

Jefferson was co-opting, or adopting for his own purposes, the plans of the Federalist secessionist movement. Burr and Wilkinson could revolutionize Mexico with a filibuster army including Burr's civilian volunteers, western militias, and the U. S. Army. It would be an "undeclared war"with Spain if it happened—a spontaneous eruption against the hated Spanish Dons. The result would be an independent republic established by Americans, which would then become a United States territory.

Burr was to rehabilitate himself by becoming a senator from a western territory or state, and rehabilitate his finances by land speculation. He was uninterested in this solution. Burr went on a western exploratory tour in 1805, meeting with his close friend, Edward Livingston, the former Mayor of New York City, in New Orleans. Livingston was the leader of the Mexican Association, a group of 300 Americans who had moved to New Orleans after the Louisiana Purchase for the purpose of revolutionizing Mexico and other Spanish colonies in the Americas.

After visiting New Orleans during the summer, Burr met with Wilkinson in St. Louis in September, 1805 and revealed his larger goals. This is when the general pulled away from him, but pretended to remain on his side. Secretary of War Henry Dearborn warned Wilkinson about his association with Burr, saying, "You ought to keep every suspicious person at arm's length, and be as wise as a serpent and as harmless as a dove." Dearborn wrote this letter to the general on August 25th. He had received word of Burr's dealings in New Orleans.

So, at this point, what was Jefferson's position on the proposed invasion of Mexico? He opposed it in 1804 when it was a plan of the Northern Confederacy. He supported it in 1805 when he placed Wilkinson and Burr's people in power positions. He opposed it again in 1806, when Burr began meeting with the British and Spanish ambassadors and French revolutionary agents in what would become a filibuster to start a new country.

As plans for the invasion progressed in 1806, Jefferson and Wilkinson kept silent. However, there must have been secret instructions to Wilkinson to arrange for a Neutral Ground Agreement with Mexican officials, and secret money was sent to him for this purpose. The silence of Jefferson and Wilkinson was to entrap Burr and his followers. Jefferson wanted to make certain there was enough evidence to convict Burr on treason charges. Was Jefferson worried about the loyalty of his general? It was only natural for him to be worried. But Wilkinson came through. He didn't start a war; he betrayed Burr; he provided a strong defense of New Orleans; and he escaped an indictment for treason.

Wilkinson by this time had a long history of being involved in conspiracies. Sometimes he started them, and sometimes he busted them. Like Burr and Jefferson, he operated behind the scenes, and tried to have as much distance as possible from any events for which he might later be questioned. All three were masters of "deniability."

In the 1790's Wilkinson was a "conspiracy-buster," defeating attempts by the British, French, and disaffected Mexican officials to organize filibusters to invade Mexico with the cooperation of Americans living in the west. The European wars; the existence of Indian lands yet to be acquired by the United States; the Spanish territories on the gulf; and the location of a cosmopolitan city, New Orleans, at the mouth of the Mississippi River were all reasons to organize revolutions. Above all else, there was the lure of the silver mines of Mexico.

There was no concept of "the United States." Instead, it was "these United States," or "the U. States"—as if united was a questionable concept. The states had been independent colonies of Great Britain less than a generation before. No one exactly knew what the authority of the federal government was. Jefferson and Madison had first questioned federal authority with the Kentucky and Virginia Resolutions. The northern states were organizing a new confederacy. It was natural to think the lands between the Atlantic and Pacific Oceans would be divided into independent countries, which could then be more easily governed.

The first Burr and Wilkinson Conspiracy to invade Mexico started with President John Adams's decision in 1799 to seek a peace with France rather than continue the Quasi-War. Without Adams being aware of it, Wilkinson established the military cantonment near the junction of the Ohio and Mississippi Rivers called "Wilkinsonville." During the hotly contested election of 1800, between Burr and Jefferson for the presidency, troops were stationed at Wilkinsonville. If Burr had won the presidency, the troops would have invaded Mexico. Philip Nolan and his men, waiting in Texas in 1800–01, were their advance agents.

When the Louisiana Purchase unexpectedly happened in 1803, the northern states were bitterly opposed because it permanently established the slave-holding states's control of Congress. Louisiana Territory was cotton country and would eventually enter the union as slave states. The 3/5's compromise of the U. S. Constitution counted slaves as "3/5's" of a person in determining the number of congressional seats in the House of Representatives.

The northern states were ready to secede and start their own country. This new country would include not only the northern states and British provinces of Canada, but also the Spanish territories of the two Floridas, Texas, and Mexico with its silver mines. Wilkinson was ready to join with his Federalist friends to establish a new country. There was always a good chance they would later reunite with the United States.

The basic difference between Burr and Wilkinson is Burr was working with international associates in an open-ended scheme of revolutionizing Spanish colonies in the Americas and starting a new country with himself as the head of it. They were empire builders. After the treason trial Burr would head for Europe and continue planning another filibuster.

Wilkinson would continue to focus on a filibuster led by Americans that would revolutionize the Spanish territories. Burr and Wilkinson were now working independently of each other.

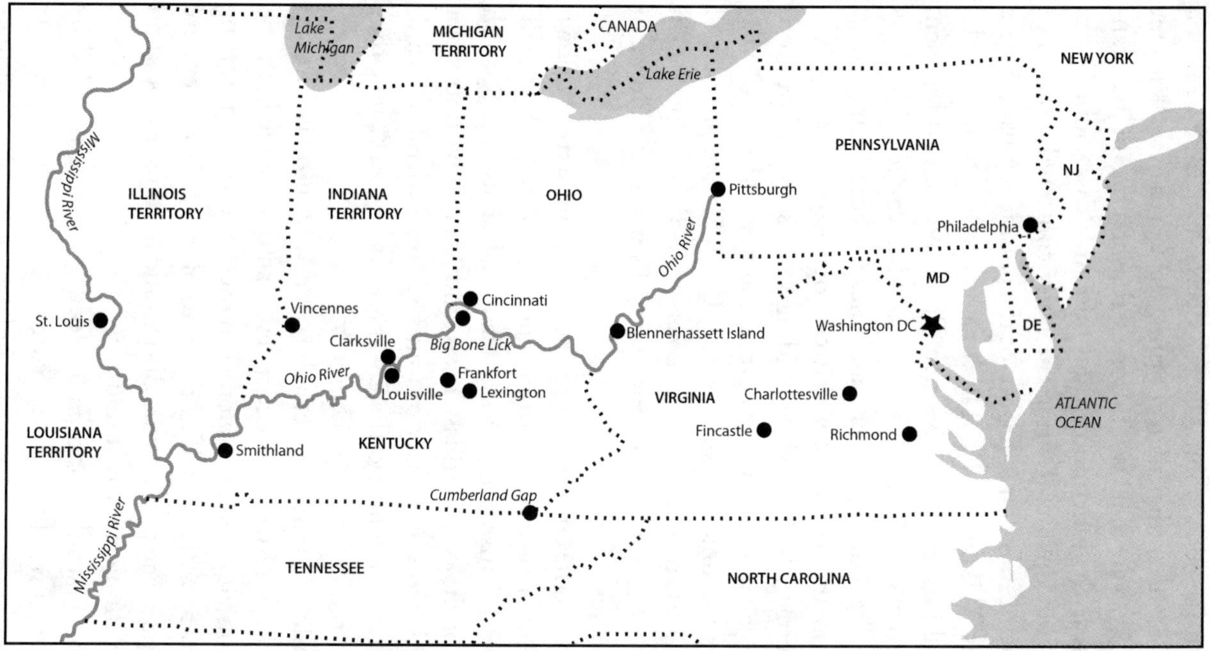

St. Louis to Washington, circa 1806–09, via the Ohio River and overland travel.

Chapter 12

Governor of Louisiana Territory: 1807-09

September 23, 1806 was the day that the Lewis and Clark Expedition arrived back in St. Louis to a heroes' welcome. Sergeant Ordway wrote when they had arrived at the settlements two days earlier, the people "were Surprized to See us as they Said we have given out for dead above a year ago." A St. Louis resident said they "really had the appearance of Robinson Crusoes—dressed entirely in buckskins." Upon arriving at St. Louis they got quarters in town and began celebrating. The captains ordered new clothes from a tailor and began to pay formal calls. Dressed in buckskins, they were entertained at a dinner and a ball that evening.[584]

Meriwether Lewis had no idea he would be returning to St. Louis as Governor of Louisiana Territory in 1808. St. Louis was in the midst of the Burr Conspiracy and the threatening war with Spain in September, 1806. The Governor of Louisiana Territory, General Wilkinson, had gone with his troops to meet the Mexican army on the Sabine. Captain Zebulon Pike and the general's son, Lieutenant James Biddle Wilkinson, were exploring the route to Mexico. Aaron Burr's brother-in-law, territorial secretary Dr. Joseph Browne, was serving as acting governor, while the conspirators in St. Louis were preparing to bring 12,000 pounds of lead down the Mississippi to Burr. The arrival of the Lewis and Clark Expedition only added to the excitement.

The captains wrote their letters the next day. Lewis wrote a long letter to the president, and Clark copied a draft of Lewis's letter to send to his brother, General Jonathan Clark, in Louisville. The letter to his brother would serve as the public account of their journey and safe return, and it would be reprinted widely in newspapers across the country. Lewis would spend the next weeks writing out pay warrants and discharging the men.

The captains were asked to examine evidence of land claim frauds by Rufus Easton, the U. S. Postmaster and U. S. Attorney for St. Louis. Easton wrote to Jefferson on December 1, 1806:

> . . . the Books & Records of Soulard have been personally inspected by Captains Lewis & Clark, who will be the faithful messengers of their state & condition, and that they contain innumerable *alterarations & forgeries!* [585]

Silas Bent, the U. S. Deputy Surveyor in St. Louis, who had replaced Antoine Soulard in mid September, wrote to the Surveyor General on October 5th that the records had pages cut out and others inserted, names erased and others inserted, and that—

> . . . the Conduct of some of the commissioners appears rather extraordinary—they do but little Bussiness lately—the large Claims of from a League square to one million of acres I believe are generally fraudulent, Yet the Combination in favor of them is strong and Powerful and by their Opulence and influence may perhaps be able to bear down all those who dare to come forward and state facts respecting them.[586]

The three federally appointed land claims commissioners had begun meeting in December, 1805 to establish the legitimacy of Spanish land titles. Two of the commissioners were Wilkinson's men: Clement Biddle Penrose, a nephew of General Wilkinson, and James Lowry Donaldson, a member of a prominent Baltimore family. The third was Judge J. B. Lucas, a French emigrant and friend of Albert Gallatin, Secretary of the Treasury. Lucas opposed the efforts of Wilkinson's two commissioners to confirm the suspect titles held by the large land claimants. A great deal of money was at stake, and Wilkinson had favored the wealthy French families of St. Louis.

The Americans arriving in St. Louis resented the old residents' preferential treatment and fraudulent claims, and wanted

The Lead Belt District in southeastern Missouri contains the highest concentration of lead in the world. The area was also a center of gun powder manufacture, which is made from saltpetre (potassium nitrate), sulfur, and charcoal. Lead is used to make bullets.

to invest in these lands themselves. On January 4th, 1807, after Meriwether Lewis arrived in Washington with evidence of the commissioners' misconduct, President Jefferson wrote to Secretary Gallatin, "I have never seen such a perversion of duty as by Donaldson & Penrose." He fired Donaldson and replaced him with Frederick Bates.[587]

Many of the disputed land claims centered on the incredibly rich lead mine district. All land containing lead, salt, and salt petre was reserved for the federal government unless it had been previously acquired by a valid Spanish land grant. Moses Austin—the Virginia entrepenur who brought new mining techniques to the lead mine district in 1798—had become a Spanish citizen in order to acquire title to one league of land at Mine à Breton (Potosi). Moses Austin was the father of Stephen Austin, one of the founders of the Texas Republic.

When the United States established Louisiana Territory, John Smith T moved to Ste. (Sainte) Genevieve to enter into competition with Moses Austin in the lead mine district. John Smith had added a "T" to his name to distinguish himself from other John Smiths. The "T" stood for Tennessee. In Kentucky, Smith T had established Smithland near the junction of the Cumberland and Ohio Rivers in a partnership with Zachariah Cox in the 1790's. At one time, Smith T claimed to own the "north half of the state of Alabama" in the Muscle Shoals area of the Great Bend of the Tennessee River, in what became known as the Yazoo land fraud.

Smith T was the most prominent "Burrite" in the St. Louis area. He was born in Essex County Virginia around 1770. His mother's maiden name was Lucy Wilkinson. Smith family tradition is their family is related to General Wilkinson, and the general and the Smith brothers thought they were related.

Smith T's brothers, Thomas and Reuben, were both West Point graduates. Thomas Smith became a respected brigadier general in the U. S. Army. For many years he and Smith T were not on speaking terms, as Tom Smith was the courier used by General Wilkinson to deliver Aaron Burr's cipher message and other letters to President Jefferson in Washington. The papers were concealed in his boot.

Two days after reviewing the contents of Lieutenant Smith's boot, on November 27, 1806, Jefferson issued his *Proclamation on Spanish Territory* forbidding citizens to plot against foreign governments, and calling for all participants to be arrested and their supplies confiscated. This directly affected John Smith T, who would soon be going down the Mississippi River with 12,000 pounds of lead for Aaron Burr.[588]

When Burr arrived at Smithland on the Cumberland River in late December, he sent word to Smith T at Ste. Genevieve. Smith T had purchased some $5200 worth of supplies from Andrew Jackson which were stored at Smithland for the filibuster. Jackson later sued him to recover the money he was owed.[589]

In January, 1807, Smith T set off from Ste. Genevieve with

a flotilla of canoes to bring men, lead, munitions, and other supplies down the Mississippi to Burr at Natchez. His companions were Andrew Steele, Henry Dodge, and Robert Westcott. Westcott was the son-in-law of Dr. Joseph Browne, territorial secretary and acting governor in the absence of Wilkinson. Browne was the brother-in-law of Aaron Burr; he had married Theodosia's half sister in a double wedding ceremony with Burr and Theodosia in 1782.

When Smith T and his associates arrived at New Madrid near the junction of the Ohio and Mississippi they learned of President Jefferson's proclamation against the filibuster. Smith T sold the boats, lead, and supplies, and they returned on horseback to Ste. Genevieve. When they got back, they discovered a grand jury in Ste. Genevieve had indicted Dodge and Smith T for treason. After he posted bail, the 24 year old Henry Dodge, who was over six feet tall and strong, "whipped nine of the grand jurors," according to John F. Darby, and "the rest ran away."

Darby, who was Mayor of St. Louis in 1835, did law work for John Smith T. Darby wrote about him—

> Col. Jack Smith T always went armed. He had two pistols under his coat in a belt around his body, generally two pocket pistols in his side-coat pocket, and a dirk in his bosom.

Darby reported that when the coroner Otho Schrader came with a writ to arrest Smith T at his house, Smith T said to him:

> Mr. Schrader, dinner is just ready—get down and come in and take dinner; but, mark, if you attempt to move a finger, or make a motion to arrest me, you are a dead man.

They ate dinner together, with Smith T's pistol cocked and loaded, laid on the table beside his plate pointing at Schrader. He was never arrested on the indictment.[590]

The travel writer Henry Marie Brackenridge visited Smith T and described him as a small man, of delicate frame, and somewhat effeminate in his appearance. Brackenridge called his gunsmith shop an "arsenal." Smith T employed two slaves full-time to make and repair guns. His guns were famous for their accuracy. John F. Darby wrote—

> It was said he killed fifteen men, mostly in duels, where his own life was in danger. . . . He was as polished and courteous a gentleman as ever lived in the State of Missouri, and "as mild-mannered man as ever put a bullet into the human body." [591]

Smith T had a fearsome reputation as a duelist, and employed a private army in his Mineral Wars with Moses Austin over the choicest claims in the lead mine district. The Mineral Wars raged for years in the Lead Mine District.

Lewis and Clark set out for the east coast on October 21st, amidst rumors of war with Spain and the Burr filibuster in progress. They had waited at St. Louis for Indian Agent Pierre Chouteau to arrive with a group of six Arkansas Big Osage chiefs who were going to visit the president. The travel group consisted of Lewis and his dog Seaman; Clark and his slave York; the Mandan chief Big White, his wife Yellow Corn, and their small son; the interpreter René Jessaume, his wife and their two small children; Pierre Chouteau and the six Osage chiefs; and former corps members, the mulatto François Labiche, who would serve as the French interpreter for Big White's Mandan interpreter, and John Ordway, who was traveling with them on his way home to New Hampshire. Pierre Provenchere accompanied Chouteau.[592]

They traveled overland by way of Cahokia to Indiana Territory, where Governor William Henry Harrison entertained them at his Grouseland home in Vincennes. Lewis wrote a letter from Vincennes on October 30th. They arrived at the Falls of the Ohio on November 5th. They must have had a great time telling George Rogers Clark about their adventures. He was still living at his home overlooking the Falls in Clarksville, Indiana.

Clark's sister, Lucy, who was married to William Croghan, gave a party at their home, Locust Grove, in Louisville on November 8th. William Clark Kennerly, a nephew of William Clark, recounted Clark family stories in *Persimmon Hill.* First of all, he says Clark said Big White was really big—that he was six foot ten inches tall. He related—

> At Louisville, bonfires burned in their honor, and banquets and balls were given. . . . The whole family gathered at Locust Grove, his sister's house. . . York was in the quarters unpacking his Indian trophies to the "oh's" and "ah's" and prideful joy of his parents Old York and Nancy [actually his sister who was Croghan's slave], the cook. . . . York recited his adventures with dramatic pose. He took much pleasure, too, in the fact of the buckskins' being abolished and in seeing his master again in ruffled shirt, silken hose and buckled pumps.[593]

The travelers split into three groups. Chouteau and the Osage delegation and Lewis and Big White's entourage went to Frankfort, Kentucky, arriving there on November 13th. After that, Chouteau and the Osages went east from Lexington to Pittsburgh, and from there to Washington, while Lewis and Big White's group went south to the Cumberland Gap.[594]

Clark remained in Louisville until December 15th. Clark's arrival home coincided with major events in the Burr Conspiracy. The leader of the Burrites at the Falls of the Ohio, Davis Floyd, was the older brother of Sergeant Charles Floyd—who was the only man to die on the expedition. Davis Floyd was Burr's quartermaster, and he had men armed with muskets and ammunition and boats loaded with supplies ready to join Burr.

Aaron Burr visited the Falls on November 27th, but there is no record of Burr meeting with Clark. Undoubtedly, some members of Clark's family were involved in the conspiracy. Thirteen years earlier, in 1793–94, George Rogers Clark, as commander of the French Revolutionary Army, had recruited over 1,000 Kentuckians to join him in a similiar, failed invasion of Spanish territory during President Washington's administration.

Later, Davis Floyd was one of the seven Burr associates who were indicted for treason and violating the Neutrality Act by the grand jury in Richmond in June of 1807. In fact, Floyd became the *only* person in the conspiracy ever to be convicted of anything—he was convicted of violating the Neutrality Act by an Indiana court, and jailed for three hours. Soon after that, he was appointed Clerk of the Indiana Territorial Assembly.[595]

On December 16th, Harmon Blennerhassett and his men arrived at the Falls after fleeing from Blennerhassett Island on December 11th. William Clark may have still be in the area, as he wrote on December 14th to his brother-in-law William Croghan at Locust Grove that he was going to visit Colonel Richard C. Anderson, who lived east of Louisville. When Blennerhassett arrived, Davis Floyd and his men joined him, and they left on the same day to join Burr at their meeting place on the Cumberland River near Smithland, Kentucky. About 100 men in ten boats arrived in Natchez where the filibuster was eventually stopped. When and where these boats joined together is uncertain.[596]

During November and early December, 1806, Burr had been involved in legal proceedings in Kentucky. He stood trial for treason in Lexington on December 5th, defended by Henry Clay. The charges were brought by U. S. District Attorney Joseph Hamilton Daviess, who had been sending warnings all year to President Jefferson about Burr's activities.

It was fortunate for Burr he was brought to trial in Kentucky, as it ultimately allowed him to escape conviction for treason in Richmond, Virginia in the fall of 1807. The grand jury indictment at Richmond charged that Burr was the leader of the group of men who left Blennerhassett Island on December 11th, but Burr was in Kentucky with his court case. The charge of treason in the Kentucky court was dismissed for lack of evidence, and Burr and John Adair rode south to Nashville around December 10th. Andrew Jackson—who had changed his mind about Burr—gave him a cool reception, but he did supply him with two of the five boats he had ordered.

When William Clark left Louisville, York accompanied him on his journey east. York's family was in Louisville. His father and mother, Old York and Rose, were owned by William Clark's parents. Apparently York's mother had died, but his sister Nancy was an enslaved cook of the Croghan family. York was married and his wife was owned by someone not related to the Clark family. It is not known whether they had children.[597]

William Clark had no official responsibilities for the expedition after it ended. He could do as he wished, and he went to Fincastle, Virginia to visit his prospective bride, Julia Hancock, and her family. He took the same route Lewis had taken a few weeks earlier through the Cumberland Gap in the Appalachian Mountains, and then rode north on the Great Valley Road.

Lewis and his party had stopped at the Cumberland Gap in order for Lewis to make a survey of the boundary line between Virginia and North Carolina. The original survey, Walker's Line, had been done by Thomas Walker, who was married to Lewis's great aunt. Around November 20th, Lewis spent two days taking new measurements, which resulted in the state of Virginia gaining an additional nine miles and 1,077 yards of land along its border with North Carolina. It must have given Lewis a great deal of satisfaction, no matter how eager he was to return home.[598]

Lewis and the Mandan party arrived at his mother's home, Locust Hill, in Albemarle County on December 13th. Jefferson had written to Lewis urging him to take Big White over to Monticello to see the Indian artifacts displayed in his Great Hall—some of which the Mandans had given to the president. After spending Christmas in Albemarle, they arrived at the President's House in Washington on December 28th. Chouteau and the Osage chiefs had arrived a week earlier, on December 20th.[599]

Clark arrived at Fincastle in late December. He may have attended the wedding of Harriet Kennerly on December 23rd. The 18 year old Harriet married John Radford. She was a cousin of 15 year old Julia Hancock, the girl Clark would marry a year later in 1808. Clark named the Judith River in Montana for Julia—Judith was her given name, but she was always called Julia. Clark had met Julia and Harriet in 1803 when Julia was 11 years old. He was 21 years older. Years later, in 1821, William Clark and Harriet would marry, after both of their spouses had died.

On January 8th, the citizens of Fincastle honored Captains Lewis and Clark.[600] Clark then went to Washington, arriving on January 21st. He was too late to attend an affair held in their honor on January 14th, which had already been postponed once, waiting for his arrival. Clark's position was now of an ambiguous nature. Even though the country referred to it as the Lewis and Clark Expedition, it was Lewis's responsibility and he still had many matters to take care of—the financial accounting; managing the distribution of scientific and geographical information; and writing an account of their travel and discoveries for publication, which would be published under both of their names.

"Donation Land" was given to members of the Corps of Discovery as a reward for their services. Secretary of War Dearborn wanted to give 1,500 acres to Captain Lewis, 1,000 acres to Lieutenant Clark, and 320 acres to the non-commissioned officers and privates. Everyone on the expedition was to get double pay. However, Captain Lewis requested equal grants of land be made to himself and Lieutenant Clark, as he didn't want to call attention to the distinction in rank between them.[601]

On March 3th, 1807, Congress authorized warrants for 1,600 acres of land each to be given to Lewis and Clark, and warrants for 320 acres to the others. Double pay was authorized. By this accounting, Lewis received $2,776.22 and Clark received $2,113.74. The men averaged $333.33 each. Drouillard earned $1,666.66 and Charbonneau, $818.32 as civilians.

Historian Arlen Large, however, discovered that later in the same year, Lewis and Clark received ample bonuses, resulting in a total pay of $4,932 to Lewis, and $4,538 to Clark. Part of Clark's pay was a subsistence ration for his " black waiter." Officers were entitled to have subsistence rations for at least one servant—slave or free. York had been Clark's slave since their early boyhood together.[602]

Clark turned back his army commission to Secretary Dearborn after he arrived in St. Louis, saying it had "answered the purpose for which it was intended." [603] He was still upset about not getting the captain's commission promised by Jefferson. On February 28th Jefferson nominated Lewis as Governor of Louisiana Territory and Clark for promotion from first lieutenant to lieutenant colonel of the 2nd Infantry Regiment. The Senate approved Lewis's nomination and rejected Clark's promotion. The promotion was denied on grounds that several officers had greater seniority.

On March 12th, Secretary Dearborn nominated Clark to become the Brigadier General of the Territorial Militia in Louisiana, an appointment which was confirmed by the Senate. Clark had already been appointed "Agent for Indian Affairs to the Several Nations of Indians within the Territory of Louisiana excepting the Great and little Osages and their several divisions, and detachments." The Chouteau family had been trading with the Osage for years, and after the U. S. acquired Louisiana, Pierre Chouteau was appointed U. S. Agent for the Osages. The Osage lived in western Missouri, Arkansas and Kansas.

Lewis was appointed superintendent of Indian affairs, as well as governor. After Lewis's death Clark became the superintendent in 1813, a position he held until his death in 1838.[604]

William Clark, the day after he arrived in Washington, on January 22nd, wrote to his brother Jonathan that he had shared Jonathan's recent letter reporting on Burr's activities with the president. He wrote:

> . . . the Expedition of Mr. B. has excted the greatest allarm and his views as are maid known the Exctve (and which will be laid before Congress tomorrow or the next day) are of the most desperate and violent ~~nature~~ aspect . . . Genl. Wilkinson has been giveing information to the Government for Some time . . .[605]

Clark would soon be directly involved with the Burrites, as he was going to St. Louis in his capacity as Indian Agent. Lewis would remain in the east for another year taking care of responsibilities regarding the expedition, and on special assignment for the president during Aaron Burr's trial for treason.

Lewis wrote Clark a letter on March 13th, after Clark had set out from Washington to St. Louis via Fincastle:

My dear Friend,

Robert Frazier arrived here yesterday and from information which he brings with respect to Burrism in Louisiana the government have thought proper to avail themselves of your services should it be necessary for you to act by placing you at the head of the Militia of that Territory, and for that purpose have placed in my hands a commission for you as Brigadier General. . . . I shall write to Secretary Bates and request him to consult with you and take such measures in relation to the territory as will be best calculated to destroy the influence and wily machinations of the adherents of Col. Burr. It is my wish that every person who holds an appointment of profit or honor in that territory and against whom sufficient proof of the infection of Burrites can be aduced, should be immediately dismissed from office, without partiality, favor, or affection, as *I can never make any terms with traitors.* Mr. Robert Waistcoat [Wescott] son in law to Secretary Brown, Col. John T Smith and Mr. Dodge sherif of the district of St. Genevieve are highly implicated and there is good reason for believing that Mr. Brown himself might be sensured without injustice—Frazier will give you a more detailed account of those transactions than leisure permits me to give you at this moment.

I must subjoin a wish that you would make your dissertations on the subject of _____ to Miss ______ as short as is consonate with your amorous desires—for god's sake do not whisper my attach to Miss ____or I am undone.[606]

Adieu and believe me your sincere friend,

Meriwether Lewis

Governor of the Territory of Louisiana

Robert Frazier was bringing Lewis's letter and his commission as brigadier general to Clark. Frazier, a former member of the expedition to the Pacific, was going to testify against Wescott in a trial for treason scheduled to be held in St. Louis in May.[607] Robert Frazier himself wrote to President Jefferson on April 16th from Henderson County, Kentucky. In this letter he states that he is worried he will be assassinated by Smith T and his "junto" while traveling to St. Louis. Frazier was bringing incriminating documents with him that apparently would implicate them in the Burr Conspiracy.

> . . . At Breckinridge court house I was informed of a number of inquiries that some of the party (dispached to overtake & wrest from me my papers) had been making relative to my business at Washington. I shall take proper measures to bring these sub-traitors if not to condign punishment, at least to a full exposure at the tribunal of my country. I also learned from a gentleman of high respectability, directly from St. Louis that Col. John Smith (T) will not suffer himself to be taken by the civil authority; but has threatened and reviled me with the harshest and most bitter epithets.
>
> From this man's character as a desperado & from the servility of a vile and desperate junto of which he is the head, I really think I am in no small danger of assassination, or some other means of taking me off. If they would face me fairly and openly, I would boldly confront them, and set them at defiance; but this is not their method of attack. I advised with several gentlemen on the subject of my personal safety before I arrived here, whom I knew were friends of the government, and who referred me to Gen. Hopkins. He advises me to travel by way of Vincennes; there being but few houses between there and St. Louis and the country chiefly prairie. That it is the safest I have not a doubt & have consequently adopted it.[608]

Frazier reached St. Louis safely, and both Smith T and Westcott were removed from the offices they held by Frederick Bates, the new territorial secretary and acting governor.

Frederick Bates (1777–1825) became the second governor of the State of Missouri in 1824. His home, Thornhill, is part of a historic village located in Faust Park, Chesterfield, MO, which is open for group tours. www.stlouisco.com (Wikimedia Commons)

Frederick Bates arrived in St. Louis on April 1st, replacing Joseph Browne, Burr's brother in law, as acting governor and territorial secretary. Browne and his son, Robert, joined John Smith T and W. H. Ashley in the lead mining business.[609] On April 28th, Bates wrote to Governor Lewis:

> Nothing but the actual experience could have fully informed me of the extent of those duties and responsibilities, which in your absence, it will be necessary for me to take upon myself. . . . I have not yet experienced so much ill natured opposition as I had expected: Yet the minds of the factions are by no means tranquil. . . . all from contrary motives of hostility and friendship anxious for your arrival: For contrary to my first expectation you must expect to have some enemies.[610]

Bates made a tour of the lead mine district, visiting Moses Austin, during his first month. He wrote to his uncle he was planning to buy a farm in the lead mine district along with a few slaves, whom he would employ in digging for lead on the off season. A slave with a pick and shovel could make back his purchase price, $400, in a season. He couldn't buy land immediately because land titles and lead mine leases were still uncertain.[610]

Bates received two additional federal appointments as the recorder of land titles, and as a member of the three man board of land commissioners. He was doing five jobs at once in Louisiana

John Smith T (c. 1770–1836) in a miniature portrait. (Missouri History Museum, St. Louis)

Territory—serving as territorial secretary, acting governor and superintendent of Indian affairs, recorder of land titles, and land claims commissioner.

The brother in law of Moses Austin was Moses Bates, a relative of Frederick Bates. The two Moses's— Austin and Bates—had come from Connecticut to Spanish Louisiana in 1798 as partners in the lead mining business.[611] Frederick Bates became associated with the Austin-Bates faction in the Mineral Wars, just as Wilkinson joined with Smith T's faction. There were family connections in both cases.

Lead was found over a region of hundreds of square miles.The lead mining district had been declared public land in an 1804 act of Congress, but the law was not enforced. On March 3rd, 1807, Congress passed the Lead and Salt Leasing Act. Conflicts over this policy developed immediately. Existing large mining claims, such as Austin's, would continue until the land commissioners decided their fate at a later time. All new mines opened since 1804 and all those digging for lead under the traditional rights of usage ("prescriptive rights") were to cease operation and obtain leases from the federal government.

Leasing on public land was allowed for small tracts of land, from 40 to 320 acres, for up to three years. Bates, as recorder of land titles, was put in charge of granting leases. The newly appointed federal land agent, Will C. Carr, took over this responsibility in 1808. Those who got leases found their lands were still being illegally mined by others. John Smith T was notorious.

He had purchased an original Spanish "floating land concession" from a French resident in 1806, giving him the supposed right to claim land wherever he wanted. It was a "wild card," which he enforced with his private army. Austin had a private army of his own to battle him, but small mine operators were simply driven off their land.

A second act, also passed by Congress on March 3rd, 1807, instructed the commissioners to make no decisions about land containing a lead mine or salt springs which had no prior claims on it. The government was stalling for time as it tried to figure out what to do with the valuable mineral lands. The commissioners had to first determine whether the land in question did contain lead or salt, in order to exclude it from sale. It was obviously to the benefit of land purchasers to have their titles approved and any minerals on their lands ignored.[613]

The situation led to highly charged conflicts of interest, involving large potential profits in land speculation and mining. Since lead was used for making bullets, the lead lands were the focus of both the government and private filibusterers. Salt petre in caves also fell under this act. Salt petre (the excrement of bats, also known as potassium nitrate or fertilizer) was used for making gun powder. The lead district contained the highest concentration of lead in the world, and was also a center of gun powder production.

The land commissioners began the task of sorting out the complicated land claims of the original French inhabitants. The residents followed the traditional French Canadian custom of living in villages with homes and vegetable gardens surrounded by high fences to keep the pigs out. They farmed on long, narrow plots of land located next to each other in an agricultural commons area, and shared a common pasture for livestock. Land was shared in common for timber, maple sugar, and lead. In the off season for agriculture, they dug for lead ore and smelted it in open fire pits to pay their bills at the local store. This system had worked well for generations, and they settled any issues among themselves. The Spanish government had seen little need to sort out their land claims.[614]

William Clark, after spending time in Fincastle and obtaining permission to marry his young sweetheart, arrived in St. Louis in May, 1807. He would spend the summer in St. Louis performing his duties as Indian agent, and then take on a new assignment for Jefferson on his return trip east—excavating mastodon fossil bones from Big Bone Lick near Cincinnati for the American Philosophical Society. In January, Clark and Julia would marry at Fincastle. They would make St. Louis their new home in July, 1808.

During the summer of 1807—following the return of Lewis and Clark's expedition the previous year with its news of the rich fur country on the Upper Missouri—several expeditions left St. Louis bound for the headwaters. Two of these expeditions employed former members of the Corps of Discovery.

The first was a joint military-trading expedition authorized by Secretary of War Henry Dearborn, who ordered Nathaniel Pryor to escort Big White and his Mandan party back home to their villages on the Knife River. Pryor, now an ensign (the lowest rank of commissioned officer), was to have "one careful, sober searjeant and ten good sober privates" assigned to him from Camp Bellefontaine. A private trading expedition of the Chouteau family was accompanying Pryor, which was authorized by Dearborn to trade with Indians from the Arikara Indian villages on the Missouri upwards. The U. S. would supply ammunition to the trading expedition.

On May 18th, Clark wrote to Dearborn that Pryor's party would have 14 men, and young Auguste Chouteau was bringing 22 men to trade with the Mandan. However, a group of fifteen Sioux chiefs and warriors, from the "most numerous and vicious bands" had arrived in St. Louis recently, escorted by Pierre Dorion, who was intending to bring them to Washington. After several days of deliberation, the Sioux chiefs decided not to go to Washington. They wanted to return home with the nearly $1,500 worth of merchandise Clark had given them.[615]

Big White suggested the Sioux delegation should accompany them back up river. The combined expeditions now numbered either 106 or 108 people. There were 18 Indian men, 8 Indian women, and 6 children in the party, including Big White and his wife and child, and the wife and two children of interpreter René Jessaume.[616]

On June 1st, Clark reported to Dearborn that the expedition had left St. Louis and started up river. After the Sioux left them and the soldiers escorting them started back, Clark wrote there would still be 48 men in Ensign Pryor's party, "fully sufficient to pass any hostile band which he may meet with." [617]

Clark's optimism proved false—when on September 9th, Pryor and Chouteau faced some very angry Arikara Indians. The Sioux chiefs and their military escort had already left them. The Arikara had recently learned that their beloved chief had died the year before while on his visit to Washington, and they were also at war with the Mandan Indians.

The other expedition going up the Missouri with members of the Corps of Discovery was under the command of St. Louis trader Manuel Lisa. Lisa had started up river ahead of Pryor and Chouteau in April. When Lisa's expedition of 42 men encountered the Arikara, they were allowed to continue only after they were forced to surrender half their trade goods.

When the Pryor-Chouteau expedition reached the Arikara villages, a Mandan woman, held captive by the Arikara, told Big White that Manuel Lisa had informed the Arikara that two boats were following his boat bringing the Mandan chief back home, and they were loaded with merchandise for the Arikara. The woman captive also told them that the Arikara intended to kill Big White and his escort, and planned to kill Manuel Lisa and his men on their return journey.

There were 650 Arikara warriors on the shore as the two boats of Pryor and Chouteau tried to pass by the Arikara's two stockaded villages on opposite sides of the Missouri River. Dorion and Jessaume were on shore, essentially held hostage by the Arikara, as the boats traveled up to the second village. Pryor told Big White to barricade himself and his party with trunks and boxes in the boat cabin. Negotiations with a chief who had been friendly with Lewis and Clark proved futile.

The Arikara seized the cable of Auguste Chouteau's boat, which was loaded with merchandise and had no soldiers on it. They indicated to Pryor that he could continue up river, but he didn't believe them. The commander of the trading expedition, young Auguste Chouteau, was only fifteen years old. He and several of his men had gone on shore to negotiate with the Indians. Pryor told Auguste to "make them an offer." Pryor wrote:

> He at last *did* make them an offer, which, had they not been determined on plunder and blood, ought to have satisfied them. He proposed to leave half his goods with a man to trade them—

A battle then took place in which they were forced to retreat, and as they retreated down river they encountered gun fire from both sides of the river for about an hour. Sioux warriors were fighting alongside the Arikara. None of Pryor's men were killed, but three were wounded. Four men of Chouteau's party were killed and six were badly wounded. One of Pryor's men was 22 year old George Shannon, the youngest member of the Corps of Discovery. Shannon, wounded in the battle, would eventually have his leg amputated in 1816. He used a peg leg for the rest of his life and became known as "Peg-Leg Shannon."

Pryor told Big White he would escort him by land to his village, but Big White refused, because of the women and children accompanying him, and because René Jessaume had been badly wounded and needed medical attention in St. Louis.

The expedition returned to St. Louis in defeat. Nathaniel Pryor concluded at the end of his report to William Clark—

> If my opinion were asked 'what number of men would be necessary to escort this unhappy Chief to his nation,' I would be compelled to say, from my own knowledge of the association of the upper band of the Sioux with the Ricaras that a force of not less than 400 men ought not to attempt such an enterprize—and surely it is possible than even One Thousand men might fail in the attempt.[618]

Meriwether Lewis spent the entire year of 1807 out east. He was a returning hero, and yet—in thinking about the remaining 34 months of his life—what about Seaman, his dog? Where was he? He was an important part of Lewis's life. Seaman was a very special dog, a "sagacious dog," the size of a black bear, who was always following Lewis around, according to the Nez Perce traditional stories. During the expedition there is no question where Seaman was—he was traveling with Lewis.

We know that he survived the expedition and wore a collar inscribed with these words:

> The greatest traveller of my species. My name is SEAMAN, the dog of captain Meriwether Lewis, whom I accompanied to the Pacifick ocean through the interior of the continent of North America.

The inscription was included in *American Epitaphs and Inscriptions*, a book written by Congregational minister and historian Timothy Alden, and published in 1814. Alden copied the inscription from Seaman's collar on display at the Masonic Museum in Alexandria, Virginia. Alden wrote that Seaman died of grief on Lewis's grave in 1809, establishing that the dog continued to be Lewis's companion.[619]

Meriwether Lewis traveled by horseback—Seaman could have accompanied him wherever he went. Did he go everywhere with him? It's possible, but there is another possibility, because Lewis added a servant to his household when he moved to St. Louis, a servant from the President's House named John Pernier.

We know nothing of John Pernier's background, but his last name is associated with the beginnings of the Haitian revolution. The wealthy Pernier plantation north of Port au Prince was the scene of an early battle in which black fighters for independence achieved a victory in 1791.[620] John Pernier could have been a "free man of color," who came to the United States after that time. He was working as a servant for Jefferson in 1804–05, and paid wages as a "free mulatto." [621] Since Jefferson employed only very highly qualified people at the President's House, it seems likely Pernier could have been one of the many refugees fleeing Saint Domingue, who would have been employed as a servant at a plantation on the island. He spoke French. His language skills would have been helpful to Governor Lewis in St. Louis.

Lewis lived at the President's House during his time in Washington. Did Seaman travel with Lewis to Philadelphia? It doesn't seem likely he would have gone with him to Richmond for the Aaron Burr treason trial. Perhaps the dog was sometimes left in Pernier's, or another servant's care, in Washington.

On December 30th, the Mandan chief Big White was formally welcomed by President Jefferson. On New Year's Day the Indians attended the president's Open House. In the evening of January 1st, the Indians themselves were a featured attraction at The Theatre in Washington. Big White and his wife Yellow Corn remained in the audience for most of the evening, viewing tumblers, acrobats and comedians. But when the Indians performed onstage, Big White, as "King of the Mandans," sat on a throne and watched the Osage chiefs dance. He also attended the dinner given in Lewis and Clark's honor on January 14th. After that, the Mandan chief's party left with Pierre Chouteau and the Osage chiefs to return to Louisiana Territory. They visited Philadelphia where Big White and Yellow Corn had their portraits done by St. Memin before returning to St. Louis, and they probably visited Baltimore and New York.

There is no record about what Jefferson—who was of Welsh descent—thought about the Mandan-Welsh connection.[622]

Lewis was planning to write a series of books called *Lewis and Clark's Tour to the Pacific Ocean Through the Interior of the Continent of North America.* A wall map of North America would be sold separately. The first part of the publication would consist of two volumes at a combined price of $10; the second part, one volume at $11; and the map, at $10. In April, 1807 his publisher in Philadelphia, John Conrad, issued a prospectus seeking subscribers to Lewis's publications.

The prospectus announced that Part One's first volume—

> WILL contain a narrative of the voyage, with a description of some of the most remarkable places in those hiterto unknown wilds of America, accompanied by a map of good size, a large chart of the entrance of the Columbia river, embracing the adjacent country, coasts and harbours, and embellished with views of two beautiful cataracts of the Missouri; the plan, on a large scale, of the connected falls of the river, and also of those falls, narrows, and great rapids of the Columbia, with their several portages. For the information of future voyagers, there will be added in the sequel to this volume, some observations and remarks on the navigation of the Columbia and Missouri rivers, pointing out the precautions which necessarily must be taken, in order to ensure success, together with an itinerary of the most direct route across the continent of North America, from the confluence of the Missouri and Mississippi rivers to the discharge of the Columbia into the Pacific Ocean.

Part One's second volume would contain—

> WHATEVER properly appertains to geography, embracing a description of the rivers, mountains, climate, soil and face of the country, a view of the Indian nations distributed over that vast region, showing their traditions, habits, manners, customs, national character, stature, complexions, dress, dwellings, arms, and domestic utensils, with many other interesting particulars in relation to them: also observations and reflections on the subjects of civilizing, governing and maintaining a friendly intercourse with those nations. A view of the fur trade of North America, setting forth a plan for its extension, and showing the immense advantages which would accrue to the mercantile interests of the United States, by combining the same with a direct trade to the East Indies through the

> continent of North America. This volume will be embellished with twenty plates illustrative of the dress and general appearance of such Indian nations as differ materially from each other; of their habitations; their weapons and habiliments used in war; their hunting and fishing apparatus; domestic utensils, &c. In an appendix there will also be given a diary of the weather, kept with great attention throughout the whole of the voyage, showing also the daily rise and fall of the principal water courses which were navigated in the course of the same.

Part Two's description of the third volume stated—

> THIS part of the work will be confined exclusively to scientific research, and principally to the natural history of those hitherto unknown regions. It will contain a full dissertation on such subjects as have fallen within the notice of the author, and which may properly be distributed under the heads of Botany, Mineralogy, and Zoology, together with some strictures on the origin of Prairies, the cause of the muddiness of the Missouri, of volcanic appearances, and other natural phenomena which were met with in the course of this interesting tour. This volume will also contain a comparative view of twenty three vocabularies of distinct Indian languages, procured by Captains Lewis and Clark on the voyage, and will be ornamented and embellished with a much greater number of plates than will be bestowed on the first part of the work, as it is intended that every subject of natural history which is entirely new, and of which there are a considerable number, shall be accompanied by an appropriate engraving.

Lewis and Clark's Map of North America would measure 5 feet 8 inches by 3 feet 10 inches, and be published separately. The three books would be 400–500 pages each, and along with the map, would be printed in Philadelphia and available for sale around the country at specified locations.

Meriwether Lewis informed his potential subscribers that no payment would be required of them until the books and map were available for delivery, and explained that he was setting a moderate price. He was asking for subscriptions mainly so he could make an estimate of how many copies to print. He cautioned the price of the second part might have to be increased, depending on the number of engravings. Subscribers soon began responding from around the country. [623]

Rumors of the impending publication of Sergeant Patrick Gass's unauthorized account of the journey infuriated Lewis and Jefferson. Gass's book was going to sell for $1. Gass's text was highly edited, or "expurgated" by his publisher, and his original journal has never been found. The woodcuts which appeared in its second edition were almost certainly created by Gass.[624]

Lewis wrote an open letter which appeared in a Washington newspaper on March 14th. Historian Paul Cutright in his *History of the Lewis and Clark Journals* says Jefferson had a hand in writing the letter, as some words were not part of Lewis's vocabulary (i.e., "promulgation" and "subjoined") and there were other indications of Jefferson's style.[625] Unfortunately, in his public letter, Lewis promised to have his first volume ready for sale in January, 1808, and the large map ready for sale by the end of October, 1807. It was a highly unrealistic plan, and the projected dates were undoubtedly influenced by the publication of Gass's book, which did come out in the summer of 1807.

Meriwether Lewis had been elected to membership in the American Philosophical Society before setting out on the expedition. Thomas Jefferson was the president of the American Philosophical Society from 1797 to 1814. There is no question that both of them considered the announced publication would become the premiere work of American scholarship, and that they were very proud of it. On March 22nd, Jefferson wrote to a fellow botanist, praising Lewis's accomplishments in the field of botany, and sharing some of the seeds that Lewis had brought back for the development of "useful or agreeable varieties." He wrote about Lewis's other accomplishments—

> He will equally add to the Natural history of our country. On the whole, the result confirms me in my first opinion that he was the fittest person in the world for such an expedition.[626]

In October, 1806, Lewis had authorized the publication of Robert Frazier's journal. But upon examining Frazier's proposal, he took back his promise that he could include certain information about natural history. In his letter to the public, Lewis said Frazier could not possibly give accurate information about the celestial observations, mineralogy, botany, or zoology of the expedition, and what could be expected from his journal was "merely a limited detail of our daily transactions." [627] Lewis's criticism must have persuaded Frazier to drop the project, because his journal was never published, and is lost to history.

Although Lewis was going to author the three volumes, he was going to share equally in its "honors and rewards" with Clark. Clark would handle the making of the large map. The two men split the cost of buying Sergeant John Ordway's journal from him for $300, and intended to split the profits. Ordway's journal was an important reference, as he was the only writer to make entries for every one of the 863 days of the expedition.[628]

As time has gone by—and as one great editor after another has tackled the task of working on the materials of the expedition—it is easy to understand Lewis's reluctance to work on it part time. When he began assembling materials, he must have realized it would require his full attention for a number of years and he would need to live in Philadelphia. It was very different in concept than simply publishing the journal entries.

Jefferson had put an impossible obstacle in Lewis's way by appointing him governor of Louisiana Territory, superintendent of Indian affairs, and commander of the militia in St. Louis. When he finally did depart for the east coast in 1809, after his 18 month stay in St. Louis, he intended to stay out east and focus on getting the books and map ready for publication.

Nicholas Biddle in Philadelphia took over the task of preparing an account of their travels after Lewis's death, consulting with William Clark and George Shannon as he worked. Biddle used Gass's model of daily entries, with no attempt to present a larger overview. The Biddle edition(1814) and Gass's journal, (1807), remained the only published accounts of the expedition until Elliot Coues's edition was published in 1893. Coues was the first editor to include natural history information.

William Clark had completed the large map by December, 1810, which was also published in 1814. Clark offered Biddle one half share of the profits, and later arranged that Lewis's share (now one quarter) would go to his mother. The first publisher, Conrad, went out of business. Nicholas Biddle, who was beginning a political career, turned over the project to another editor, Paul Allen, and paid him $500 for his work. Biddle told Clark he would not accept pay for his own work. Less than 1,500 copies of the two volume *History of the Expedition Under the Command of Captains Lewis and Clark* were printed and they sold poorly in 1814. The second publisher, who also went out of business, explained that there were no profits to be had from the sale of the books.[629]

The scientific information was omitted by Nicholas Biddle, because it was going to be published in a separate volume authored by Philadelphia physician, Dr. Benjamin Smith Barton, who had trained Lewis in botany before the expedition. Lewis discussed his natural history material with Barton after the expedition, and intended for Barton to assist him in writing the third volume devoted to scientific research. Dr. Barton, who was in failing health, died in 1815.

The science material was never was fully published until the Moulton edition was published almost 200 years later. Editor Gary Moulton has written that Biddle's omission of the scientific data "established a view of the expedition, common to this day, as primarily a romantic adventure." It is a great popular story, but the lack of understanding of its scientific mission is another factor affecting the reputation of Meriwether Lewis.

Many suicide theorists point to Lewis's supposed inability to publish his account of their travels as an indication of depression, and as a reason why they believe that he committed suicide. But they are overlooking the complex nature of his project and how long it would take to complete it. It took over twenty years for Moulton and his team of research associates to do justice to the entire body of material, including all of the known journals. *The Journals of the Lewis and Clark Expedition*, were published in a series of thirteen volumes in 1982–2001 by the University of Nebraska Press.[630]

Meriwether Lewis set off for Philadelphia in April of 1807. Over the next few months he must have enjoyed himself greatly. He sat for his portraits with Charles Willson Peale and Saint-Mémin. He attended meetings of the American Philosophical Society with his friend Mahlon Dickerson. He made a new friend, the bird artist, Alexander Wilson, and met with him and other members of the scientific and artistic community in arranging for the publication of his materials.[631]

He tried to find a woman who would marry him and move to St. Louis. Lewis was 32 years old. He needed a wife who would be comfortable in the social circles in which he moved, a good Republican wife. He was an extraordinary young man—a celebrity, with a bright future ahead of him. He had just returned from three years of exploration, and was absorbed in preparing an account of his experiences for publication.

What kind of future could a young woman expect with him? In a few months time he would be assuming the responsibilities of governing Louisiana Territory. St. Louis was very far away. The Aaron Burr Conspiracy was just coming to its end, and threats of war with Britain and Indian attacks were a constant worry. Philadelphia was one of the most sophisticated cities in the United States. It would have taken an equally extraordinary woman to marry him under these circumstances, based on an acquaintance of a few months. It was much more reasonable to wait until his life had settled down before he found a wife.

In any event, Jefferson assigned him another responsibility—attending the Aaron Burr treason trial in Richmond, Virginia. Lewis had expected to be in St. Louis by October 1st. He had written to Auguste Chouteau to have Gratiot's house prepared for his arrival by that date, or to help him find another house to rent if he couldn't afford Gratiot's house.[632] Under that timeline, Lewis would have been traveling to St. Louis in early August.

On July 31st, the accountant for the war department, William Simmons, asked Lewis to explain in writing if any of the provisions for the expedition should be charged personally to Lewis or to Clark. Simmons asked for an explanation of more than 30 ledgers of expense records. On August 5th, Lewis submitted his account of expedition expenses totalling $38,722.25. These records place Lewis in Washington at this time. [633]

Lewis attended the treason trial of Aaron Burr in Richmond, which began on August 3rd. He must have left immediately after submitting his financial accounting to Simmons. He stayed in Richmond until Burr's second trial for high misdemeanor ended on September 14th.

The trial had several phases. It was presided over by Jefferson's enemy—who was also his distant cousin—the Chief Justice of the Supreme Court, John Marshall. Burr arrived at Richmond, under arrest, on March 26th, 1807. On April 1st, Judge Marshall, after hearing evidence, scheduled a trial for a crime of "high misdeameanor" rather than "treason." Jefferson took offense at this, and sent inquiries around the country, asking people to supply evidence of treason.

The initial evidence of treason was based on Wilkinson's account of Burr's plotting western secession and civil war, and Jefferson's own declaration that Burr had committed treason. The critical point was whether there was an "open assemblage of men" for the purpose of moving against a territory of the United States. Filibusters against a foreign power did not constitute treason. The charge of "high misdemeanor" was for organizing a military action against Spain in violation of the Neutrality Act of 1794—that is, the filibuster planned against Mexico.

Enough of Wilkinson's actions were questionable that the charges against Bollman and Swartwout—the conspirators he had arrested in New Orleans—had been dismissed in previous hearings held by the high court. Justice Marshall was now acting in his capacity as a Virginia circuit court judge when a grand jury was convened in Richmond. A Virginia court was necessary in order to establish a location for where the act of treason had occurred, in this case, Blennerhassett Island on the Ohio River.

The grand jury, on June 24th, indicted Aaron Burr for both treason and high misdemeanor, setting the framework for one of the most famous trials in American history. The jury interrogated 29 witnesses, including the main government witness, General Wilkinson. The general himself was almost indicted, for misprision (concealment) of treason and high misdemeanor. The vote was 7 for his indictment and 9 against. Jefferson was subpoenaed to appear in court and turn over letters in his possession, but he refused on the grounds of separation of powers—the court could not compel a sitting president to appear.

Jefferson remained at Monticello in Albemarle County. The population of Richmond, 6,000, nearly doubled in size during the trial. According to one estimate there were "no less than two hundred persons in Richmond"—witnesses, jurymen, and lawyers—who had some official connection to the trial.[634] The trial for treason began on August 3rd. The charge of treason was limited to what occurred on December 11th on Blennerhassett Island, and as to whether its participants were *overtly* acting to start war against the United States. The other issue was whether Burr, who was not present on the island, could be convicted of treason. Justice Marshall delivered an opinion that what happened on the island was not an overt act of war against the United States.

This was an important legal opinion, because it established that under American law, a *conspiracy to commit treason* was not enough to bring charges. There had to be an actual *overt act*. On September 1st, the jury after deliberating 25 minutes, delivered its qualified verdict—

> We of the jury say that Aaron Burr is *not proved to be guilty under this indictment by any evidence submitted to us*. We therefore find him not guilty.

Aaron Burr was outraged and demanded he be judged "guilty" or "not guilty." However, Marshall let the jury's verdict remain on record, while entering it as "not guilty."

Jefferson was even more outraged. He instructed his lawyers that no witnesses be paid or allowed to depart until their testimony was put into writing. A second trial for Burr, on the charge of violating the Neutrality Act began on September 9th, with a new jury. The trial fizzled out, with a verdict of "not guilty" delivered within a week. The government tried once more, accusing Burr of treason at the Cumberland River, and asked Marshall, as chief justice to arraign Burr and others for a new trial in a western circuit court. Burr complained this was the sixth trial he had had to endure on the same charge (two in Kentucky, one in Mississippi, and three in Virginia). The trials at Richmond ended on October 30th. The trial in a federal court in Ohio was finally abandoned by the government.

In June of 1808 Aaron Burr left the country bound for Europe, to try his luck again.[635]

Sometime during this period of time, Lewis wrote a lengthy paper he entitled, *Observations and reflections on the present and future state of Upper Louisiana, in relation to the government of the Indian nations inhabiting that country, and the trade and intercourse with same. By Captain Lewis.*

Observations and Reflections has the same name and content that he proposed to publish in his second volume. There is no manuscript known to exist. It was published by Nicholas Biddle as an appendix to the *Lewis and Clark Journals* in 1814. Lewis used some of its material in a letter he wrote to Secretary of War Dearborn in 1808 when he was in St. Louis, and used parts of it in a newspaper article that appeared in the *Missouri Gazette* in August, 1808 signed "Clatsop." [636]

At the end of the second trial, Lewis left for Albemarle. General Wilkinson wrote a letter to Jefferson on September 15th, saying he had intended to send Captain Zebulon Pike's report to the president by Lewis, but he had been "too occupied" to get it ready.[637]

By October 8th, Lewis had returned again to Richmond. The hearing, as to whether there should be a trial in a western circuit court, was still going on. Lewis stayed in Richmond through October 20th, settling financial matters, which were negotiated by William Wirt, the government's chief prosecutor. Lewis's account book is his private record of money, dates of transactions, and names. The reason for the transactions is rarely stated. It seems likely he was settling accounts for the president. He then returned to Albemarle. It was only two or three days journey from Richmond to Albemarle on Three Notched Road.[638]

Lewis was able to support the education of his 22 year old stepbrother, John Hastings Marks, who was going to Philadelphia to study medicine. On November 3rd, Lewis wrote to his friend Mahlon Dickerson in Philadelphia, asking him to cover John Marks's expenses as Lewis was temporarily short of funds. He would pay Dickerson back, and reimburse him for any future expenses for Marks.

Lewis then gossiped about his love life, writing—

> . . . [I] am now a *perfect widower with rispect to love.* I feel all that restlessness, that inquietude, that indiscribable something common to old bachelors . . . Whence it comes from I know not, but certain it is, that I never felt less like a heroe than at the present moment. What may be my next adventure god knows, but on this I am determined, *to get a wife.*[639]

Lewis and his 30 year old brother Reuben Lewis set off for St. Louis soon after making arrangements for John Marks. Reuben was relocating to St. Louis, where he would enter the fur trade. The two brothers were accompanied by John Pernier, who was entering employment with Meriwether Lewis as his servant, and Lewis's dog, Seaman.

The Lewis brothers stopped in Fincastle in November. Clark had written to Lewis in April about a "most lovly girl Butifule rich possessing those accomplishments which is calculated to make a man happy." Her name was Letitia Baldridge, but she was already in love with another man and she left Fincastle two days after the Lewis brothers arrived, not wanting to give Meriwether false hopes. In July of 1808, Lewis learned from his friend William Preston that Letitia was "off the hooks" and had married Robert Gamble, Jr., in Richmond. Lewis had known Gamble since he was a boy. Lewis wrote that her new husband, was—

> a good tempered, easy honest fellow . . . both his means and his disposition well fit him for sluming away life with his fair one in the fashionable rounds of a large City. such is the life she has celected.[640]

On February 15th, 1808, Lewis wrote to their mother from Louisville, Kentucky. He said he had been "near two months traversing this state" locating and surveying the land claims belonging to himself, his stepbrother John Marks and stepsister Mary Marks. On February 10th, Reuben had set out by flatboat with Pernier, bringing Lewis's baggage and his carriage to St. Louis. Reuben and Pernier would go part of the way by boat and then go overland. Lewis himself was setting out the next day and expected to arrive in St. Louis about the same time.[641]

While Lewis was in Louisville, he engaged the services of Joseph Charless, a printer. After they both arrived in St. Louis, on April 29th, he loaned Charless $100, and advanced him another $125 he had obtained as newspaper subscriptions for the soon to be published *Missouri Gazette.*

The printing press arrived with William and Julia Clark in early July. Meriwether Lewis had almost certainly bought the press in Philadelphia from Adam Ramage, the leading manufacturer of printing presses in America.[642]

Charless set up the printing press, and published the first issue of the *Gazette* on July 12th. On July 22nd, Lewis, who had appointed Charless the Printer to the Territory, made out a draft on the Secretary of State for $500 to purchase paper for printing the *Laws of the Territory of Louisiana* (250 copies in English and 100 copies in French). Charless returned to Louisville to buy the paper and look after his printing business there. His young assistant, Jacob Hinkle, did the printing in St. Louis.

The first fifteen issues of the *Missouri Gazette* list Joseph Charless as the printer—not the publisher—of the newspaper. On September 21st, issue number 16, listed him as publisher. Meriwether Lewis was actually serving, anonymously, as the editor and publisher of the first newspaper published in Louisiana Territory. Two issues were combined in September, due to the "serious illness of its editor." Lewis was suffering from malaria.

A strong indication of Lewis's silent partnership in the newspaper business is that Charless changed the name of the newspaper to the *Louisiana Gazette*, shortly after learning of Meriwether Lewis's death. It began publication under its new name on November 30th, 1809.[643]

For several years Charless's newspaper in Louisville ran an ad seeking subscribers for the *Missouri Gazette*. The ad stated:

> A capable Editor is employed, and a number of Gentlemen have volunteered to devote their leasure hours in writing on such subjects as will enrich its columns. Essays on indian antiquity, Mines, Minerals, and an account of the Fur-trade, with Topographical Scetches will be diligently sought after.[644]

Jacob Hinkle supposedly edited and published the issues of the *Missouri Gazette* during this time. In November, 1808, one week after Charless had returned to St. Louis, Hinkle ran off with $200 worth of goods and $600 collected for newspaper subscriptions. To aid in his capture, Charless wrote the following description of his young printer—

> This hopeful *Sprig* is about 24 years old, 5 feet 9 or 10 inches high; he shuffles as he walks; has a shallow complexion, black and curled hair, is cross eyed and very near sighted; in short, Dame Nature never formed his countenance to *deceive* [tho' sad experience proves that he has outwitted even her in that respect' for she plainly stamped *The Villain* in every lineament thereof. [645]

Lewis had rented one of the nicest houses in St. Louis, which he offered to share with William and Julia Clark. Their first child, Meriwether Lewis Clark, was born here on January 10, 1809.

The house was located at South Main and Spruce Streets across the street from the Government House. It was quite a large and modern residence for its time, 40 x 20 feet, having two chimneys, glass windows, and rooms decorated with wall paper and oil paint—the first residence in St. Louis known to use wallpaper. It had four rooms on the ground floor, a large cellar, and an attic room for slaves to sleep in. There was a kitchen building and a smokehouse in the back yard. The yard had an excellent water well, fruit trees, and a vegetable garden. The well was a special feature. Most residents hauled water from the river because the town was built on limestone rock, and large cellars were a rarity.

On the double lot there was a stone warehouse, 16 x 60 feet, which became the first Masonic Lodge in St. Louis. The lodge, founded and presided over by Meriwether Lewis, met here from 1808–1811. The warehouse would have also been used as a gathering place for the militia and as an Indian Council House. Clark stored mercantile goods here.[646]

William and Julia had brought a number of slaves with them. He wrote to his brother Jonathan about his troubles with them on July 21st, 1808, three weeks after arriving in St. Louis:

> I have hired out Sillo, nancy, Aleck, Tenar & Juba,—Ben is makeing hay, York employed in prunng wood, attendng the garden, Horses &c. &c. Venus the Cook and a very good wench Since She had about fifty [lashes], indeed I have been obliged [to] whip almost all my people, and they are now beginning to think that it is best to do better and not Cry hard when I am compelled to use the whip. they have been troublesom but are not ~~at~~ all so now—[647]

The Clarks were accompanied by his 18 year old niece, Ann Anderson, when they arrived in St. Louis. The bachelors in St.

Louis flocked around Ann, but by October, she wanted to return home to Louisville. Clark was upset about her going because 16 year old Julia was pregnant and homesick herself. He wrote to his brother Jonathan on November 9th that Nancy [Ann] was about to return to Louisville.[648]

He also wrote—

> Govr. Lewis leaves us Shortly and we Shall then be alone, without any white person in the house except our Selves. I shall Send york with nancy, and promit him to Stay a fiew weeks with his wife, he wishes to Stay there altogether and hire himself which I have refused. he prefurs being Sold to return [ing] here, he is Serviceable to me at this place, and I am deturmined not to Sell him, to gratify him . . . if any attempt is made by york to run off, or refuse to provorm his duty as a Slave, I wish him Sent to New Orleans and Sold, or hired out to Some Severe master until he thinks better of such Conduct I do not wish him to know my deturmination if he conduct[s] [himself] well This Choice I must request you to make if his Conduct deservs Suverity.[649]

On November 24th, he wrote to Jonathan:

> Govr. Lewis is here and talks of going to philadelphia to finish our books this winter, he has put it off So long that I fear they will [not] bring [?] us much. I wrote to you the other day about york, I do not wish him Sold if he behaves himself well . . . —he does not like to Stay here on account of his wife being there. he is Serviceable to me here and perhaps he will See his Situation there more unfavourable than he expected & will after a while prefur returning to this place. . . .
>
> This morning Charless Comes with a Sour Countinonce and tells me Hinkle has run off in his debt $240. . . .[he] has contrived to clear out about $1000 in debt. . . . I am a little Supprised that this fellow Should have been So Suckcessfull.[650]

York statue by sculptor Ed Hamilton is located near 5th & Main Sts. in Louisville, Kentucky. www.edhamiltonworks.com (Photo by Rick Blond)

On December 10th, 1808 he wrote:

> I wrote to you in both of my last letters about York, I did wish to do well by him—but has he has got Such a notion about freedom and his emence Services, that I do not expect he will be of much Service to me again; I do not think with him that his Services has been So great (or my Situation would promit me to liberate him). . . he could if he would be of Service to me and Save me money, but I do not expect much from him as long as he has a wife in Kenty. [651]

In 1810 York was hired out as a wagon driver in Louisiville, with Clark receiving payment for his services. In November, 1815, he was working for William Clark and one of Clark's nephews driving a wagon in Louisville. The only other mention of York is an account by the writer Washington Irving who visited William Clark in 1832. Clark told Irving that, after he finally freed him, York became an independent wagon driver in Nashville, Tennessee. He said York had failed in his business, and was planning to return to St. Louis to work for Clark when he died of cholera in Tennessee. Historian James Holmberg speculates York may have

Lewis's Great Mound is seen on the right. It was 130 x 120 feet on its flat top and about 12 feet high.

gone to Nashville because his wife was taken there by her owner, and he may have died around 1820–26 when cholera epidemics were prevalent in the region.[652]

It seems unlikely that Lewis took separate quarters from the Clarks, as he was still living with them in late November, 1808. The Government House and the printing office were across the street, and the Masonic Lodge was next door.[653]

In 1808, there were about 1200–1400 people living in St. Louis. It was called "Mound City" because of its old Indian mounds. Cahokia Mounds, across the river in Illinois, is one of eight World Heritage Sites in the U. S. In 1050–1200 A. D., Cahokia was the center of one the largest cities in the world, with a population of 10,000–20,000. Over 120 mounds were built at Cahokia by the residents who hauled dirt in baskets to build them. They lived on the mounds and on the surrounding land. The largest mound, the 100 foot high Monks Mound, was named for Trappist monks who lived there in 1809–13. Mounds were a solution to the problem of annual river flooding.

Meriwether Lewis, as governor, was required to own at least 1,000 acres of land in the territory. Within two months of his arrival he had purchased the Great Mound in St. Louis and several more mounds lying near it, totalling 42 acres. Historian Grace Lewis Miller says Lewis probably intended to build a museum on the Great Mound. All of the mounds in St. Louis were eventually leveled.

Before the end of the year, on December 1st, 1808 he wrote to his mother he had purchased a total of 5,700 acres north of town.[654] One tract was 3,000 acres and she would have 1,000 acres of it. Her tract had a comfortable three room house, an enclosed garden, a good well, horse stables and other improvements, a 40 acre enclosed field, and three springs. He proposed to give her the 1,000 acres in either a life estate or fee simple, in return for relinquishing her Dower Rights to the Ivy Creek property in Albemarle County. Lewis, as the oldest son, had inherited the property after his father's death, but his mother had the right to use it and receive income from it during her lifetime.

Lewis had paid about $3,000 for the new property. An additonal payment of $1,500 was due in May, 1809, and a final payment of $1,200 on May 1, 1810. For this reason, he had to sell the property on Ivy Creek. Lewis was intending to put down roots in St. Louis and was planning that their mother—and perhaps their siblings—would join him and Reuben in their new home in Louisiana Territory.[654]

Actual hard currency was in very short supply in St. Louis. Most transactions consisted of handwritten "drafts" in which it was stated the amount one person owed to another person for supplying goods or services. These drafts would change hands several times, with stores and businesses accepting them in lieu of money. The drafts could be cashed for bank notes in other cities. St Louis's first bank was established in 1813, with lead, fur, and skins being accepted as deposits.

Lewis's salary as governor was $2,000 a year. When he paid for goods or services on behalf of the government, he wrote a "bill of exchange" and sent it to Washington. If the government refused to pay the debt, he was personally liable for the amount.

Fortunately, Lewis was a "master of mathematical detail," as historian Thomas Danisi has commented in reviewing his fiscal records, because his transactions, both personal and as governor of the territory, were very complex. The penny-pinching conduct of William Simmons, the government's chief accountant made his financial affairs even more difficult. Simmons, for example, had examined Lewis's total of $38,722.25 in expenses for the expedition, and decided Lewis owed the government $9,685.77 because of inadequate records. Lewis was unware that a suit would be filed against him in three years time by the government. He would have discovered it when he arrived at Washington and began dealing with Simmons. After his death, his so-called debt for expedition expenses was expunged from the record in 1812.[655]

Simmons was challenging the expenses of all federal officials on the western frontier because the government was broke. In December, 1807, President Jefferson had imposed an embargo forbidding American ships to trade in foreign ports. His purpose was to avoid declaring war against Great Britain. Americans were outraged because British ships were impressing (taking) sailors off American ships to serve on their ships. Jefferson thought the embargo was the best alternative to war. An embargo would bring economic pressure to bear against Britain, and allow time for American ships to return home before any war might start.

The result of the embargo, however, was that smuggling by American ships flourished in 1808. It emboldened the Northern Confederacy to continue their plot to secede from the federal union. The embargo directly affected Massachusetts and the other northern states dependent upon commercial shipping. When James Madison became the fourth president of the United States in March of 1809, he sought to conciliate the factions in the northeast by appointing them to positions of power.

It was against this background of escalating tensions—and a change in administrations—that Governor Lewis was charting the destiny of his distant territory. He set logical goals, and was accomplishing them. They were based on the principles he shared with Jefferson and other founders of the Early Republic.

He brought a printing press and a printer to St. Louis so the *Laws of the Territory* could be printed and a newspaper could be established. He edited the paper and wrote news and analysis under assumed names (*Clatsop, Enquirer, Reporter*), which was standard practice in those days. It was a four page weekly. His militia orders, proclamations, and other state papers were published in the newspaper and signed by him as governor. News of foreign wars raging on the continent were reported, and articles and essays on various topics reflected his interests. Other contributors also wrote articles.

He founded the first Masonic Lodge in St. Louis in August, 1808 and served as its first grand master.

He presided over two sessions of the territorial legislature in 1808, for ten days in June and twenty days in November. The legislature consisted of four men: Governor Lewis, Judge J. B. C. Lucas, Judge Otto Schrader, and Secretary Edward Hempstead. They passed twenty acts in 1808.

Among the most important were those acts for laying out and constructing an all-weather road, capable of handling a four wheeled carriage, going south to Ste. Genevieve, Cape Girardeau, and New Madrid. The current route south of Ste. Genevieve was barely passable by horseback or foot. The only four wheeled vehicle road in St. Louis went north to Camp Bellefontaine. The road to St. Charles was for two wheeled vehicles.

Other acts concerned the incorporation of towns; dower rights of widows; and establishment of corporate bodies. Fiscal acts concerned punishment of insolvent, absent, and absconding debtors; mortgages; recovery of debts; regulating interest on money; penalty for passing counterfeit money; bail; expenses of maintaining a jail; and paying Edward Hempstead for preparing and supervising the printing of the territorial laws.

Lewis issued a proclamation dividing the district of New Madrid in half, establishing the district of Arkansas. He was beginning to outline the boundaries of Missouri, which became a territory in 1812, and the State of Missouri in 1821.[656]

The Secretary of War had ordered the establishment of two trading forts for the Indians, one on the Mississippi for the Sac and Fox, and the other on the Missouri for the Osage. Both forts were built in 1808. Fort Madison on the Mississippi was under almost constant Indian attacks—attacks instigated by British traders who had dominated the trade of the Upper Mississippi for years.

Lewis wrote a letter over 4,000 words in length to Secretary Dearborn on July 1st, 1808. The next day, on July 2nd, Dearborn wrote to him complaining he had received no communications from Lewis for many months, except drafts for money. Lewis replied on August 20th. He explained Dearborn's letter had taken 42 days to reach him during "this most favourable season of the year." He noted he had sent him a letter on May 3rd, which apparently had not reached him at all, and he had written him another letter on July 16th. Since Lewis had arrived in St. Louis in April, it seems a reasonable schedule of letter writing.

Dearborn told him that he had exceeded his authority by using regular troops to provide an escort for Nicholas Boilvin, the newly appointed Indian subagent to the Sac and Fox at Prairie du Chien on the Upper Mississippi. Dearborn had learned about this from another source, and was worried his troops might become involved in warfare. Lewis replied he would be very careful in the future to not use regular troops. He was at a loss to determine under what circumstances it was permissible, if the escort for Boilvin wasn't deemed necessary. If he wasn't given discretionary powers, he would defend the territory as best he could using his militia, and observed many cases would arise that would require his immediate action.

At Monticello, President Jefferson replied to Lewis's letter of July 1st, saying General Dearborn was on his annual retreat to Maine. Most officials left Washington during the "sickly season" of August and September. He approved of Lewis's handling of the Osage Indians. He urged military action be avoided with Indians. He discussed the trial of four Ioway Indians who had murdered a white man, which was happening in St. Louis at the time and urged leniency. It was the beginning of establishing an American system of justice with the tribes.[657]

Lewis wrote two more lengthy letters with enclosures to the Secretary of War—who was also in charge of Indian Affairs—a 50 page letter on December 15th, 1808 and a 42 page letter on August 27, 1809.[658] Anything he wrote to the Secretary of War was read by Jefferson and by his successor, James Madison.

Jefferson, in the fall of 1808, was reluctant to believe Indians on the frontier needed anything more than government trading houses, and he thought Lewis was "too much committed to war." [659] A few months later in April, 1809, William Clark was warning the new Secretary of War, William Eustis, that the Shawnee Prophet and his brother Tecumseh were recruiting allies to go to war against the United States.[660]

The Osage had raided white settlements in the spring, stealing and destroying property. In response, Lewis had issued an order declaring them "out of the protection of the United States," and forbidding all commercial trade with them. Anticipating that an Indian war was about to begin, a new relationship with the Osage needed to be established.

William Clark built a trading fort for the Osage Indians near present day Kansas City, Missouri. The fort—first called "Fort Clark" and later "Fort Osage"—was built in August and September of 1808 with the aid of a militia unit. Clark appointed Reuben Lewis, Meriwether's brother, subagent for the Osage. In September, Clark made a treaty with the Osage, even though he was not their agent. Lewis and Clark had decided to bypass Pierre Chouteau, the longtime trader who had been appointed their agent by the government.

Clark acquired "near Thirty Million of acres of excellent country" for the United States—50,000 square miles of territory which the government could sell for $1.25 to $2.50 per acre. The ceded land extended from Fort Clark on the Missouri River south to the Arkansas River, and east to the Mississippi. It established a claim to all the land in modern day Missouri and half of Arkansas. Clark bragged to his brother that he had gotten the land "fer a verry Small Sum," but, much later, as a seasoned Indian diplomat, he apologized for the Osage treaty, saying—

> ... [it was] the hardest treaty on the Indians he ever made and that if he was to be damned hereafter it would be for making that treaty.

When Clark returned to St. Louis, he discovered that other Osage chiefs were complaining about a treaty being signed without their consent. They were also insisting that Pierre Chouteau receive a gift of land from them. Lewis renegotiated the treaty, as the Superintendent of Indian Affairs. The Indians called St. Louis, "Red Head's town" because of Clark's red hair. Over the course of his long career as an Indian diplomat in St. Louis from 1807–1838, William Clark signed more treaties with Indian nations than anyone else in America, and tried to help them during a time of tremendous upheaval.[661]

In August, 1808, Manuel Lisa returned to St. Louis from the post he had established on the Bighorn River with $9,000 worth of furs and skins. Most of his men remained in the field. Over the winter, many leaders of the business community became partners with Lisa in the new St. Louis Missouri Fur Company. Members in St. Louis were: Manuel Lisa; Benjamin Wilkinson, the general's nephew; Pierre Chouteau and his son Auguste Chouteau; Reuben Lewis; William Clark; and Sylvestre Labadie. Other members were Pierre Menard and William Morrison of Kaskaskia; Andrew Henry of Potosi; and Clark's brother-in-law, Dennis Fitzhugh, of Louisville. Wilkinson, Morrison, and Fitzhugh were storekeepers. William Clark became the company's director in St. Louis. The other partners, or their representatives, pledged to participate in field operations.[662]

Their first expedition would be combined with a military expedition escorting Big White and his family, and René Jessaume and his wife and daughter, back to the Mandan villages. The Jessaumes' son, 13 year old Toussaint, wanted to stay in St. Louis. Meriwether Lewis agreed to be responsible for his welfare, and signed a five year indenture bond with René Jessaume on May 13th, 1809. He promised to send Toussaint to school to learn English and French, and to provide him with a general education and means of earning a living. He would feed and clothe him and pay all of his expenses.[663]

When Lewis was eleven years old, his cousin William Meriwether became his guardian so that Meriwether could stay in Virginia and attend school after his mother, stepfather, and younger brother Reuben moved to Georgia. Now, Lewis was expecting that his mother would be joining him and Reuben within the next year or two, to live on the property he had acquired for her. It was a hopeful time for Governor Lewis, with few signs of the troubles he would soon encounter.

Meriwether Lewis had enemies—multiple enemies. After Frederick Bates arrived in St. Louis to become acting governor in April, 1807, he warned Lewis—

> . . . all from contrary motives of hostility and friendship [are] anxious for your arrival: For contrary to my first expectation you must expect to have some enemies.

The first group who hated Lewis before he arrived were the Burrites, John Smith T and his associates. Jefferson had instructed Lewis to remove them from "positions of power and influence." Robert Frazier had feared assassination by Smith T in April, 1807 when he was bringing documents from Washington to St. Louis concerning the Burrites.

The second group were the land claimants whose claims were included in the altered and forged documents Lewis brought to Washington in December of 1807. Wilkinson, as the first governor, had favored the French. His land claims commission had been corrupt, with President Jefferson saying he had never seen such "a perversion of duty" as two of the three commissioners had committed.

The third land claims commissioner, J. B. C. Lucas, was a friend of Secretary of the Treasury Albert Gallatin. Lucas was still on the land claims board, and he was also a territorial judge. Lucas wrote to Gallatin on October 19th, 1809—a week after Lewis's death, which was then not known in St. Louis—

> SIR the hatred of certain species of Land Claimants against me which appeared for som time past to have relaxed, or Lie dormant is once more revived, and at present is acting with

> greater vigor than ever ... I am assailed by a powerfull host, I mean *powerfull* in the territory ...

Judge Lucas enclosed a notice which had been posted in public places urging that Lucas be removed from office by hanging—

> People of Louisiana are you slaves how long will you suffer the vile & infamous John Lucas to keep you in the chains of oppression voilence & injustice is not 4 years of Slavery sufficient to rouse your indignation againest that little Robispiere & his adherents . . . the power of Govern [ment] has raised this enemy of your country above you & to a station he does not Merit—& your power with a little hemp [a hangman's rope] may raise him still higher & to a Station to which he is really entitled.[654]

On August 2, 1809, a group of land claimants met at Ste. Genevieve to start a petition drive urging the federal government to not reappoint Lucas to the Land Claims Commission, and to remove him from office as judge. John Smith T carried the petition to Congress, where it was received on January 3, 1810 by the House Committee on Public Lands.

A letter writer to the *Missouri Gazette* on October 16, 1809 advised his fellow land claimants to: *"Look before your leap," and think twice, before you sign once.* He pointed out if they wanted to change the laws governing land claims, they should ask the members of the claims commission to represent their views in Washington, not John Smith T. He had no doubt but that—

> col. John Smith T, and other gentlemen who claim the right of representing you, would be willing to mix their large or elastic 10,000 or 20,000 acre claims with your *Bona fide* claims of moderate size . . .[666]

Moses Austin wrote to Bates on August 27th, 1809 saying, even if he was considered a "troublesome meddler," he had to warn him in "strictest confidence" about what he had learned from a member of the "Grand Committee" (the land claimants committee). They were planning to remove extracts from land commission books to bring to Washington which would injure the reputation of Bates and Lucas. Saying he knew how hard it was to undo first impressions, Austin told Bates he could share this information with Lucas, but not with anyone else. Austin wrote:

> I need not tell you how much I have suffered by this same party. You know what I should suffer was it known that I made a communication to either of you. I therefore pray you, as you value my safety and the peace of my family who have already been sufficiently oppressed, not to . . . seen or spoken of . . .

He was obviously referring to Smith T, with whom he had been in a running battle in the mineral district for years.

What Austin didn't understand was that Bates didn't like Lucas, so it's doubtful he passed the information along. Bates wrote to his brother that he was feeling "triumphant" about the current circumstances. Bates wrote he had been asked to support the petition, but it would place him in the "most dangerous & delicate of all imaginable situations" if he did. The situation required all the "wariness and circumspection of which I am master. . . ."[667]

Frederick Bates was also an enemy of Lewis's. He wrote to his brother on November 9th, after learning of Lewis's death that he had "no personal regard for him." They had quarreled over the conduct of business between them as governor and secretary of the territory. He wrote when they both attended a ball, Lewis had sat down next to him, but Bates had refused to speak to him and had gotten up and left. Lewis was angered, and said Bates had treated him with "contempt & insult." Bates said Lewis should have known he would refuse to speak to him except on matters of business. Clark tried to mend the rift between them, but Bates refused to apologize, saying he had to "step a *high* and a *proud* path."

Bates said he wouldn't speak of these things except that

"Governor Lewis on his way to Washington became *insane*." He reported that Clement Biddle Penrose, a colleague on the land commissions board, had told several people —

> . . . that the mental derangement of the Governor ought not to be imputed to his political miscarriages; but rather to the *barberous conduct of the Secretary*. That Mr. Bates had been determined to tear down Gov Lewis, at all events, with the hope of supplanting him in the Executive Office with a great deal of scandal equally false and malicious.

Bates wound up his letter to his brother with these comments:

> Richard, this is a strange world in which we live! I had thought my habits were pacific; yet I have had acrimonious differences with almost every person with whom I have been associated in public business. . . before God, I cannot acknowledge that I have been blamable in any one instance. My passions blind me I suppose.

He said he didn't want the governor's office. where he "might have three years greatness and sink into oblivion." He could make more money and have better job security by continuing to serve as territorial secretary and recorder of land titles.[668]

There was another powerful enemy in the St. Louis area—John Hay, the postmaster at Cahokia, Illinois. Historian Thomas Danisi has analyzed the delivery time of official correspondence from St. Louis to Washington during the first years of the territory. Governor Lewis's mail took longer than mail sent by other senders. Two letters Lewis sent in April, 1808 with drafts and a bill of exchange on the government took 81 days and 98 days to be delivered. In August, 1808, a letter from Clark with a bill took 190 days, and one from Lewis with a bill took 225 days to be delivered.

From October through December, five of Clark's letters with bills or drafts averaged 105 days to reach Washington. Three of Lewis's letters with vouchers took an average of 128 days in 1808.

Lewis sent a draft of $500 that was to be taken from his salary to pay a store in St. Louis. When the store sent in its voucher, the government wouldn't pay it. It took at least six months for Lewis to become aware of the missing draft.

Danisi has found from January to August, 1809 that letters written by Lewis and by Clark, containing drafts or bills of exchange, took an average of 84 days to reach Washington. The average delivery time for their correspondence not related to payments was 33 days.

Lewis wrote to Clark in April, 1808, when Clark was still in Fincastle, that he had reason to believe that "letters and valuable papers" remained at Cahokia for several days. Although Danisi won't accuse postmaster John Hay without definitive proof, he is the only candidate.[669]

John Hay was the son of Jehu Hay, who was the assistant to the "hair buyer" general, Henry Hamilton. Jehu Hay was arrested along with General Hamilton in 1779 by George Rogers Clark in his surprise raid on the British post at Vincennes. Hamilton was paying Indians for scalps of American settlers. Both Hay and Hamilton were jailed as war criminals by Thomas Jefferson, who was then the Governor of Virginia. There was plenty of motive for John Hay to seek revenge on both Lewis and Clark. John Hay was Lewis's host in Cahokia in 1803 before the expedition set out, but that doesn't mean he wasn't duplicitous.

These delays obviously wrecked havoc with both Lewis and Clark's personal finances. Both men made plans to go to Washington to straighten out their financial accounts and have face to face meetings with federal officials.

How many letters never reached Washington at all? Lewis had a letterbook in his possession at his death. A letterbook is a record of all correspondence sent and received. The small letterbook was listed in an inventory of Lewis's possessions sent to Washington after his death. It was noted as being received by William Clark.[670] One of the last letters Lewis wrote to the Secretary of War discussed the problem of missing letters. On August 18th, 1809, he wrote:

> I have reason to believe that sundry of my Letters have been lost, as there remain several important Subjects on which I have not yet received an Answer.

Danisi points out that Donald Jackson, editor of the *Letters of the Lewis and Clark Expedition*, omitted the words, "I have reason to believe that sundry of my Letters have been lost" (while continuing the rest of the sentence) in his transcription of the letter. It is an inexplicable omission on the part of Jackson, unless it shows a bias towards promoting the suicide argument.[671]

Jackson comments in a footnote to another letter, "Lewis well deserved a scolding for his failure to write." The footnote refers to a letter Jefferson wrote to Lewis on July 17, 1808 saying he had not heard from him since they met at Albemarle in September, 1807. Lewis wrote to the government on May 3rd, 1808, one month after arriving in St. Louis, but the letter never reached Washington. Jackson must have seen the letter of August 20, 1808 in which Lewis tells Dearborn about the missing letter.

Jackson includes no letters, treaties, legislative acts, newspaper editorials, or proclamations written by Lewis for a period of 15 months (December,1807–February, 1809), creating the false impression Lewis was doing nothing of importance and suffering from depression. Jackson omits lengthy letters written by Lewis, on July 1, 1808, December 15, 1808 and August 27, 1809 reporting on Indian affairs. All of Lewis's letters sent to Washington were shared with Jefferson and Madison.

Jackson cannot be trusted on matters relating to Lewis. His collection doesn't include important documents relating to Lewis's death, and he published for the first time a forgery concerning Lewis's death—matters which are dealt with in the next chapter. In addition, Jackson failed to include in his second edition letters on the death of John Pernier which were found by the prominent historian Julian Boyd. Jackson published these letters in an academic journal, but not in his own book. Jackson's *Letters* is a valuable source of documents, but—like other material over the last 200 plus years—it has been shaped to a predetermined conclusion supporting the suicide argument.

It is conceivable that John Hay acted entirely on his own in delaying or destroying correspondence, but—because there was a flourishing conspiracy by Canadian officials and leaders of the Northern Confederacy to separate the northeastern states from the union in 1808–09—it is more likely he was acting under British orders. Hay's father became the British Lieutenant Governor of Detroit in 1782–85 after being released from prison.

Historians Donald Jackson and Dale Morgan speculated John Hay copied a letter Lewis wrote to Governor William Henry Harrison about the expedition, and sent it to David Thompson, the Canadian explorer. It is the only explanation for Thompson having a copy of Lewis's 70 page report in his possession. The unsigned report, dated September 29th, 1806, was accompanied by an unsigned note, dated Cahokia, October 10th, 1806.[673]

George Prevost, a nephew of Aaron Burr's late wife, was the Lieutenant Governor of Nova Scotia in 1808. Prevost set up "free ports" in Nova Scotia and New Brunswick, encouraging smuggling by American ships during the embargo against foreign commerce. The ports provided free access to West Indies trade. On his way to Europe, Burr visited Governor Prevost in Halifax in June, 1808 and obtained letters of introduction from him. Prevost served as the Commander-in-Chief of British North America during the War of 1812.

In 1807–08 Prevost was working with members of the Northern Confederacy in plotting secession. John Quincy Adams visited President Jefferson in March of 1808 to warn him about Prevost. Two months later, Adams resigned his seat in the U. S. Senate as the Senator from Massachusetts because of opposition within his own Federalist Party to Adams's support of the Embargo Act.

When James Madison became president in 1809—worried about the secessionists and their opposition to the embargo—he substituted a "Non-Intercourse Act" with Great Britain and France. War with Great Britain threatened during 1809–1810, and if war came, the northeastern states planned to break away and join with Britain. A secessionist convention in Massachusetts was being planned for 1809.[674]

Madison chose his cabinet members based on sectional politics. Robert Smith, the brother of Senator Samuel Smith, became the Secretary of State. The Smith brothers were Baltimore shipping merchants who engaged in extensive international trade. They were enemies of Albert Gallatin, and blocked his appointment as Secretary of State. Much to his displeasure, Gallatin remained Secretary of Treasury. Senator Smith was a leader of the "Invisibles," a group of Republican senators secretly working against Madison's leadership.[675] When Madison replaced Robert Smith with James Monroe as Secretary of State in 1811, Robert Smith wrote a pamphlet attacking Madison, which backfired and damaged his own reputation.

William Eustis, a medical doctor from Massachusetts, became the Secretary of War. Dr. Eustis had no military qualifications other than serving as a surgeon in the Revolutionary War. He was a Republican from the Federalist stronghold of Massachusetts, and he was Aaron Burr's closest personal friend.[676]

Both Secretary of War Eustis and Secretary of State Smith refused to reimburse Meriwether Lewis for bills he had drawn on the federal government. This was the cause of the financial collapse that threatened to ruin him. It would seem—either working separately or together—Eustis and Smith wanted to drive Governor Lewis out of office.

Before the story of Lewis's financial problems continues, there is one more enemy in St. Louis, to be identified—Rudolph Tillier, stepfather of Clement Biddle Penrose. Tillier had married Sarah, the older sister of Ann Biddle Wilkinson. Tillier was the former president of the Swiss Confederacy. He received his appointment as factor at Camp Bellefontaine through his connection to Wilkinson. Tillier was in charge of the factory at the army post where goods were stored to be distributed or sold to Indians and soldiers.

In December, 1808 Tillier learned he was being replaced by William Clark, and the factory at Bellefontaine would be shut down. Clark would receive additional pay in St. Louis to handle the distribution of Indian goods to the government's trading forts. In retaliation, Tillier sent three letters of

Big White, the Mandan chief was 6 foot 10 inches tall, and the "whitest man of his nation." His Indian name was Sheheke or White Coyote. The portrait was made by St. Memin in 1806.[678]

(American Philosophical Society, FAP 144).

complaint to Washington regarding the expedition that was being assembled to deliver the Mandan chief back to his home. The expedition was a combined military and mercantile operation, with the militia under the command of Pierre Chouteau.

The St. Louis Missouri Fur Company signed a contract with Governor Lewis to deliver Big White to his people. The company would be paid $7,000 for safely delivering him to the Mandan villages. William Clark was the St. Louis director of the company, and Reuben Lewis was a director. Tillier asked whether it was proper for Meriwether Lewis and William Clark to "take any share in Mercantile and Private concerns." When Tillier's complaints and misleading accusations reached President Madison, he authorized Secretary of War William Eustis to reject any payments beyond the original $7,000.

Eustis wrote to Governor Lewis on July 15, 1809 that he would not honor the bill for $500 Lewis had drawn on May 13th for additional Indian gifts for Pierre Chouteau to secure a safe passage through Indian territory. Lewis, after learning the Cheyenne, as well as the Sioux and Arikara, might attack the expedition, determined that Chouteau would need extra gifts to obtain the services of Indians as auxilary fighters with the militia, or to buy his way through hostile territory. If the goods and extra ammunition were not needed, they would be returned to the U. S. storehouse in St. Louis.[679]

The 1807 Pryor-Chouteau expedition to return Big White had ended in defeat with four men killed and nine wounded by the Arikara. Nathaniel Pryor, said no less than 400 men would be needed to successfully pass through Arikara and Sioux territory, and that 1,000 might not succeed.

The 1809 contract stated the company would provide 120 men—40 of them had to be American expert riflemen—as a military escort. The volunteer militia would return to St. Louis after reaching the Mandan villages. The expedition departed in three groups between May 17–June 17, 1809. It numbered about 350 men in thirteen keelboats and barges. Half were Americans and half were French engagés in the fur trade. Manuel Lisa was in charge, and Lisa was accompanied by most of the company's partners, including Reuben Lewis.

They passed safely by Yankton and Teton Sioux villages. Chouteau tried to recruit 250 Teton warriors to fight the Arikara, but the chiefs refused to go to war. Six Teton chiefs accompanied them up river. When they arrived at the Arikara villages, the militia marched along the shore with a cannon. A Mandan delegation was waiting at the Arikara villages to welcome their chief back home. News of Big White's return had traveled far in advance of the boats.

At a council, the Arikara chiefs blamed the death of their head chief, and a French trader living among them for their attack on the 1807 Pryor-Chouteau expedition. The trader had told them the boats were loaded with goods for the Arikara but instead the goods were destined for the Mandan with whom they were at war. He also told them to open fire on the Americans. Chouteau tried to pursue the matter, but the Arikara refused to take them to where the trader had relocated.

Chouteau did not carry out the punishment ordered by Lewis, who wanted either military action; the surrender of those who had killed Americans; or the execution of four warriors and surrender of part of their horse herd in return for their attack on the 1807 expedition.[680] Since Chouteau did not carry out his orders, they escaped punishment. Peace prevailed over war.

Lewis understood the Arikara could block Americans from the Upper Missouri. At the time, the river was the only known route to the west. In 1823, the Arikara attacked the Ashley and Henry expedition, which resulted in the Arikara War and ended fur trade traffic on the Missouri. After that, the fur trade switched to a system of wagons hauling goods along the Platte River and through the South Pass of the Rockies to annual mountain men rendezvous (trading fairs). Mountain men stayed in the mountains rather than going back and forth on the river.

On September 14th the Mandan party finally arrived home, five years after they had set out from the Knife River to meet Thomas Jefferson.Thirteen year old Toussaint Jessaume remained in St. Louis as a ward of Meriwether Lewis.

Sacagawea and Toussaint Charbonneau and 4½ year old Jean Baptiste (Pompey) accompanied Pierre Chouteau back to St. Louis. Jean Baptiste was baptized in St. Louis on December 28th, 1809 by Trappist monks. His godfather was Auguste Chouteau, Pierre's brother. The Charbonneaus arrived in St. Louis after both Lewis and Clark had left for Washington.

William Clark provided a small farm for them in the Florissant area, where they lived for the next year and a half, until Sacagawea and Toussaint went home with Manuel Lisa's 1811 expedition, leaving Jean Baptiste in the care of Clark. Jean Baptiste never lived with the Clark family. He graduated from a school taught by a Baptist missionary. In 1824, at the age of 20, he accompanied the German Prince Paul, Duke of Württemberg, to Europe where he lived for seven years before he returned and became known as a famous guide and mountain man.[681]

Sacagawea died in 1812 at Manuel Lisa's fort at Kenel, South Dakota. Their infant daughter Lisette was brought to St. Louis to be adopted by William Clark along with her brother in 1813. Toussaint Charbonneau remained close to William Clark, and served as a guide to the German Prince Maximilian and artist Karl Bodmer on their western tour in 1833-34.

In 1809—before he left for Washington in September—Lewis was dealing with several issues. The St. Louis Missouri Fur Company was established in March. He had no financial interest in the company, although William Clark and his brother Reuben were directors.[682] The company was the instrument of his strategy for gaining control of the Upper Missouri, and he used it to return Big White to his people.

In local Indian affairs, Lewis was keeping careful watch on British intrigue on the Upper Mississippi. He called out the militia, in spite of public opposition to fulfilling the quota demanded by the federal government. He wrestled with the legal issues of indicting Indians for the murder of white citizens in a territory where legal boundaries had not yet been established. One of his last actions before leaving St. Louis was to write a letter to Thomas Jefferson on August 27th, discussing jurisdiction issues in a territory without boundaries.

The federal government had little interest in the problems of establishing a government on the western frontier. Secretary of War Eustis complained about the mounting expense of Indian relations. Eustis objected to Clark's appointments of agents, and other expenditures. After Clark went to Washington in October, the government repaid him $1,069.91 for money Clark had spent in paying agents' salaries.[683]

On July 8th, 1809 Secretary of State Robert Smith refused to pay an $18.70 draft by Lewis, dated February 6th, saying the bill had not been authorized by his department. The bill was for translation of the *Territorial Laws* into French.[684] Altogether, Lewis had paid $1,596.17 of his own money for printing and translating the laws. Publishing the laws of the territory came under the jurisdiction of the State Department, after Treasury refused to pay it.

On July 15th, 1809 Secretary of State Eustis denied payment to Lewis for the $500 he had paid for additional Indian gifts. Altogether Lewis paid $940 in additional expenses for the expedition. When Lewis received the July 15th letter, he realized his financial affairs were in serious jeopardy. It appeared that both the expense of printing the laws, and the extra ammunition and

Indian goods for the expedition were going to be rejected by the government. He determined to leave for Washington as soon as he could.

Thomas Danisi has analyzed Lewis's financial records. After Lewis's death, the government reimbursed Lewis's estate $5,257.46. Danisi says additional expenditures of about $3,800 were never reimbursed to his estate. In other words, Lewis spent about $9,000 of his own money in paying government expenses. His annual salary was $2,000.[685]

Lewis was on his way to Washington to clear up these matters. Logically—because his estate received this amount—he would have received at least $5,200 from the government. If he had reached Washington, he would have sought the full amount. Is it any wonder that Meriwether Lewis was worried in the summer of 1809 about the possibility of financial ruin?

Danisi blames treasury department accountant William Simmons for the government's refusal to reimburse Lewis and Clark. Albert Gallatin, the Secretary of the Treasury, was focused on reducing the national debt and approved of Simmons's actions. In the opinion of historians, Madison had a weak and mediocre cabinet. Simmons, with his penny pinching, arbitrary decisions may have—unwittingly—been serving larger conspiracies.

The motive of Secretaries Eustis and Smith would have been to force Lewis out of his governorship of Louisiana Territory. Dr. Eustis—Aaron Burr's close friend—was from Massachusetts, the center of the secessionist movement. Senator Samuel Smith, Robert Smith's brother, was General Wilkinson's close friend. Both Burr and Wilkinson were plotting new, separate, conspiracies in 1809-10 to invade Mexico.

Lewis's term of office was ending in 1809. His appointment would be up for renewal by Congress in January of 1810. He intended to go to Philadelphia to work on publication of his three volumes. Perhaps he was planning to give up his post as governor, or perhaps, he was hoping to repeat the acting governor arrangment of 1807. These were matters that would be settled in Washington. Most importantly, Lewis needed to get rereimbursed for his government expenses. On August 18th, 1809 Lewis wrote to Secretary of War Eustis:

> Be assured Sir, that my Country can never make "A Burr" of me—She may reduce me to Poverty; but she can never sever my Attachment from her.
>
> Those protested Bills from the Departments of War and Treasury, have effectively sunk my Credit; brought in all my private debts amounting to about $4,000, which has compelled me, in order to do justice to my Creditors, to deposit with them, the landed property which I had purchased in this Country, as Security.
>
> The best proof which I can give of my Integrity, as to the use or expenditure of public Monies, the Government will find at a future day, by the poverty to which they had now reduced me—still, I shall do no more than appeal to the Generosity of the Government by exposing my claims.
>
> I had sooner bear any pecuniary embarassment, than attempt, in any manner to wound the Feelings, or injure in the public Opinion, the present Executive [Madison], or either of the Heads of Departments [Gallatin and Eustis], by complaining of Injuries done me, other than in friendly Expostulations.[687]

Lewis owned seventeen land parcels, which he put up as security for his $4,000 dollars of private debt in St. Louis. The private debt was related to additional land purchases. He had a good eye for property and many in St. Louis were hoping he would be forced to liquidate his holdings.[688] When the government repaid him, he could easily manage his finances.

After he finished writing his books in Philadelphia, Lewis would have returned to St. Louis and brought his mother back with him. His future was in the west. After Lewis's death, Clark was appointed Superintendent of Indian Affairs, and in 1813,

William Clark became the first Governor of Missouri Territory. As Brigadier General of the Militia, he skillfully managed the defense of the western frontier during the War of 1812. He served as teritorial governor until Missouri became a state in 1821, and held the office of Superintendent of Indian Affairs until his death in 1838. After Lewis died, Clark wrote to his brother, on January 12, 1810, that the post of governor was his, if he wanted it, but he wrote: "I do not think mySelf Calculated to meet the Storms which might be expected." [689]

On August 18th, 1809, his 35th birthday, Lewis was suffering with malarial fevers, and and he dictated his letter to Eustis. The sheriff of St. Louis, Jeremiah Connor, served as his scribe. Lewis signed the letter with a postscript. Lewis wrote his own letter on August 27th about Indian affairs to Jefferson, indicating his health was better. On August 29th he held a council with the Osage Indians of Arkansas and took his own notes.[690]

Thomas Danisi believes that Lewis committed suicide six weeks later, in a delusional state brought on by malarial fevers. His theory is contradicted by the common sense observation that Lewis suffered periodically from malarial attacks for many years—as did many, if not most, residents of the Ohio and Mississippi Valleys. It obviously did not impair his thinking as he dictated the letter to Eustis when he was too ill to write. Malaria is a disease brought on by the bite of an infected mosquito, and it can kill people. Its symptoms are high fevers, headaches, and violent shaking. The treatment was "Peruvian Bark," which contained quinine. The bark was so bitter it was mixed with wine or brandy.

By September 2nd, Lewis had concluded his treaty making with the Arkansas Osage, who signed the treaty he had renogiated with the Great and Little Osages of the Osage River in November,1808. He recommended Pierre Chouteau continue in his office as agent to the Osage, and not be removed. Lewis was ready to leave the territory, and as he wrote to Eustis—

> I shall leave the Territory in the most perfect state of Tranquility which I believe, it has ever experienced.

He was, however, not thinking of the petition drive started in Ste. Genevieve to remove Lucas from his offices as judge and land commissioner. General Wilkinson also hated John Lucas,[691] and it is my theory that John Smith T and Wilkinson not only wanted Lucas removed from office, but they planned to assassinate Governor Meriwether Lewis while he was on his way to Washington.

Their motive would have been to prevent Lewis from bringing incriminating papers to the government concerning the large land claimants and the lead mine district. The land commissioners had finished dealing with the small land claimants and would deal with the large land claimants in 1810. Lucas was up for reappointment, just as Lewis was.

The hidden agenda was they were planning to mount another filibuster expedition to invade Mexico in 1810. The evidence for this is that—as soon as Lewis's death was known—an advance reconnaissance party set out from Ste. Genevieve in November. John Smith T would have set out for Washington with the petition to remove Judge Lucas probably sometime after William Clark left for Washingon in October.

Benjamin Wilkinson, one of the directors of the Missouri Fur Company, didn't go on the expedition to the Upper Missouri. Instead he remained in St. Louis. Wilkinson was the general's nephew, and he operated a store in St. Louis. He must have been a passenger on the boat with Lewis, his servant John Pernier, and Seaman the dog, because he was at Fort Pickering with them a few weeks later. He was on his way to New Orleans, and from there, to Baltimore by ocean voyage.

What kind of boat they traveled on is unknown. It left St. Louis on September 4th. Among the other passengers or crew, there must have been at least one other assassin working with the general's nephew. The next chapter will present my case for assassination rather than suicide.

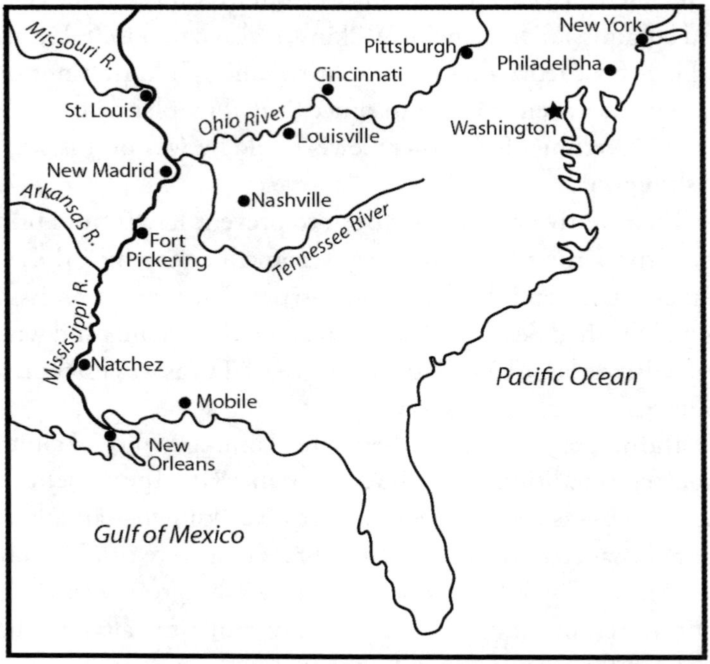

St. Louis to Washington via New Orleans: Meriwether Lewis planned to travel by boat from St. Louis to New Orleans, and from there, on board a sailing ship to Washington D. C. He changed his plans at Fort Pickering (today's Memphis, Tennessee). From Fort Pickering he traveled by horseback on the Natchez Trace, intending to go to Nashville, Tennessee. However, he died at Grinder's Stand, some 70 miles southwest of Nashville, on October 11, 1809.

Chapter 13

The Assassination

Meriwether Lewis left St. Louis on September 4th, 1809. His departure had been delayed by treaty making with the Osage Indians. William Clark wrote to his brother Jonathan on August 26th, that he expected Lewis to set out the next day for Philadelphia to "write our book." The Indians must have shown up unexpectedly. Clark wrote that he was concerned about Lewis—

> Govr. L. I may Say is r—d by Some of his Bills being protested for a Considerable Sum, which was for moneys paid for Printing the Laws and expences in Carrying the mandan home all of which he has vouchers for. if they Serve me in this way what______ I have Sent on my accounts to the 1st of July with vouchers for all to about $1800 . . . .
>
> . . . his Crediters all flocking in near the time of his Setting out distressed him much, which he expressed to me in Such terms as to Cause a Cempothy which is not yet off—I do not believe there was ever an honester man in Louisiana nor one who has purer motives than Govr. Lewis. if his mind had been at ease I should have parted Cherefully. he has given all his landed property into the hands of Judge Steward Mr. Carre & myself to pay his debts. with which we have Settled the most

> of them at this place. Some yet remains and Some property yet remains—his property in this Country will—pay his—by a Considerable amount, tho' I think all will be right and he will return with flying Colours to this Country . . .[692]

The source of written information about Lewis's death has to be carefully analyzed. Many false stories and documents were created to support the death by suicide story. False documents for the historic record were a trademark of General Wilkinson, who understood the value of a paper trail. However, cover ups can go both ways. They are sometimes the only evidence linking someone to a crime. In the case of Meriwether Lewis, there are not only historic cover ups, there are also contemporary distortions of the historic record to support the suicide theory. Of necessity, this chapter will be dealing with both historic and contemporary inaccuracies in trying to puzzle out what might have really happened.

On September 11th, Lewis was at New Madrid. He went to the courthouse with John Pernier to make out a will. The will was later recorded at the Albemarle courthouse. Lewis wrote:

> I bequeath all my estate real and personal to my Mother Lucy Marks of the County of Albemarle and the State of Virginia after my private debts are paid of which a statement will be found in a small minute book deposited with Pernier my Servant.

By September 15th, they had arrived at Fort Pickering at Chickasaw Bluffs. Lewis was sick with malarial fevers, and stayed at the fort for two weeks, leaving on September 29th. While he was ill, Lewis was cared for by Captain Gilbert C. Russell, the commander of the fort. Russell wrote two letters to Jefferson reporting on Lewis's 14 day stay at the fort, and what he knew about the 14 days after Lewis left the fort before his death. Russell's letters were in the nature of an official report to the president and Lewis's family and friends. There never was an investigation of his death.

According to a false story—that was told by an unnamed person to Captain Amos Stoddard, who then reported it to Captain James House—Lewis tried to commit suicide several times while he was on the boat. Captain House, the former commander of Camp Belle Fontaine in St. Louis, wrote to Frederick Bates from Nashville on September 28th:

> I arrived here two days ago on my way to Maryland—yesterday Maj. Stoddart of the army arrived here from Fort Adams, and informs me that his passage through the indian nation, in the vicinity of Chickasaw Bluffs he saw a person, immediately from the Bluffs, who informed him, that the Governor had arrived there (sometime previous to his leaving it) in a state of mental derangement, that he had made several attempts to put an end to his own existence, which this person had prevented, and that Capt. Russell, the commanding officer at the Bluffs had taken him into his own quarters where he was obliged to keep strict watch over him to prevent his committing violence on himself and has caused his boat to be unloaded and the key to be secured in his stores.[693]

The most interesting point about this letter is that Captain House doesn't *identify* the person who told Major Stoddard he had "several" times prevented Lewis from committing suicide while traveling on the boat between St. Louis and Fort Pickering. This is the first evidence of a conspiracy with "talking points." It appears that House and Bates were at least on the edges of the conspiracy. Captain House is the first person to use the attempted suicide story. If it was only gossip, House would have identified the person who saved the mentally deranged Lewis from killing himself. Instead, he is laying the paper trail for the suicide story. "Mr. X," (see note 699, page 508) who told Amos Stoddard about saving Lewis's life, was undoubtedly traveling in the vicinity of Chickasaw Bluffs to make new arrangements for murdering Lewis on the Natchez Trace—the assassination which happened at Grinder's Stand on October 11th, 1809.

Captain Gilbert C. Russell wrote two letters to Jefferson, on January 4th and January 31st, 1810. There are also two false documents written in his name, one of which William Clark reported receiving shortly after Lewis's death, which is lost, and another created during General Wilkinson's court martial in 1811, known as the "Russell Statement." Donald Jackson published the "Russell Statement" in his *Letters of the Lewis and Clark Expedition* for the first time in 1962. The "Russell Statement" was examined by two handwriting analysts testifying at the Coroner's Inquest held at Lewis County, Tennessee in 1996. They both concluded that Russell did not write and he did not sign the so called "Russell Statement." [694]

The day after Lewis arrived at Fort Pickering, on September 16th, he wrote a letter to President Madison:

> I arrived here yesterday about ~~2 Ock~~ P. M. very much exhausted from the heat of the climate, but having taken medicine feel much better this morning. My apprehension from the heat of the lower country and my fear of the original papers of the voyage falling into the hands of the British has induced me to change my rout and proceed by land through the state of Tennisee to the city of Washington. I bring with me duplicates of my vouchers for public expenditures &c. which when fully explained, or rather the general view of the circumstances under which they were made I flatter myself ~~that~~ they will receive both sanction & approbation ~~and~~ sanction.
>
> Provided my health permits no time shall be lost in reaching Washington. My anxiety to pursue and to fulfill the duties incedent to the international arrangements incedent to the government of Louisiana has prevented my writing you ~~as~~ more frequently. ~~Mr. Bates is left in charge.~~ Inclosed I herewith transmit you a copy of the laws of the territory of Louisiana. I have the honour to be with the most sincere esteem your Obt. ~~and very humble~~ Obt. and very humble servt.
>
> Meriwether Lewis [695]

This signed letter appears to have been a draft. A final version of the letter could have accompanied the book of *Territorial Laws* sent to Madison. This is the only known version, in the collection of the Missouri History Museum. The letter is obviously written in poor health, but its meaning is entirely clear. Suicide theorists view it as evidence of Lewis's mental instability.

Captain Russell wrote on January 4th, 1810 to Jefferson:

> Considering it a duty incumbant upon me to give the friends of the late Meriwether Lewis such information relative to his arrival here, his stay and departure, and also of his pecuniary matters as came within my knowledge which they otherwise might not obtain, and presuming that as you were once his patron, you still remain'd his friend, I beg leave to communicate it to you and thro' you to his mother and such other of his friends as may be interested.
>
> He came here on the 15th of September last from whence he set off intending to go to Washington by way of New Orleans. His situation rendered it necessary that he should be stoped until he would recover, which I done & in a short time by proper attention a change was perceptible and in about six days he was perfectly restored in every respect & able to travel. Being placed then myself in a similar situation with him by having my Bills protested to a considerable amount I had made application to the General [Wilkinson] & expected leave of absence every day to go to Washington on the same business with Governor Lewis. In consequence of which he waited six or eight days expecting I would go with him, but in this we were disappointed & he set off with a Major Neely who was going to Nashville.
>
> At the request of Governor Lewis I enclosed the land warrant granted to him in consideration of his services to the Pacific Ocean to Bowling Robinson Esq Sec'y of the T'y [Trea-

sury] of Orleans with instructions to dispose of it at any price above two dollars per acre & to lodge the money in the Bank of the United States or any of the branch banks subject to his order.

He left with me two Trunks a case and a bundle which will now remain here subject at any time subject to your order or that of his legal representatives. *Enclosed is his memo respecting them but before the Boat in which he directed they might be sent got to this place I rec'd a verbal message from him after he left here to keep them until I should hear from him again.* [italics emphasis added].

He set off with two Trunks which contain'd all his papers relative to his expedition to the Pacific Ocean. Gen'l Clark's Land Warrant, a Port-Folio, pocket Book, and Note Book together with many other papers of both a public & private nature, two horses, two saddles & bridles, a Rifle, gun, pistols, pipe, tomahawk & dirks, all ellegant & perhaps about two hundred and twenty dollars, of which $99.58 was a Treasury Check on the U. S. branch Bank of Orleans endorsed by me. The horses, one saddle, and the check I let him have. Where or what has become of his effects I do not know but presume they must be in the care of Major Neely near Nashville.

As an individual I verry much regret the untimely death of Governor Lewis whose loss will be great to his country & surely felt by his friends. . . .[696]

Russell's letter of January 4th is important for several reasons. If Lewis had tried to commit suicide before he got to the fort, Russell would have said so. If he had written a second will at the fort, Russell would have said so. William Clark believed that he received a letter from Russell shortly after Lewis's death telling him Lewis had made a second will at the fort naming William Meriwether and himself as the executors of the will, and saying that Clark was to take charge of his papers. Clark expected to find this will, and have it sent to him, but it never materialized. The letter Clark received has also never been found. This is one of the tricks of which General Wilkinson was a master. The letter was written in the name of Captain Russell to make certain William Clark believed the suicide story and the story was reinforced with the mention of a new will. The general had played the same kind of tricks twenty years earlier, when he destroyed the reputation of Clark's older brother, George Rogers Clark in Kentucky politics.[697]

In Russell's authentic letter to Jefferson he states that Lewis was fully restored to health in six days time and was waiting at the fort for another 6–8 days hoping Russell would receive permission to go to Washington, so that they could travel together. Russell must have requested a leave of absence from General Wilkinson weeks earlier. It is the only reason Lewis would have delayed his travels.

The other important detail is that Russell said he received a "verbal message" from Lewis to hold Lewis's trunks and other items at the fort rather than follow the instructions in the memorandum that Lewis had Russell sign. The memo strongly suggests Lewis feared assassination.There are two other possible indications that he was worried—making out his will at New Madrid, and changing his plans to go by boat to New Orleans. Both of those actions could have been just sensible precautions, but the memorandum is a different story—

Fort Pickering, Chickasaw Bluffs
September 28th 1809

Captain Russell will much oblige his friend Meriwether Lewis by forwarding to the care of William Brown Collector of the port of New Orleans, a Trunk belonging to Capt. James Huse addressed to McDonald and Ridgely Merchants in Baltimore. Mr. Brown will be requested to forward this trunk to its place of destination. *Captain R. will also send two trunks a package and a case addressed to Mr. William C. Carr of St. Louis unless otherwise instructed by M. L. by letter from Nashville.*

> *M. Lewis would thank Capt. R. to be particular to whom he confides these trunks &c a Mr. Cabboni* [Jeane Pierre Cabanné] *of St. Louis may be expected to pass this place in the course of the next month, to him they may be safely confided.* [italics emphasis added]
>
> [on reverse side] Sent Capt. Hous's Trunk by Benjamin Wilkinson on the 29th Sept. 1809. Gov'r Lewis left here on the morning of the 29th Sept. Russell [698]

The two trunks are what his assassins wanted to get possession of. This is shown by the so-called "verbal message" from Lewis to hold the trunks at the fort. Will Carr was the federal land agent in St. Louis. The trunk's contents must have been land records. Lewis tried to protect the contents by leaving the trunks at the fort, and saying they should be sent to him in Nashville, only *if he wrote a letter* requesting they be sent. It mattered enough for him to leave a written memorandum with Russell.

If he didn't write a letter, the meaning seems clear—Lewis would no longer be alive and the trunks should be sent back to Will Carr in St. Louis. Russell should be "particular" about who took them to Carr, and they could "safely be confided" in the care of Jean Pierre Cabanné, a wealthy St. Louis merchant. The contents were something Lewis wanted to protect. Russell was only to follow his written instructions, not any verbal message.

Who brought the "verbal message" from Lewis to Captain Russell saying Russell should hold the trunks at the fort? Was it Mr. X, the man Major Stoddard encountered around the Chickasaw Bluffs, who said he had saved Lewis from several attempts at suicide on the boat? Or was it someone else?

The original plan of the conspirators must have been to kill Lewis on the boat and blame it on suicide. Lewis knew by September 16th, the day after his arrival at the Bluffs, that he was going to travel on the Natchez Trace. He probably decided that while he was still on the river. Was he consciously trying to get away from his assassins? How did the conspirators change their plans in order to assassinate him, and still call it a suicide?

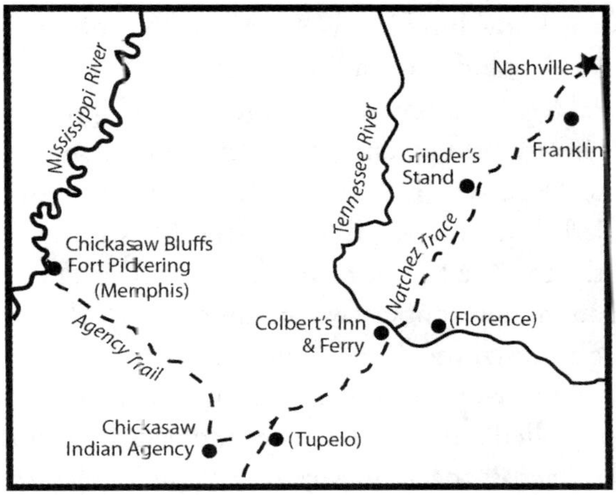

Fort Pickering to Chickasaw Agency to Nashville.

The premise of this book is that General Wilkinson and Smith T planned to have Lewis assassinated. Benjamin Wilkinson, the general's nephew, was a storekeeper in St. Louis. He traveled on the same boat with Lewis, and was going to Baltimore from New Orleans. He took charge of Captain House's trunk, the one Lewis had agreed to bring to New Orleans for Captain House. Wilkinson left the fort by boat on September 29th for New Orleans, on the same day after Lewis left for Nashville. Benjamin Wilkinson must have been a conspirator.

Making the simplest assumption, there had to be at least one more member of the conspiracy. Benjamin Wilkinson stayed at the fort watching Lewis, while someone else—probably the unknown man Major Stoddard met in the vicinity of Chickasaw Bluffs, "Mr. X"—left immediately to recruit Indian Agent James Neelly at his Indian agency near Pontotoc; tavern owner and inn keeper Robert Grinder at Grinder's Stand on the Natchez Trace; and army officer Captain John Brahan in Nashville. All three men play leading roles in the mysterious circumstances surrounding Lewis's death.

It seems likely Mr. X would have been a trusted friend of General Wilkinson, and that he knew both James Neelly and Captain Brahan. Robert Grinder was just the means of committing the murder. Neelly would have suggested Grinder, if Mr. X didn't already know him. It was Agent Neelly who told Lewis to go to Grinder's Stand and wait for him there.[699]

General Wilkinson was in charge of developing the old Indian Trace into a military and postal service road in 1801–1803. The Natchez Trace was heavily traveled. It was used by boatmen who traveled down to Natchez with the Mississippi River current, and then walked back to Nashville on the Trace. Because travelers had money, robbery was commonplace. Slaves in chains were brought down the Trace to be sold at auction in Natchez. There were slaves working in the adjoining fields. The Chickasaw Indians controlled much of the Trace and had the exclusive right to run a ferry across the Tennessee River, and to operate traveler's inns (stands) in Chickasaw territory.

The War Department was in charge of Indian Affairs. The Chickasaw Indian agent Neelly would have owed his job to Wilkinson, or would at least have known him. Mr. X was probably an army officer who served with Wilkinson during his tours of duty in the southwest frontier, and knew the country.

The unknown conspirator, "Mr. X," would have left Fort Pickering as soon as they arrived and gone down the Agency Trail to confer with James Neelly, the newly appointed Indian Agent. Lewis wrote that his boat had arrived at Fort Pickering in the afternoon of September 15th. The distance was about 120 miles from the fort to the agency. "Mr. X" could have reached the Indian agency on the 18th. Neelly could have started to Fort Pickering on September 19th and arrived there late on the 21st. The timeline could be extended by several days. All we know is that the letter which forms the basis of the suicide story says that Neelly arrived at the fort "on or about" the 18th of September.

Agent James Neelly is a key player in both the suicide and the assassination theories. His account of the death of Lewis by suicide has formed the basis of the suicide theory. However, his so-called letter to Jefferson, describing events which he had not-witnessed, was neither written nor signed by him, so all information contained in the letter is part of the cover up.

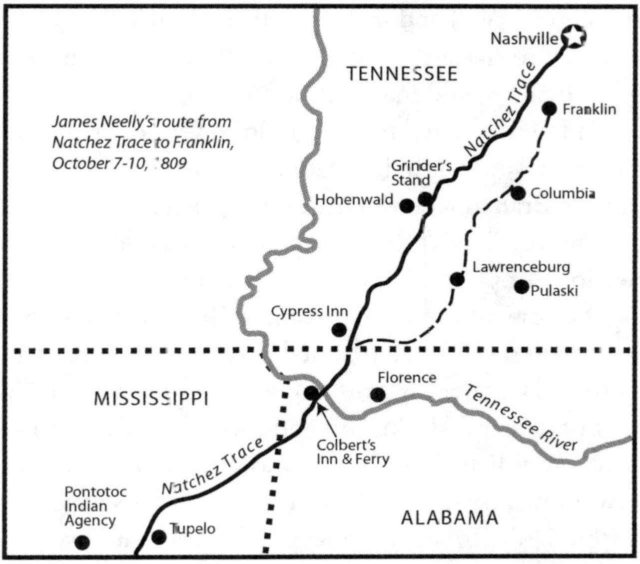

After leaving the Chickasaw Agency, Lewis and Neelly traveled to Cypress Inn on the Natchez Trace, where Neelly took the road to Franklin. Lewis went on to Grinder's Stand.

When "Mr. X" arrived at Neelly's Indian Agency at Pontotoc, he found Neelly willing to cooperate, but Neelly had a highly important obligation. Neelly had to appear in court in Franklin, Tennessee on October 11th, where he was being sued for a debt of $103.44. This court appearance had been arranged a year earlier, and if he didn't appear he would forfeit his bond and face jail as a debtor. Neelly, in fact, did appear in court. So October 11th became a convenient alibi date. If he was ever accused of the murder of Governor Lewis, he had the best of all possible alibis—he was in court on that date.[700]

It was arranged that Neelly would go to Fort Pickering and volunteer to escort Lewis on the Natchez Trace to Nashville, but Neelly would leave the party after they crossed the Tennessee River and take the alternate shorter route to Franklin. He would tell Lewis to go on ahead and wait for him at Grinder's Stand, the first traveler's inn and tavern owned by a white man after leaving Chickasaw territory. Robert Grinder would arrange the murder of Meriwether Lewis, and make it appear a suicide. Agent Neelly would return to his agency at Pontotoc, and Captain John Brahan in Nashville would manage the cover up. Neelly's job was to travel with Lewis and tell him to go on to Grinder's Stand alone, saying that he would join him later.

Captain Brahan would write and sign a letter in the name of Neelly to report Lewis's death by suicide to Thomas Jefferson at Monticello. "Mr. X" would be in the vicinity of Grinder's Stand to bring the news of the assassination to Brahan at Nashville. The conspiracy had several well planned parts. James Neelly and John Brahan would be far away from the scene of the murder. "Mr. X," the agent of General Wilkinson and John Smith T, was not connected to it publicly. The general was at Natchez, and Smith T was at St. Louis. Benjamin Wilkinson was on his way down the Mississippi. Only Robert Grinder and his associates were vulnerable, and local residents could be intimidated into silence.

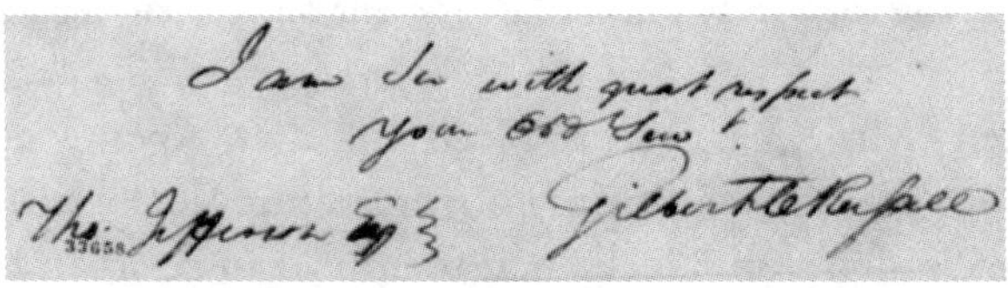
I am Sir with great respect
Your Obt Servt
Thos. Jefferson Esqr
Gilbert C. Russell

Gilbert C. Russell wrote a second letter to Jefferson describing what he knew of events after Lewis left the fort. This is his signature from his letter of January 31, 1810. What is important about his signature is that he invariably wrote it as one consecutive name. The C stands for Christian and it joins together the names of Gilbert and Russell. That is why his contemporaries often referred to him as "Gilbert C. Russell." The Russell Statement which General Wilkinson created in 1811 forgot that important detail and his forged signature simply reads "Gilbert Russell." On January 31st, Captain Russell wrote to Jefferson about James Neelly and the role he played in Lewis's suicide:

Sir,

I have lately been informed that James Neely the Agt. to the Chickasaws with whom Gov Lewis set off from this place has detained his pistols & perhaps some other of his effects for some claims he pretends to have on his estate. He can have no more just claim for any thing more than the expenses of his interment unless he makes a charge for packing his two Trunks from the nation. And for that he cannot have the audicity to make a charge after tendering the use of a loose horse or two which he said he had to take from the nation & also the aid of his servant. He seemed happy to have it in his power to serve the Govr & but for making the offer which I accepted I should have employ'd the man who packed the trunks to the Nation to have them taken to Nashville & accompany the Govr. Unfortunately for him this arrangement did not take place, or I hesitate not to say he would this day be living. The fact is that his untimely death may be attributed to the free use of liquor which after he recovered & expressed a firm determination to never drink any more spirits or use snuff again both of which I deprived him of for several days & confined him to claret & a little white wine. But after leaving this place by some means or other his resolution left him & this Agt. being extremely fond of liquor, instead or preventing the Govr from drinking or keeping him under any restraint advised him to it & from everything I can learn gave the man every chance to seek an opportunity to destroy himself. Also from the statement of Grinder's wife where he killed himself I cannot help to believe that Purney was rather aiding & abeting the murder than otherwise.

This Neely says he lent the Govr money which cannot be so as he had none himself & the Govr had more than one hund. $ in notes & species besides a check which I let him have of 99.58 none of which it is said could be found. I have wrote the Cashier of the branch bank at Orleans as to whom the check was drawn in favor of myself on order to to stop payt. when presented. I have this day authorized a gentlemean to pay the pretended claim of s & take the pistols which will be held sacred to the order of any of the friends of M. Lewis free of encumbrance.

I am Sir with great respect
Your Obt Servt [Obedient Servant]
Gilbert C. Russell

Tho. Jefferson Esq. [Esquire] [701]

From the letter we learn Captain Russell sent his own man with Lewis to manage the two pack horses carrying Lewis's trunks with the expedition papers and other items. Neelly insisted that after they arrived at the Pontotoc Agency, he would employ his own servant and use horses taken from the Chickasaw nation to transport Lewis's trunks to Nashville. Captain Russell blamed Neelly for causing Lewis's death, saying that Neelly encouraged his drinking, and that it was the cause of Lewis committing suicide.

When they left Fort Pickering on September 29th, the party would have consisted of Lewis and John Pernier, riding on two horses that Russell had provided, Seaman the dog, James Neelly and one man, probably an enlisted army man, from the fort. Russell would have learned about the drinking when his man returned to the fort, and he believed that his man would have prevented Lewis from committing suicide. He also somehow thought Pernier was involved in the suicide.

Russell says Neelly was lying when Neelly said he lent Lewis money. In addition, Neelly wanted Captain Russell to pay him money before he would relinquish Lewis's two pistols and other possessions. Neelly told Russell Lewis had no money, so the more than $100 Lewis was carrying had disappeared. Russell stopped payment on the endorsed check for $99.58 he had given Lewis. Russell was going to pay Neelly the money so that he could get back Lewis's pistols. However, that didn't happen.

In December, 1811, Lewis's stepbrother John Marks went to Tennessee to retrieve Lewis's two pistols, rifle, horse, dirk [a small sharp weapon] and watch from Neelly. He wrote to his stepbrother Reuben Lewis, that he visited Neelly's farm at the Duck River (near Franklin) and had gotten Lewis's rifle and horse from Mrs. Neelly. Neelly was living at the Chickasaw Agency and Marks was "informed that Nealy carries the dirk and pistols constantly with him." Neelly had Lewis's gold watch as well as his pistols.[702]

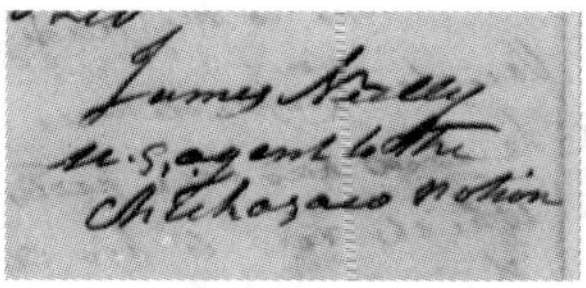

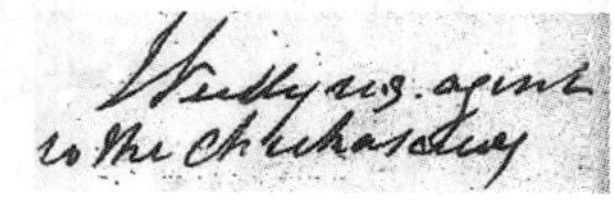

Suicide theorists depend on Neelly's letter reporting Lewis's death to Thomas Jefferson. Thomas Danisi acknowledges the Neelly letter was written by John Brahan at Nashville, as Brahan wrote an almost identical letter to Jefferson signed by himself. Both letters are dated October 18, 1809 and were delivered to Jefferson by Pernier.[702] When he signed for Neelly, Brahan didn't get Neelly's signature block correct. Neelly always signed "J Neelly, u.s. agent" with the J and N joined together on one line and on the second line, "to the Chickasaws." Brahan made it a three line signature block, used Neelly's first name, "James" and put "u.s. agent to the" on the second line and "Chickasaw Nation" on the third line. The signature on the right is from the real letter that James Neelly wrote on the same day, October 18, 1809 to Secretary of War Eustis from the Chickasaw Agency near Pontotoc.[703]

Neelly wrote to Eustis on October 18th requesting reimbursement for the $90 he had given to Jeremiah K. Love for transporting a prisoner from the Indian agency to jail at Nashville. $90 was a month's salary for Neelly. The prisoner, George

Lanehart had stolen two saddlebags containing $621 silver dollars belonging to Joseph Van Meter, a guest at James Colbert's Inn at Pontotoc. The saddlebags of money were in the safekeeping of Colbert. The theft occurred on September 20th, 1809.

The theft happened while Neelly was at Fort Pickering, but he would have been quickly informed about it, because it was a major crime committed in Chickasaw territory. It was a very large sum of money, and Colonel Van Meter was a personal friend and neighbor of President James Madison. It almost certainly had to do with the public lands sale of former Chickasaw territory at the Great Bend of the Tennessee River (near today's Huntsville, Alabama). Investors were swarming in from all over the country, to purchase the prime land for growing cotton.

When Neelly and Lewis arrived at the Chickasaw Agency, Neelly took extreme measures in dealing with the crisis. It was vital to restore public confidence in the Colbert family, the half Scotch-half Chickasaw family, which ran the Chickasaw Nation for many years.

On October 3rd, Neelly assembled eight leading men of the area to sign a statement with him that announced:

> James Colbert, at whose house the money was lost . . . has always conducted himself, and has been reputed an honest man, and we conceive no blame is, or ought to be attached to the character of Colbert or his family—let malicious characters report what they may.

This statement was published in the same issue of the Nashville *Democratic-Clarion* newspaper which reported the death of Governor Lewis on October 20th. Neelly at the bottom of the paid announcement, added the statement: "the printers at Natchez and New Orleans will insert this and forward their accounts to me." Publication in the three leading newspapers of the south shows how important the matter was to Neelly. The fact that he waited at the fort to escort Lewis rather than returning to the agency to take care of it sooner demonstrates his determination to be Lewis's escort.

Timeline of Events

Sept. 15: Lewis arrives at Fort Pickering.

September 20: Saddlebags of Colonel Joseph Van Meter of Virginia are stolen at James Colbert's Inn at the Tennessee River containing $621 silver dollars.

September 29: Lewis and Neelly leave Fort Pickering.

October 3: Neelly assembles local leaders at the Chickasaw Agency to endorse honest reputation of the Colbert family.

October 11: Lewis dies at Grinder's Stand.

October 11: Neelly appears in court at Franklin.

October 18: John Brahan at Nashville writes two letters to Jefferson—one in his name, and the other in the name of James Neelly—reporting the death of Lewis by suicide.

October 18: Neelly at the Chickasaw Agency at Pontotoc writes to Eustis requesting $90 reimbursement for paying Jeremiah K. Love to deliver the prisoner to Nashville. Delivering the prisoner would have been Neelly's duty.

October 20: The *Nashville Democratic Clarion* publishes an article on the death of Governor Lewis by suicide, and the paid public announcement by Neelly.

November 30: George Laneheart is convicted of stealing Colonel Van Meter's saddlebags and money and is sentenced to two months in the debtors room of the jail in Nashville.[704]

When Neelly, Lewis, Pernier, Neelly's servant, and Seaman left the agency, they crossed the Tennessee River on George Colbert's Ferry. They would have stayed at Chief Tuscomby's Stand, sixteen miles from the river, near the junction of the trail to Lawrenceburg. The stand later became known as Cypress Inn.[705]

This is where the so-called letter from Neelly to Jefferson claimed that at "one day's journey" from the river they lost two of their horses, and Neelly stayed behind to look for them, telling Lewis to go on ahead and wait for him at the first house inhabited by white people, Robert Grinder's Stand.

Lewis, Pernier, Neelly's servant and Seaman traveled on the Natchez Trace, and Neelly went to Lawrenceburg and from there he went north to Franklin. Neelly owned a large farm outside of Franklin. On October 10th, Lewis and his party reached Grinder's stand, which was just off the Natchez Trace. It was a two room cabin, with one room reserved for travelers. Robert Grinder supposedly was absent from the stand. His wife Priscilla was there. Their household consisted of three children, two slaves and a cook. The stand had a bad reputation because Grinder sold whiskey to the Indians.[706]

Mrs. Grinder stories play an important role in the account of Lewis's death by suicide. The first account is found in the so called Neelly letter to Jefferson written by John Brahan in Nashville. It is entirely crafted by the conspirators to support the suicide story.

There are two more Mrs. Grinder stories. The second story she told to Lewis's friend, Alexander Wilson, on May 10th, 1810. Wilson was the ornithologist from Philadelphia who was going to provide bird illustrations for Lewis's books. Wilson was traveling on the Natchez Trace in search of more birds.

The third story Mrs. Grinder told to a schoolteacher who taught in the Cherokee nation, first printed in the *North Arkansas* newspaper, and then reprinted in the New York *Dispatch* newspaper in 1845. Robert Grinder died in 1827, so the third Mrs. Grinder story may be more truthful.

Mrs. Grinder's husband is the prime suspect in the murder of Meriwether Lewis, and all of her stories may be considered cover ups, while perhaps containing some elements of truth. Many people have puzzled over these stories. The first story was supposedly told to Brahan by Neelly, who wasn't present at Grinder's Stand at the time of Lewis's death, and who, in fact, wasn't present at Nashville to tell Brahan the story of what Mrs. Grinder says happened. It might have been the mysterious "Mr. X" who was near the Stand at the time of Lewis's death, saw him buried, and then went to Nashville to meet Brahan where they concocted the letters to Jefferson. In any event, this is the version that has come down in history. My earlier book, *The Death of Meriwether Lewis: A Historic Crime Scene Investigation,* contains all three versions of the story, and many other documents.

Any lawyer looking at this account would declare it inadmissible as evidence. The Neelly/Brahan letter says:

> . . . he reached the house of a Mr. Grinder about sunset, the man of the house being from home, and no person there but a woman who discovering the governor to be deranged gave him up the house & slept herself in one near it. His servant and mine slept in the stable loft some distance from the other houses. The woman reports that about three o'clock she heard two pistols fire off in the Governors Room; the servants being awakined by her, came in too late to save him. He had shot himself in the head with one pistol & a little below the breast with another. when his servant came in he says: I have done the business my good servant give me some water. He gave him some water, he survived but a short time, I came up some time after, & had him as decently Buried as I could in that place. if there is anything wished by his friends to be done to his grave I will attend to their Instructions.[707]

The Alexander Wilson account says:

> I took down from Mrs. Grinder the particulars of that melancholy event, which affected me extremely. This house or cabin is seventy-two miles from Nashville, and is the last white man's as you enter Indian country. Governor Lewis, she said, came hither about sunset, alone, and inquired if he could stay for the night; and alighting brought the saddle into the house. He was dressed in a loose gown, white striped with blue. On being asked if he came alone, he replied there were two servants behind, who would soon be up. He called for some spirits and drank very little. When the servants arrived one of whom was

a negro, he inquired for his powder, saying he was sure he had some powder in a cannister. The servant gave no distinct reply, and Lewis, in the meanwhile, walked backwards and forwards before the door talking to himself. Sometimes, she said, he would seem as if he were walking up to her; and would suddenly wheel round, and walk back as fast as he could. Supper being ready he sat down, but had eaten only a few mouthfuls when he started up, speaking to himself in a violent manner. At these times, she says, she observed his face to flush as if it had come on him in a fit. He lighted his pipe, and drawing a chair to the door sat down, saying to Mrs. Grinder, in a kind tone of voice, "Madame this is a very pleasant evening." He smoked for some time, but quitted his seat and traversed the yard as before. He again sat down to his pipe, seemed again composed, and casting his eyes wistfully towards the west, observed what a sweet evening it was. Mrs. Grinder was preparing a bed for him; but he said he would sleep on the floor, at dusk, the woman went off to the kitchen, and the two men to the barn, which stands about two hundred yards off. The kitchen is only a few paces from the room where Lewis was, and the woman being considerably alarmed by the behaviour of her guest could not sleep, but listened to him walking backwards and forwards, she thinks , for several hours, and talking aloud, as she said, "like a lawyer." She then heard him at her door calling out "Oh madam! Give me some water, and heal my wounds." The logs being open, and unplastered, she saw him stagger back and fall against a stump that stands between the kitchen and room. He crawled for some distance, and raised himself by the side of a tree, where he sat about a minute. He once more got to the room; afterwards he came to the kitchen-door, but did not speak; she then heard him scraping the bucket with a gourd for water; but it soon appeared that this cooling element was denied the dying man! As soon as the

> day broke and not before—the terror of the woman having permitted him to remain for two hours in this most deplorable situation—she sent two of her children to the barn, her husband not being home, to bring the servants; and on going in they found him lying on the bed; he uncovered his side, and showed them where the bullet had entered; a piece of the forehead had blown off, and had exposed the brains, without having bled much. He begged they would take his rifle and blow out his brains, and he would give them all the money he had in his trunk. He often said, "I am no coward; but I am so strong, so hard to die." He begged the servant not to be afraid of him, for that he would not hurt him. He expired in about two hours, or just as the sun rose above the trees. He lies buried close by the common path, with a few loose rails thrown over his grave. I gave Grinder money to put a post fence round it to shelter it from the hogs, and from the wolves; and he gave me his written promise he would do it. I left this place in a very melancholy mood, which was not much allayed by the prospect of the gloomy and savage wilderness which I was just entering alone. . . . [708]

The 1845 account by the schoolteacher says:

> The writer says the remains of Captain Lewis are "deposited in the southwest corner of Maury county, Tennessee, near Grinder's old stand, on the Natchez trace where Lawrence, Maury, and Hickman counties corner together." He visited the grave in 1838, found it almost concealed by brambles, without a stone or monument of any kind, and several miles from any house. An old tavern stand, known as Grinder's, once stood near by, but was long since burned. The writer gave the following narrative of the incidents attending the death of Capt. Lewis, as he received them from Mrs. Grinder, the landlady of the house where he died in so savage a manner. She

said that Mr. Lewis was on his way to the city of Washington, accompanied by a Mr. Pyrna and a servant belonging to a Mr. Neely. One evening, a little before sundown, Mr. Lewis called at her house and asked for lodgings. Mr. Grinder not being at home, she hesitated to take him in. Mr. Lewis informed her two other men would be along presently, who also wished to spend the night at her house, and as they were all civil men, he did not think there would be any impropriety in her giving them accommodations for the night. Mr. Lewis dismounted, fastened his horses, took a seat by the side of the house, and appeared quite sociable. In a few minutes Mr. Pyrna and the servants rode up, and seeing Mr. Lewis, they also dismounted and put up their horses. About dark two or three other men rode up and called for lodging. Mr. Lewis immediately drew a brace of pistols, stepped towards them and challenged them to fight a duel. They not liking this salutation, rode on to the next house, five miles. This alarmed Mrs. Grinder. Supper, however, was ready in a few minutes. Mr. Lewis ate but little. He would stop eating, and sit as if in a deep study, and several times exclaimed, "If they do prove any thing on me, they will have to do it by letter." Supper being over, and Mrs. Grinder seeing that Mr. Lewis was mentally deranged, requested Mr. Pyrna to get his pistols from him. Mr. P. replied, "He has no ammunition, and if he does any mischief it will be to himself, and not to you or any body else." In a short time all retired to bed; the travellers in one room as Mrs. G. thought, and she and her children in another. Two or three hours before day, Mrs. G. was alarmed by the report of a pistol, and quickly after two other reports, in the room where the travellers were. At the report of the third, she heard some one fall and exclaim, "O Lord! Congress relieve me!" In a few minutes she heard some person at the door of the room where she lay. She inquired,, "Who is there?" Mr. Lewis spoke and said, "Dear Madam, be

> so good as to give me a little water." Being afraid to open the door, she did not give him any. Presently she heard him fall, and soon after, looking through a crack in the wall, she saw him scrambling across the road on his hands and knees.
>
> After daylight Mr. Pyrna and the servant made their appearance, and it appeared they had not slept in the house, but in the stable. Mr. P. had on the clothes Mr. L. wore when they came to Mrs. Grinder's the evening before, and Mr. L.'s gold watch in his pocket. Mrs. G. asked what he was doing with Mr. L.'s clothes on; Mr. P. replied "He gave them to me." Mr. P. and the servant then searched for Mr. L., found him and brought him to the house, and though he had on a full suit of clothes, they were old and tattered, and not the same as he had on the evening before, and though Mr. P. had said that Lewis had no ammunition, Mrs. G. found several balls and a considerable quantity of powder scattered over the floor of the room occupied by Lewis; also a canister with several pounds in it. [709]

My own preference is to believe some of the third account. It seems probable that three men rode up to the stand and that Lewis drove them off with his pistols. On the assumption that Lewis's assassins wanted to prevent him from bringing incrimating documents to Washington, they would have offered him an opportunity to save himself by surrendering the documents, but he refused to deal with them.

It seems probable that Pernier switched clothes with Lewis and was going to ride off as a decoy at sunrise. Mrs. Grinder blamed Pernier for the gold watch that Neelly took when he arrived. Pernier was made to appear a suspect, with robbery as a motive. It seems likely Lewis was dressed in old and tattered clothes when he was found on the road. It seems likely Pernier, "Mr. P.," appeared like a white person to her, and the negro accompanying Lewis was Neelly's slave.

A local account may be found on the Tennessee Geneaology website about the burial of Lewis and a coroner's inquest that was held in 1809. Robert Smith was a postal employee, who discovered Lewis's body on the road in the early morning of October 11th. Since Lewis was a very big man, they made his coffin from a huge chestnut oak, and forged large square iron nails to close it.[710]

It was those nails that established the identity of Lewis's remains when his burial site at Grinder's Stand was developed into a Tennessee State Monument in 1850. The Tennessee State Monument Committee had a medical doctor on its commission. When they identified Lewis's remains, they declared:

> The impression has long prevailed that under the influence of disease of body and mind—of hopes based upon long and valuable services—not merely deferred but wholly disappointed—Governor Lewis perished by his own hands. It seems to be more probable that he died at the hands of an assassin.[711]

Local residents believed that Robert Grinder and his accomplices killed Lewis. There was a coroner's inquest, and the record was kept in a docket book by the jury foreman, Samuel Whiteside. The docket book was known to exist in the early 1900's, but has since vanished. It was said that all six members of the coroner's jury believed Grinder was guilty but were afraid to convict him. It was said Thomas Runnions was an accomplice in the murder, and that his mocassin imprint was seen around the murder scene, which took place near the stables.[712]

There is also the question of the bloodstained Masonic Apron known to have belonged to Meriwether Lewis, which remains a mystery. Why didn't Lewis receive a proper Masonic burial at the time of his death? The gravesite is still located on the property that once was Grinder's Stand, and is now a National Park Service site. Two hundred years later, in 2009, his National Monument & Gravesite was the scene of a public and Masonic ceremony honoring Lewis, attended by over 2,000 people.[713]

Historian Jay Buckley points out the similarity of the Freemason story of Hiram Abiff, builder of Solomon's temple and a widow's son, who was killed by three men for refusing to reveal

Meriwether Lewis Gravesite at National Monument, Hohenwald, Tennessee.

Grinder's Stand replica cabin at Meriwether Lewis National Monument & Gravesite.

Lewis's bloodstained Masonic Apron is on display at the Masonic Lodge in Helena, Montana.
(Photos by Rick Blond)

masonic secrets. Like Hiram Abiff, Lewis was buried in a shallow grave, and later reburied with a broken shaft monument.[714]

Why was his body buried in a primitive inn and tavern yard, without even a fence to guard it from the pigs and wolves? Why weren't his remains returned to the family graveyard in Charlottesville? Why was Alexander Wilson the only one of his friends and family to visit his gravesite? Why didn't Jefferson investigate the death of his friend? The answer has to be that if his remains were moved, or his death was investigated, it would become obvious he was assassinated. Suicides didn't have to be investigated. An assassination couldn't be proven, and nothing was going to bring Lewis back. There were powerful people behind Lewis's assassination, and the deed was done. Jefferson kept silent.

What happened to Seaman? It is more likely that Seaman was shot and died defending Lewis, rather than the story he died of grief at his gravesite. What happened to Pernier, who witnessed what really happened that night?

Did James Neelly witness Lewis's burial? It was a day and half's travel from Franklin to Grinder's Stand. He might have been there, but it is likely he arrived soon after the burial. "Mr. X" would have conferred with Neelly at Grinder's Stand, and then gone north with Pernier to Nashville, bringing Lewis's possessions with them.

"Mr. X" and Captain John Brahan composed the letters to Jefferson. It would have made a much better cover up if Brahan hadn't sent his own letter to Jefferson, duplicating the content of the so-called Neelly letter. There was no need for it. It was an odd admission that he had actually written the Neelly letter. Crimes get revealed by subsequent cover ups. The letters to Jefferson were dated October 18, 1809, one week after Lewis's death. The news story and Neelly's paid announcement defending the reputation of the Colbert family were published in the Nashville newspaper on October 20th.

On November 23rd, 1809, Lewis's possessions, inventoried and neatly bundled, were to be "safely conveyed to Washington City" by Thomas Freeman. They were received by Isaac Coles, who had replaced Lewis as private secretary to both Jefferson and

Madison. Isaac Coles wrote on January 10, 1810:

> The bundles of Papers referred to in the above memorandum were so badly assorted that no idea could be given them by any terms of general description. Many of the bundles containing at once Papers of a public nature—Papers intirely private, some important & some otherwise, with accts. Receipts, &c. They were all carefully looked over and put in separate bundles.

Coles attempted to reassemble the jumbled materials to match the descriptions of the list, and he and William Clark noted on the list the persons or departments receiving the materials. W. C. means William Clark.

One small bundle of Letters & Vouchers of consequence (Depts.)
One Plan & View of Fort Madison (Dept. War)
Two books of Laws of Upper Louisiana (Dept. War)
One Book an Estimate of the western Indians (W. C.)
One small package containing the last will & testament of Govr.
Lewis Deceas'd & one check on the Bank of New Orleans for
99 58/100 dollars (T. Jefferson)
One memorandum book (Th. Jefferson)
A transcript of Records &c. (President)
Nine Memorandum Books (W. C.)
Sixteen notebooks bound in red morocco with clasps (W. C.)
One bundle of Miscelans. paprs. (Th. Jefferson)
No 2 Returns of Militia—papers not acted on (War Dept.)
One bundle papers indorsed—"From the Drawer of my
Poligt. [Lewis's own polygraph duplicate writing
instrument](Th. Jefferson)
One do. [do. means ditto, in this case, "bundle"] papers
relative to the Mines (P. U. S. [President of the U. S.]
One do. Public Vouchers (War Dept.)
Six notebooks unbound (W. C.)

One bundle of Indorsed Letters &c. (Th. J.)
One do. of McFarlane's acct. for the outfit of Salt Petre
expend. (War Dept.)
One journal of do. (War Dept.)
One bundle of Maps &c (W. C.)
One do. "Ideas on the Western expedition" (W. C.)
One bundle of Musterrolls (War Dept.)
One do. Vouchers for expenditures in the War Departmt.
One do. Vocabulary (W. C.)
One do. Maps & Charts (W. C.)
One bundle of papers marked A (Th. Jefferson)
One Sketch of the River St. Francis with a small
Letter Book (W. C.)
One do. Containg. Commisn. and Diploma (Th. Jefferson)
One. do. Sketches for the President of the U. States (P. U. S.)

Thomas Freeman was in charge of taking Lewis's trunks to Washington. The inventory list was signed by Freeman, Captain Boote and Captain Brahan of the U. S. Army, and William P. Anderson of Nashville on November 23, 1809.[715]

Thomas Freeman was employed by General Wilkinson to build Fort Adams in 1799, after Andrew Ellicott fired him for being an "idle, lying, troublesome, discontented, michief-making man" while making the southwest boundary survey in 1798. Freeman was then employed by Wilkinson to explore the Red River in 1806.

In 1809, he was the Deputy U. S. Surveyor in charge of surveying the public lands of former Chickasaw territory that were being offered for sale at the Great Bend of the Tennessee River. Freeman was acting as a land speculator as well as a surveyor. Freeman was the single biggest purchaser of land, acquiring almost 8,500 acres of land for $18,000 in Madison County, where Huntsville, Alabama was established in 1809. Freeman lived in Nashville. His friend Captain John Brahan was the leader of a group of influential Nashville residents who benefitted from insider knowledge in purchasing the land. Freeman was obviously acting as a purchaser for anonymous buyers. [716]

John Brahan had received a federal appointment as the Receiver of Public Money for the land sales. Six weeks of public auctions were held in Nashville to discourage purchases by the squatters who were living on the land and farming it. After the auctions in the summer and fall of 1809, private sales continued. Captain Brahan resigned from the army in January, 1810 and married the daughter of Tennessee Congressman Robert Weakley in June.

Brahan was among the Huntsville elite who became as "rich as princes" from the sale and development of the cotton lands in northern Alabama. While acting as Receiver of Public Money at the Huntsville land office in 1817–19, Brahan used federal money to speculate in land. He purchased 44,647 acres for $318,579, intending to repay the money from reselling the land. However, he came up $80,000 short. Brahan was also a director of the Huntsville Planters and Merchants Bank. He was dismissed from his federal position, and resigned as bank director in 1819.[717]

The men who assumed responsibility for notifying Jefferson of Lewis's death, and taking his possessions to Washington were both deeply involved as speculators and as federal officials in the great land sales going on in the fall of 1809. The fact that Lewis's possessions arrived in Washington in a completely disorganized condition, meant that someone went through the papers looking for material they wanted removed. Someone also wanted Lewis's two trunks held at Fort Pickering rather than being sent back to Will Carr in St. Louis. Did they find what they were looking for? We don't know. In any event, Lewis was no longer alive to disclose any information they wanted to keep hidden.

John Smith T went to Washington in the winter, and delivered his petition to Congress, asking for the removal of Judge Lucas from office and requesting second stage status in territorial government. Congress reappointed Lucas, and deferred action on the territorial status of Louisiana.

John Pernier brought the two letters written by Brahan to former president Jefferson at Monticello. First Pernier had gone to Lewis's mother, who believed Pernier was somehow responsible for Lewis's death. Undoubtedly, he was extremely nervous, and unwilling to tell her what he knew. It is my opinion that Pernier told Jefferson what really happened at Grinder's Stand. John Pernier had worked for Jefferson at the President's House in 1804–05. We don't know—but it seems likely Jefferson knew from this time on that it was an assassination, and chose to go with the suicide story.

Pernier had arrived at Monticello by November 26th, as Jefferson wrote to Madison on that date about receiving Brahan's letters concerning Lewis's goods. Jefferson suggested that Isaac Coles distribute them to the proper recipients. He informed Madison that Pernier was due $240 in wages from Lewis, and he was penniless. Jefferson told Madison he had advised Pernier to sell Captain Russell's horse that he had ridden from Tennessee and apply it to his back pay. He gave Pernier $10 towards his expenses and Pernier brought Jefferson's letter to Madison.[718]

On November 30th, Jefferson rode over to Colonel James Monroe's house in Charlottesville to talk over matters with him. He wrote to Madison:

> I had an hour or two's frank conversation with him. the catastrophe of poor Lewis served to lead us to the point intended.

Jefferson wanted James Monroe to become the Governor of Louisiana Territory. Monroe wanted a place in Madison's cabinet, not the governorship. Monroe said in regards to a military appointment, that he would "sooner be shot than take a command under Wilkinson." James Monroe replaced Robert Smith as Secretary of State in April, 1811, and then became the next President of the United States in 1817–25.[719]

Jefferson wrote to Madison using the words, "the catastrophe of poor Lewis," instead of using "suicide," or "death." Catastrophe means disaster, havoc, ruin, sudden damage, a momentous tragedy—a very good description of Lewis's situation and the circumstances of his death as he traveled towards Washington to right the wrongs inflicted on him.

William Clark visited Jefferson on the 7th of December. He wrote in his journal:

> Soon after I got to Charlotsville saw Mr. Thos. Jefferson. He invited me to go and Stay at his house, &c. I went with him and remained all night, spoke much of the af [fair]s of Gov. Lewis &c. &c. &c.

On December 13th he met with William Meriwether to express concerns about the legality of Lewis's will, as Clark was still under the illusion a second will had been written at the fort naming him and William Meriwether the executors of the estate.

This second will was never found; it was undoubtedly a falsehood create by General Wilkinson to deceive Clark about how Lewis died. Supposedly the letter was written by Captain Russell to Clark, but if Lewis had made a second will at the fort, Russell would have been a witness, and he would have informed Jefferson about it. In any event, matters were arranged so that Clark took charge of the expedition materials and John Marks became the executor of the estate.

Clark met with Secretary of War Eustis on December 17th:

> . . . went to see the Secretary of War, had a long talk abt. Gov. Lewis, pointed out his intertentions & views for his protests, [Eustis] Declaired the Govr. had not lost the Confidence of the Govt.[720]

James Neelly lost his job as Chickasaw Indian Agent in 1812. Tony Turnbow, the Franklin, Tennessee lawyer who discovered Neelly's appearance in court on October 11th, writes:

> Neelly was relieved of his duties in 1812 for questions about his own loyalty or for failing to encourage the Chickasaws to properly support the American government at a time of impending war. After losing the support of the Chickasaws, Neelly was charged with attacking a man in the Chickasaw Nation, frequently becoming intoxicated and indecently ex-

> posing himself to the Chickasaws, becoming intimate with a woman other than his wife at the Agency, and self-dealing on land transactions with the Chickasaws.

Neelly never gave back Lewis's set of pistols, his dirk, and his gold watch. It must have been a constant reminder to all who knew him of his association with Lewis's death.[721]

In 1814, Robert Grinder paid $250 for a farm of 100 acres on the Duck River and left the Natchez Trace.[722]

On February 10, 1810 John Pernier wrote to Jefferson, signing a letter written in another hand. He explained that he was in Washington City in a "distressed situation," and he had been employed by Lewis without receiving very much pay. He enclosed an account showing he was owed a balance of $271.50 from Lewis's estate, after deducting the $43.50 he had received from the sale of the horse. He said that if he didn't receive the money he would be imprisoned for debt.

Pernier was living with John Christopher Sueverman, an old, blind, and very poor former servant of Jefferson who had served in the President's House. Sueverman was "wholly or partly of Negro ancestry" according to Donald Jackson. The next letter concerning Pernier was written on May 5th, 1810 by someone who wrote the letter for the blind Sueverman, who signed it with a crude signature. It announced:

> Respectfully I wish to inform you of the Unhappy exit of Mr. Pirny. He boarded, and lodged, with us ever since his return from the Western Country. The principal part of the time he-has been confined by Sickness. I believe ariseing from uneasyness of mind, not having recd. anything for his late services to Govr. Lewis. He was wretchedly poor and destitute. . . . Last week the poor man appeared considerably better, I believe in some respects contrary to his wishes, for unfortunately on Saturday last he procured himself a quantity of Laudenam. On Sunday morning [April 29th] under the pretence of not being so well went upstairs to lay on the bed, in which situa-

> tion he was found dead, with the bottle by his Side that had contained Laudanam.[723]

John Pernier had no money, and was facing debtor's prison, and he met someone who gave him a bottle of laudanum. Laudanum is a tincture of opium, a narcotic. It was a widely used medicine until early in the 20th century. The laudanum could have easily been laced with arsenic.

General Wilkinson had arrived in Washington in April to begin to prepare his defense for congressional investigations into his conduct.[724] In the list of suspicious deaths associated with General Wilkinson, John Pernier's name should be added.

On August 18th, 1925 the Meriwether Lewis National Monument & Gravesite was dedicated at Hohenwald, Tennessee. A crowd estimated at between 7,000–10,000 attended the event. Shown here are P. E. Cox, State Archeologist, John Trotwood Moore, State Historian, Governor Austin Peay and Dr. J. N. Block. The petition to President Calvin Coolidge, requesting that it be designated a national monument, stated:

> Investigations have satisfied the public that he was murdered presumably for the purpose of robbery.[725]

(Tennessee State Library and Archives)

Chapter 14

Afterwards

General Wilkinson had an alibi. He published a warrant for his own arrest in his book, *Burr's Conspiracy Exposed and General Wilkinson Vindicated Against the Slander of His Enemies.* The warrant was issued on October 9th, 1809 in Natchez, Mississippi, two days before Lewis's death.

Wilkinson was being sued by John Adair, the former Senator from Kentucky, on charges of trespass, assault and battery, and false imprisonment, for $20,000 in damages. John Adair was one of the Burr conspirators whom Wilkinson had arrested in New Orleans in January, 1807. Wilkinson showed that he posted bail of $7,000 on October 25th, 1809. It is doubtful the general spent any time in a jail cell in Natchez, but the court records are missing. The only evidence this happened is found in footnotes in his own book, published in 1811.

Where he actually was at this time remains a mystery. There is a letter written by him dated September 25, 1809 in New Orleans, and there is a letter written by him dated November 15, 1809 in Natchez.[725] Since he was courting a young woman, Celestine Trudeau, the cousin of William Claiborne's wife, it is safe to assume he stayed in New Orleans as long as possible.

In 1809, there were two competing conspiracies to participate in the first Mexican Revolution, which was being organized by a Catholic priest, Father Manuel Hildago. The date of the first Mexican War for Independence is September 16, 1810, which is still celebrated in Mexico on September 16th every year as *El Grito Dolores*, or the "Cry of Dolores." Father Hildago was a priest in the town of Dolores at the center of the silver mine district in the Sierre Madre Mountains.

Apparently—for those who had been conspiring to invade Mexico for many years—there was advance knowledge of Hildago's plan to start a revolution. Both Aaron Burr and James Wilkinson were preparing for action and organizing separate invasion plans in 1809. Wilkinson, as usual, was doing several things at once. He was defending New Orleans; sending out his own reconnaissance mission to Mexico; and organizing the assassination of Meriwether Lewis. Burr was over in Europe trying to persuade the British government/Napoleon/a group of international associates to support his invasion plans.

In December, 1808, Jefferson ordered 2,000 American troops to New Orleans to defend the city against a potential British invasion. After making diplomatic visits to Cuba and Spanish Florida, General Wilkinson arrived at New Orleans in April, 1809 to take charge. He found the new recruits in deplorable condition. His enemies in New Orleans—Aaron Burr's allies—the Mexican Association, did their best to ridicule Wilkinson, calling him "his Serene Highness," and the "the Grand Pensioner de Godoy." Manuel Godoy was the former prime minister of Spain.

Napoleon had overthrown King Ferdinand VII in 1808 and placed his brother, Joseph Bonaparte, on the throne of Spain. The Pennisular War between the French empire and a coalition of British regular forces and Spanish and Portuguese guerilla fighters started in 1808. The Spanish War for Independence lasted from 1808–1814 on the Iberian Pennisula. With a French king on the Spanish throne, the desire for independence on both the Spanish mainland and its far flung colonial empire rapidly gained strength. In 1809–1811 there were revolutions in Bolivia, Ecuador, Argentina, Columbia, Chile, Paraguay, Uruguay, and Mexico, as various factions fought to either restore the rule of King Ferdinand or to achieve some form of democratic rule.

In 1809, the Spanish colony of Cuba, fearing a revolution, expelled the French speaking Ste. Domingue refugees who had fled to Cuba during the Haitian revolution to start new lives under Spanish rule. More than 10,000 Ste. Domingue refugees arrived in New Orleans from May through December of 1809, doubling the size of the population of New Orleans. Another 7,000 refugees would arrive in 1810. They were a mix of whites, free persons of color, and slaves.[726]

It severely strained the resources of the area. The 2,000 American troops stationed there were already living and dying in miserable conditions in the city of New Orleans. General Wilkinson arranged for the removal of the troops to farm land east of the city, *Terre aux Boeufs*, or "Land of the Cattle." It was the land where beef was raised for the city and vegetable crops were grown. It was also the site where the great Battle of New Orleans was fought in January, 1815, the Chalmette Battlefield of Saint Bernard Parish. As evident from its later fame in defending the city from British attack, Wilkinson had chosen the single best location for his troops.

However, on April 30th—less than two months after assuming his new office—Secretary of War William Eustis ordered the removal of troops from the unhealthy climate of New Orleans to high ground near the vicinity of Natchez, Mississippi, some 300 miles up river. Since General Wilkinson's orders from the previous Secretary of War, Henry Dearborn, had been to defend New Orleans, Wilkinson chose to disregard Eustis's orders. The troops arrived at *Terre aux Boeufs* on June 10th. They would leave the camp on September 14th.

William Claiborne, the territorial governor of New Orleans, wrote to his superior, Secretary of State Robert Smith on July 29th, defending Wilkinson's choice of *Terre aux Boeufs*:

> You will have heard no doubt, many rumours of the dreadful mortality among the Troops of the U. States stationed in this vicinity. But you may be assured, they are greatly exagerated.—This climate is, in truth, unfavorable to strangers, and it could not be expected, that the troops would have been exempted from the diseases common to the Country.—The number of deaths have not been considerable, and the sick

> List not unusually numerous.—I was a few days since at the *Camp* on a visit to General Wilkinson;—I found it in excellent order and every possible exertion appeared to have been used to render the men comfortable.—The position is an eligible one; it is situated about 12 miles below New Orleans and is represented by several of the old Inhabitants to be as healthy as any point on the Mississippi. . . . I have understood that an order has been issued for the immediate removal of the Troops by water from hence to Natchez or Fort Adams.—
>
> I regret the circumstance, because a voyage up river, at this time of year, will be hazardous to the health of the men, and because I consider the presence of a respectable detachment of Troops near to New Orleans as being at this period absolutely necessary.[728]

Why did Governor Claiborne write, "I consider the presence of a respectable detachment of Troops near to New Orleans as being *at this period absolutely necessary,*" in the summer of 1809? A new crisis involving Aaron Burr was happening. Aaron Burr was organizing his "X affairs"—the invasion of Mexico—in Britain when he was expelled from Britain in April, 1809. The British government arranged for Burr's passage to Helgoland, a British-held island in the North Sea which was a center of smuggling and intrigue against Napoleon. Burr's expulsion was a sham on the part of the British government, as Burr's friends told Governor Claiborne's investigator later that year.[729]

There were British ships in the Caribbean. Burr's relative, Sir George Prevost, commanded an invasion force of 10,000 British troops, which captured the French islands of Martinique and Guadelupe in January, 1809–February,1810. American officials were worried that British troops would invade New Orleans and West Florida.

American volunteers invaded Baton Rouge in 1810 and established the short lived Republic of West Florida, which was annexed by the United States in October, 1810 as part of the Territory of Orleans.

Francis Newman, a 23 year old American army officer, wrote to his wife's cousin in New Orleans on May 1st, 1809. He was no ordinary army officer. Newman was the illegitimate child of an English nobleman and an American woman—who were both married to others at the time—who later settled in the Baltimore area. The young officer had married into a distinguished Spanish Creole family in New Orleans, a leading family in the sugar cane industry.

Newman's wife was pregnant, and he wrote to her cousin Joseph Solis from Natchitoches, the American army outpost on the edge of Louisiana Territory, 100 miles west of New Orleans.

> . . . My dear cousin, I am afraid to give you the details on a plan which is not yet clear to me. I would simply advise you to tell my father-in-law, to try to sell and leave the country, and your father, too. Tell them to avoid getting caught up in the horrible disaster that I see as inevitable.

The next letter followed quickly, on May 19th. Newman told his cousin that Manuel Salcedo—the Governor of Texas from 1808 until his execution in 1813—was conspiring with American army officers to declare Mexican independence from Spain. The letter stated the French emigrés from Cuba in New Orleans were going to support the revolution.

The final letter concerning the conspiracy was dated July 20th. Newman reported he heard Governor Salcedo at a dinner say that he had 15,000 men at his disposal and more were joining the independence movement every day. Newman was worried his life was in danger if his part in revealing the conspiracy was known, and told his cousin to keep his identity hidden.

On September 3rd, 1809, Governor Wilkinson sent copies of the three letters, which had been given to him by the Mayor of New Orleans, to Secretary of War Eustis. The general said the letters were supposedly written by Francis Newman, a young artillery officer, who was "an Englishman by birth, [and who] was foisted onto the service I know not how." Wilkinson said the letters were fabricated and Newman's name was forged.[730]

Newman's father-in-law and his brother, the father of Joseph Solis, had given Newman's letters to the Mayor. The two men were arrested and jailed on a $50,000 bond. Eventually, a Grand Jury declared the letters were forgeries and they were acquitted and released from jail in December.[731]

In November, after swearing the letters were forgeries, Francis Newman received a promotion to captain. Captain Newman later commanded the strategic Fort Petit Coquilles (the site of the later Fort Pike) during the attempted British invasion of New Orleans in 1815.[732]

Governor Claiborne thought the letters were genuine, and he reported to Secretary of State Robert Smith on November 12th that he had heard Governor Salcedo had been arrested and the province of Texas was "much agitated." Claiborne was also trying to secure the release of two American schooners out of New Orleans, which were being held in Yucatan on the suspicion they were engaged in activities against the Spanish government.

Claiborne sent his Adjutant General, Colonel Henry Hopkins, out in the field to investigate. On December 31st, at the close of the tumultuous year of 1809, Governor Claiborne wrote to Secretary of State Robert Smith telling him what Colonel Hopkins had discovered:

> In a Letter which Colo Hopkins had addressed me, dated La Fourche [The Forks], November 23, 1809 he says— "I have had a nocturnal interview with two persons residing in this place, one by the name of Hopkins, a Kinsman of C. Taylor and reported to the natural son of Colonel Burr; —and the other Doctor Savage the Brother in Law of Edward Livingston [Auguste Davezac]. From these men I obtained the following information,—Burr was never ordered from England—Burr would have received an appointment in the Revolutionary Army of Spain, if his want of the knowledge of the Language had not been an obstacle;—Burr among other enterprises in person, was to encamp in the Spanish Dominions, and before the feeble Government of Spain could remove them, they

> would be joined by twenty thousand of the most brave and enterprising men of the United States.
>
> We have in this Territory many persons of similar characters to Hopkins and Savage; and whose frequent conversations as to the facility of *revolutionizing Mexico* are calculated to excite the people to Jealousy of our neighbors.

Claiborne went on to report that the Solis brothers had been acquitted of libel by a Grand Jury, and that Francis Newman had sworn his letters were forgeries.[733]

Wilkinson had sent Francis Newman's three letters, written in French, to Secretary of War Eustis on September 3rd. Dr. Eustis was not only Aaron Burr's closest friend, he was the physician to both Burr and his daughter Theodosia. The wonder is that Eustis had been appointed Secretary of War, with no military experience beyond that of serving as a physician in the Revolutionary War. President James Madison had appointed him to the cabinet in the interest of party unity.

Eustis was also from Massachusetts, the center of the secessionist movement of the Northern Confederacy. It seems reasonable to assume that Secretary of War Eustis was trying to get American troops out of New Orleans in order to facilitate the secret goals of Aaron Burr and the Northern Confederacy working with Great Britain to invade Mexico. General Wilkinson, on his way to New Orleans, had been entertained at Norfolk and Charleston, where he had publicly called Madison's new Secretary of War a "black mouthed Federalist," and had proposed a toast to—

> The New World, governed by itself and independent of the Old.[734]

This is the heart of the matter—the struggle between Burr and his British connections and Wilkinson and his American connections to revolutionize Mexico. Burr's way would lead to the establishment of a new country, and the break away of the Northern Confederacy from the United States. Wilkinson's way would lead to what eventually happened—the establishment of independent republics in Spanish territories which later joined the United States of America.

It was after the letters by Francis Newman became public knowledge that Wilkinson finally obeyed Eustis's orders to start moving the troops 300 miles up river to the Natchez area. As Governor Claiborne had predicted, it was hazardous to the health of the men. Wilkinson provided a table of official army records in his *Memoirs*. A total of 794 men died and 166 men deserted of the about 2,000 troops sent to defend New Orleans against British attack in 1809. Of those deaths, 590, or 74%, occurred after the troops left *Terre aux Boeufs*.[735] It took weeks for the troops to move upriver to Natchez, in the heat of summer during the malarial season.

General Wilkinson was relieved of his command in October and ordered to report to Washington for congressional investigations. He married the daughter of the surveyor general of Louisiana, Celestine Trudeau, on March 5th, 1810 before leaving for the federal city. Twin daughters were born to the general and his wife in 1816. In his retirement, the general and his new family lived on a plantation in Plaquemines Parish, where he grew sugar and cotton in the rich delta lands at the very tip of the New Orleans pennisula.

General Wilkinson and John Smith T had their own filibuster plans for the invasion of Mexico in 1810. This was the true reason for their assassination of Meriwether Lewis. Lewis may not have even realized they were planning another filibuster. They would have wanted Lewis killed if they believed—or knew—that he was carrying evidence to Washington implicating them in major land claims fraud. The land claims of the large land holders were going to be examined by Judge Lucas and the other two commissioners in 1810. Wilkinson had set up many of these deals when he served as first governor of the territory and Smith T was heavily invested in the lead mine district. Meriwether Lewis was simply too honest to remain the Governor of Louisiana Territory. Any resulting actions concerning land fraud would seriously interfere with their time-dependent plans to participate in the first revolution for Mexican Independence.

As soon as Lewis's death was reported in St. Louis, an expedition to Sante Fe set out on a reconnaissance mission. On December 28th, 1809, Joseph Charless, editor and publisher of the newly renamed *Louisiana Gazette,* reported that:

> We are informed that about the 20th Ult. [November 20th] Capt. R. Smith, Mr. McLanahan and a Mr. Patterson set out from the district of St. Genevieve upon a journey to St. A Fee, accompanied by Emmanuel Blanco, a native of Mexico. We presume their object are mercantile; the enterprise must be toilsome and perilous the distance computed at 5 to 800 miles, altogether thru a wilderness heretofore unexplored.[736]

"Capt. R. Smith" was Reuben Smith, the brother of John Smith T. "Mr. McLanahan" was Josiah McLanahan who served as Sheriff of St. Louis when Wilkinson was governor of the territory. "Mr. Patterson" was 25 year old James Patterson of the town of Pulaski, in Giles County, Tennessee.

Patterson went to the lead mine district of Missouri in the spring of 1809 with three slaves. His father was a prominent leader in Giles County. The Pattersons must have known Smith T. The "T" of Smith T stood for Tennessee and Smith T had once claimed ownership of the northern half of Alabama under fraudulent land titles in the massive Yazoo land scandal. John Smith T had deep connections to the area.

The Early History of Giles County, Tennessee says that James-Patterson and his associates were arrested when they reached Sante Fe on charges of being spies for the French government. They were imprisoned in Chihauhau for 9 months under heavy irons, and in jail for another 9 months after they were released from the irons. They suffered greatly. The history relates:

> He [Patterson] was a Mason and the priest who visited him was also a Mason, and through his influence, as he supposed their condition was rendered more tolerable. At the end of eighteen months they were turned out of prison but not allowed to leave the city. Smith was a gun-smith and a very ingenious mechanic. He repaired gun-locks, etc., and invented a machine to mill their silver coin in moulding the metal, which he turned to some profit. The Revolution which finally resulted in the independence of Mexico, commenced after they were arrrested.
>
> Don Miguel Hildago the first leader of the revolution, was captured with others of the insurgents and taken to Chihauhau, whilst Mr. Patterson and his party were there; and [Hildago] was there shot in July, 1811. Mr. Patterson witnessed the execution. . . .
>
> Colonel Nelson Patterson (the father of Jas. Patterson) saw in a newspaper an extract from a Vera Cruz paper, published in the City of Mexico, an account of their capture and imprisonment, and wrote to the authorities in Washington. The U. S. Government demanded their release from Spain. After having been nearly three years prisoners they were permitted to leave, the three negroes and Indians declined to return with them. . . . The Governor's Secretary was a Mason, and he told Mr. Patterson privately not to take the route the Governor designated; that the Governor had posted a band of Indians at a certain watering place, on the way where they were to stop, with strict instructions to kill all of them . . .

The Nashville *Democratic Clarion* newspaper reported their arrival in Natchez in a story dated May 6, 1812. They were reported to have said:

> That the country is in a complete state of civil war, and that nothing but arms are wanted to shake off the trammels of the

> old Spanish government. . . .The country, according to their account, abounds in gold, silver and iron ores—the climate salubrious and pleasant, and the land, considering the great droughts, that are experienced there, very productive.[737]

The next month, in June, 1812 a joint Mexican-American filibuster expedition, called the Gutiérrez-Magee Expedition, began rendezvousing near the Sabine River and Natchitoches. Bernardo Gutiérrez de Lara, a leader in Father Hildago's revolution, visited Washington in December, 1811 seeking aid from Secretary of State James Monroe. A 24 year old army officer, a protegé of General Wilkinson, Lieutenant Augustus Magee, resigned his commission in the U. S. Army in June, 1812 and joined with Gutiérrez to lead the filibuster into Mexico. James Patterson, Josiah McLanahan, and Reuben Smith were leaders of one of the three groups assembling on the Neutral Ground. The general's son, Joseph Biddle Wilkinson, participated in the expedition.[738]

The Gutiérrez-Magee Expedition ultimately ended in failure after many twists and turns and perhaps the murder of Magee by arsenic. As this book is a biography of Meriwether Lewis—and the story of the conspiracies and filibusters to invade Mexico is as lengthy and complex as any of the topics in this book—the story, regretfully, has to end here.

Mexico would finally achieve its independence in 1821. In 1823, General Wilkinson presented his Gilbert Stuart painting of President George Washington to the Congress of Mexico. The general died in 1825 in Mexico at the age of 68, while he was waiting to receive a Texas land grant from the Mexican government. He was buried in a Catholic Cemetery in Mexico City.[739]

These attempts to invade Mexico in 1809–1812 form the basis of the evidence of why I believe Meriwether Lewis was assassinated in October, 1809 by General Wilkinson and John Smith T. Their underlying motives were the lure of the fabulous wealth of the silver mines and other mineral resources of Mexico, and their desire to benefit through land speculation and commercial trade with a newly-formed independent government of Mexico.

In 1811, during his court martial at Frederick Town, Maryland, the general created a false document called the "Russell Statement." He would have been very pleased to know that his forgery would cause trouble in the Twentieth Century.

Editor Donald Jackson discovered the "Russell Statement" in the Jonathan Williams Archives of the University of Indiana's Lilly Library, and decided to publish it in the first edition of his *Letters of the Lewis and Clark Expedition*. The authenticity of the document was questioned by Vardis Fisher, author of *Suicide or Murder?: The Strange Death of Governor Meriwether Lewis.* Jackson's and Fisher's books were both published in 1962.

Fisher discussed the Russell Statement in his book, and wrote:

> But for us there is something strangely unlike Russell in his manner of writing this account of Lewis's death. In his two letters to Jefferson his friendship and his feeling for Lewis shone through at every point. Here he says that Lewis killed himself in "the most cool desperate and Barbarian-like manner"—strange words from an army man whose profession was killing with guns.[740]

In 1996, James E. Starrs organized the Coroner's Inquest which was held in Lewis County, Tennessee where the Meriwether Lewis National Monument & Gravesite is located. The Russell Statement was examined by two document examiners, Gerald B. Richards, former head of the F. B. I. Documents Operations, and Dr. Duayne Dillon, former Acting Director of the San Francisco Police Department Crime Lab. The two experts examined the Russell Statement independently of each other and concluded it was a forgery.[741]

Their testimonies are recorded in *The Death of Meriwether Lewis: A Historic Crime Scene Investigation*, which I co-authored with Professor Starrs, and which was first published in 2009, the 200th anniversary of the death of Lewis. A second "New Evidence Edition" was published in 2012. The "Russell Statement" is an interesting document for several reasons—how it was constructed as a cover up; why it was created during a court martial seemingly entirely unrelated to the question of Lewis's death, and who did the actual writing of it.

> [26 November 1811]
>
> Governor Lewis left St. Louis late in August, or early in September 1809, intending to go by the route of the Mississippi and the Ocean, to the City of Washington, taking with him all the papers relative to his expedition to the pacific Ocean, for the purpose of preparing and putting them to the press, and to have some drafts paid which had been drawn by him on the Government and protested. On the morning of the 15th of September, the Boat in which he was a passenger landed him at Fort pickering in a state of mental derangement, which appeared to have been produced as much by indisposition as other causes. The Subscriber being then the Commanding Officer of the Fort on discovering from the crew that he had made two attempts to Kill himself, in one of which he had nearly succeeded, resolved at once to take possession of him and his papers, and detain them there untill he recovered, or some friend might arrive in whose hands he could depart in Safety.
>
> In this condition he continued without any material change for five days, during which time the most proper and efficatious means that could be devised to restore him was administered, and on the sixth or seventh day all symptoms of derangement disappeared and he was completely in his senses and thus continued for ten or twelve days. On the 29th of the same month he left Bluffs, with the Chickasaw agent the interpreter and some of the Chiefs, intending to proceed the usual route thro' the Indian Country, Tennessee and Virginia to his place of distination, with his papers well secured and packed on horses. By much severe depletion during his illness he had

been considerably reduced and debilitated, from which he had not entirely recovered when he set off, and the weather in that country being yet excessively hot and the exercise of traveling too severe for him; in three or four days he was again affected with the same mental disease. He had no person with him who could manage or controul him in his propensities and he daily grew worse untill he arrived at the house of a Mr. Grinder within the Jurisdiction of Tennissee and only Seventy miles from Nashville, where in the apprehension of being destroyed by enemies which had no existence but in his wild immagination, he destroyed himself, in the most cool desperate and Barbarian-like manner, having been left in the house intirely to himself. The night preceeding this one of his Horses and one of the Chickasaw agents with whom he was traveling Strayed off from the Camp and in the Morning could not be found. The agent with some of the Indians stayed to search for the horses, and Governor Lewis with their two servants and the baggage horses proceeded to Mr. Grinders where he was to halt untill the agent got up.

After he arrived there and refreshed himself with a little Meal & drink he went to bed in a cabin by himself and ordered the servants to go to the stables and take care of the Horses, least they might loose some that night; Some time in the night he got his pistols which he loaded, after every body had retired in a Separate Building and discharged one against his forehead not making much effect—the ball not penetrating the skull but only making a furrow over it. He then discharged the other against his breast where the ball entered and passing downward thro' his body came out low down near his back bone. After some time he got up and went to the house where Mrs. Grinder and her children were lying and asked for water, but her husband being absent and having heard the report of the pistols she was greatly allarmed and made him no answer. He then in returning got his razors from a port folio which happened to contain them and Seting up in his bed was found about day light, by one of the Servants, busily engaged in cutting himself from head to foot. He again beged for water, which was given to him and as soon as he drank, he lay down and died with the declaration to the Boy that he had killed himself to deprive his enemies of the pleasure and honour of doing it. His death was greatly lamented. And that a fame so dearly earned as his should be clouded by such an act of desperation was to his friends still greater cause of regret.

(Signed) Gilbert Russell

The above was received by me from Major Gilbert Russell of the [blank] Regiment of Infantry U. S. on Tuesday the 26th of November 1811 at Frederick-town in Maryland.

J. Williams

The phrase in the document that is the most loathesome and offensive is—

> . . . he lay down and died with the declaration to the Boy that he had killed himself to deprive his enemies of the pleasure and honour of doing it.[742]

It is the wicked statement of a sociopath. Wilkinson was confessing he took pleasure in killing Meriwether Lewis and considered it an "honour" to have committed the crime.

The general had good reason to fear Lewis—in 1796, Lewis was General Wayne's trusted courier in the last days of Wayne's life; in 1801, Lewis provided Jefferson with the names of officers who were close associates of Wilkinson and who were dismissed from the service; in 1807, Lewis replaced Wilkinson as Governor of Louisiana Territory; and in 1809, when the general's enemies were clamouring for his dismissal, Lewis was bringing damaging information to Washington that undoubtedly would have brought the general's career to an end.

General Wilkinson had asked for a military court-martial to clear his name. He was being harassed by his enemies in Congress and wanted to end their investigations. The charges brought against him in 1811 were:

(1) that he received pay as a secret agent of the government of Spain from 1793-1804, with ten specific accusations.
(2) that he engaged in treasonable correspondence with Spanish officials in 1795 regarding the dismemberment of the union, with five specific accusations.
(3) that he conspired with Aaron Burr and his associates in 1805-06 to set up an independent empire of western states and territories.
(4) that while commanding the army of the United States in 1805-06, he participated in the conspiracy of Aaron Burr and his associates, and he did not make "any timely discovery of their pernicious designs."
(5) that he conspired with Aaron Burr and his associates to "set on foot a military expedition against the territories of a nation then at peace with the United States."
(6) that he disobeyed orders and instructions from the War Department, dated April 30, 1809, to remove the troops from New Orleans to the "high ground in the rear of Fort Adams and to the high ground in the rear of Natches in Mississippi Territory" and instead removed the troops to a station called *Terre aux Boeufs* below New Orleans, and remained there until September.
(7) that he had neglected his duty in regard to the supply of provisions during the summer and autumn of 1809, with three specific accusations.
(8) that there was misapplication and waste of public monies and supplies, with three specific accusations.

To all of these charges, the general pled "not guilty."

The question remains—what motivated the general to prepare a false statement in the name of Major Russell regarding the death of Lewis? It had absolutely nothing to do with the charges brought against him in 1811. It has taken seven years of research for me to construct an answer as to why the "Russell Statement" was created. First I learned from Jim Starrs that the original was in the Jonathan Williams Collection of Lilly Library at the University of Indiana, and obtained a copy of the handwritten document.

Then I realized the statement coincided with the general's court martial in Frederick-town and that it was created during the trial proceedings. I found the hand written records of the court martial in a confusing jumble on microfilm prepared by the National Archives. Eventually I got very lucky and found a typed transcript of the court-martial at the Historical Society of Frederick County, prepared by a researcher in 1976. Even in typed transcript, I didn't seriously review the proceedings.

Meanwhile, I acquired a colleague and friend in my investigations, Earl Weidner of Baton Rouge. Earl had his own interests in General Wilkinson and proved to be a diligent investigator on the trail of who actually wrote the forgery for the general. After several years of obtaining handwriting samples of various suspects, Weidner found the answer. It was a clerk in the War Department named Lewis Edwards. It makes sense because the general employed him as a scribe during the court martial and his preparations for congressional investigations. Edwards wasn't mentioned by name. It can't be proved except by the kind of professional handwriting analysis done at Coroner's Inquest in 1996, but it seems very likely.

The next stages of research took years for me to comprehend the implications of what I discovered. I found records concerning Russell at the Alabama State Archives which revealed that Major Russell—who was testifying at the court-martial—went Absent With Out Leave (AWOL) from December 3rd through December 27th, 1811. The court-martial judges delivered their verdict of "not guilty" on December 25th. The so-called "Russell Statement" was composed on November 26th.

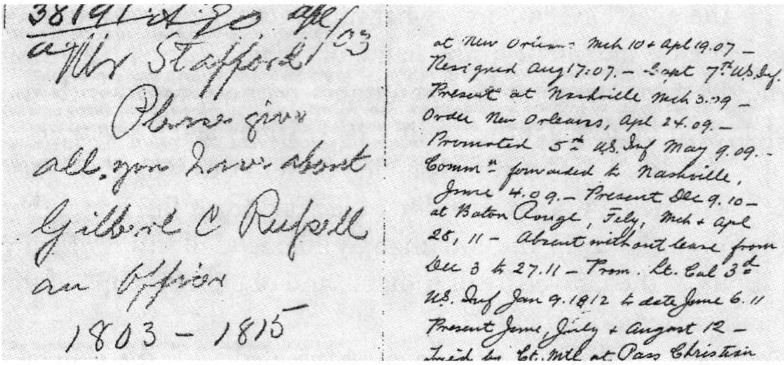

38191 A. 20 Apl/03
Mr Stafford
Please give
all you have about
Gilbert C Russell
an Officer
1803 – 1815

at New Orleans Mch 10 & Apl 19.07 –
Resigned Aug 17.07. – Capt 7th U.S. Inf.
Present at Nashville Mch 3.09 –
Order New Orleans, Apl 24.09. –
Promoted 5th U.S. Inf May 9.09. –
Commn forwarded to Nashville,
June 4.09. – Present Dec 9.10 –
at Baton Rouge, Feby, Mch & Apl
25, 11 – Absent without leave from
Dec 3 to 27.11 – Prom. Lt. Col 3d
U.S. Inf Jan 9.1812 to date June 6.11
Present June, July & August 12 –
Tried by Ct. Mtl at Pass Christian

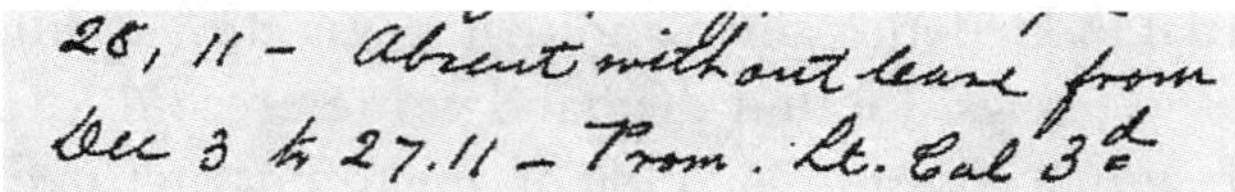

25, 11 – Absent without leave from
Dec 3 to 27.11 – Prom. Lt. Col 3d

This is the information on file at the Alabama State Archives concerning Gilbert C. Russell going AWOL and his promotion.

Major Russell disappeared shortly after the statement was written, and reappeared after the general was declared not guilty. Russell was afraid of being another victim of an assassination by the general, because the only way the statement would have any validity was if Russell was no longer around to dispute the information contained in the statement. That was the final piece of the puzzle for me—that the "Russell Statement" was prepared in advance of another planned assassination. Major Russell was promoted to Lieutenant Colonel in the 3rd U. S. Infantry two weeks after the trial ended, on January 9th, 1812. From his going AWOL and then receiving a promotion, we may safely assume that others in the government realized the general was an assassin.

What provoked the general during his court-martial? It must have been that Major Russell testified on November 5th that Major Seth Hunt told him that General Wilkinson had tried to get him [Seth Hunt] killed. In 1806 Russell and Hunt had traveled together to Washington City and Russell said Hunt was bringing evidence against Wilkinson that would "remove the General from both his military and civil offices." The general must have been afraid that, since Russell had testified about Seth Hunt's accusation, Russell was next going to accuse the general of Meriwether Lewis's assassination—which is why the general prepared the "Russell Statement" cover up.[743]

There is one final issue regarding Lewis's death by suicide—the letter written by Thomas Jefferson for the *Lewis and Clark Journals*. Jefferson was asked to contribute a biography of Meriwether Lewis for the journals, which were published in 1814. He wrote about choosing Lewis to lead the expedition:

> Captain Lewis, who had then been near two years with me as private secretary, immediately renewed his solicitations to have the direction of the party. I had now the opportunity of knowing him intimately. Of courage undaunted, possessing a firmness & perserverance of purpose which nothing but impossibilities could divert from it's direction, careful as a father of those committed to his charge, yet steady in the maintenance of order & discipline, intimate with the Indian character, customs & principles, habituated to the hunting life, guarded by exact observation of the vegetables & animals of his own country, against losing time in the description of objects already possessed, honest, disinterested, liberal, of sound understanding and a fidelity to truth so scrupulous that whatever he should report would be as certain as if seen by ourselves, with all these qualifications as if selected and implanted in nature in one body, for this express purpose, I could have no hesitation in confiding the enterprize to him.

Jefferson wrote about his time as governor—

> A considerable time intervened before the Governor's arrival at St. Louis. He found the territory distracted by feuds & contentions among the officers of the government & the people themselves divided by these into factions & parties. He deter-

> mined at once, to take no side with either; but to use every endeavor to conciliate and harmonize them. The even-handed justice he administered to all soon established a respect for his person & authority, and perseverance & time wore down animosities and reunited the citizens again into one family.

He wrote about Lewis's mental state—

> Governor Lewis had from early life been subject to hypocondriac affections. It was a constitutional disposition in all the nearer branches of the family of his name & was more immediately inherited by him from his father. They had not however been so strong as to give uneasiness to his family. While he lived with me in Washington I observed at times sensible depressions of mind, but knowing their constitutional source, I estimated their course by what I had seen in his family. During his Western expedition the constant exertion which that required of all the faculties of body & mind, suspended these distressing affections; but after his establishment in St. Louis in sedentary occupations they returned upon him with redoubled vigor, and began seriously to alarm his friends. He was in a paroxysm of one of these when his affairs rendered it necessary for him to go to Washington.[744]

This last statement should be considered as nothing more than an obligatory defense of the suicide story.

Jefferson dated the letter to the publisher, "Monticello, Aug. 18. 1813." It was the day of Meriwether Lewis's birthday. This book opened with the birth of Meriwether Lewis in 1774 at Locust Hill, just a few miles from Monticello, and now closes with this sad testimony from Jefferson 39 years later.

Meriwether Lewis was an extraordinary person, fully meriting Jefferson's praise of his character and accomplishments. His life story begins with the start of the American Revolution for Independence and ends with the start of the Mexican Revolution for Independence. It was a revolutionary era, and in telling the story of his life it has been necessary to tell the story of James Wilkinson, because I believe General Wilkinson assassinated Meriwether Lewis and arranged the story of his suicide as a cover up.

The story of Wilkinson has necessarily led to a discussion of fifty years of Wilkinson's intrigue. He was, and remains, a convenient "scape goat" for both his contemporaries and historians, who blame him for any number of sins over this period. What they can't explain is why he was retained in power by the first four presidents, and survived numerous congressional investigations and military courts. In this book I have attempted to demonstrate that Wilkinson was both a patriot and a sociopath. His contemporaries accused him of being an assassin, but historians have concentrated on his role as Spanish Agent #13 and not explored the assassination accusation.

Meriwether Lewis's achievements were many. The most well known is his successful mission to the Pacific—reinforcing the United States claim to the Pacific Northwest and a continental empire—and his governorship of Louisiana Territory during a troubled time. But perhaps his most lasting accomplishment is the contribution he made to the natural sciences and description of life as he saw it during his journey to the Pacific. As Jefferson wrote, he was—

> honest, disinterested, liberal, of sound understanding and a fidelity to truth so scrupulous that whatever he should report would be as certain as if seen by ourselves . . .

Lewis's report on Indian cultures dating back thousands of years, and his natural history observations will be remembered for all time. He also deserves to be remembered as an American hero, without the false label of suicide clouding his reputation.

1. Albemarle County

1. David Hackett Fischer and James C. Kelly, *Bound Away: Virginia and the Westward Movement* (Charlottesville and London, University Press of Virginia, 2000), pp. 95-98.

2. Thomas Perkins Abernethy, *Three Virginia Frontiers* (Gloucester MA, Peter Smith, 1962), pp. 44-45. Colonel Nicholas Meriwether obtained 13,000 acres in 1727; Charles Lewis obtained 12,000 acres in 1732.

3. Patricia L. Zontine, "Lucy Meriwether Lewis Marks (1752-1837): Her Life and Her World" (www.monticello.org, Robert H. Smith International Center for Jefferson Studies, 2009, accessed 12/20/12). Do a search on website for Lucy Marks. This section also has an interactive map of early buildings in Albemarle County and extensive biographies. Zontine writes that Lucy and William either married in 1768 or 1769, I have chosen 1769. Their first child, Jane, was born in 1770.

4. David Hackett Fisher, *Albion's Seed: Four British Folkways in America* (New York and Oxford, Oxford University Press,1989) pp. 281-286.

5. Rhyss Isaac, *The Transformation of Virginia, 1740-1790* (Williamsburg VA, Omohundro Institute of Early American History and Culture, and Chapel Hill NC, The University of North Carolina Press, 1982) p. 258.

2. George Rogers Clark

6. James Alton James, *The Life of George Rogers Clark* (Chicago, University of Chicago Press. 1928/1978) p. 31.

7. *Ibid*, p. 52-54.

8. *Ibid*, p. 69.

9. *Ibid*, p. 112 (citing Clark to Patrick Henry, *Clark Papers,* p. 30).

10 *Ibid*, p. 114. Thomas Jefferson, George Mason and George Wyethe served as advisors to Governor Henry in ordering a secret expedition to capture Kaskaskia.

11. *Ibid*, pp. 119-121 (citing, "Mason Letter," *Clark Papers,* p. 121).

12. The Liberty Bell is still on exhibit at Kaskaskia Bell State Historic Site in Kaskaskia, Illinois. Due to the shifting of the Mississippi River, the old town of Kaskaskia is located on the west bank of the river, surrounded by the state of Missouri.

13. James, *The Life of George Rogers Clark,* pp. 124-126. In January, 1792 Pollock made a claim for reimbursement from the state of Virginia for $139,739 , which he paid to support Clark and his Virginia militia; altogether he advanced $300,000 see p. 293.

14. *Ibid,* pp. 128-129 (citing "Memoir," Clark Papers, p. 244).

15. John Bakeless, *Background to Glory: The Life of George Rogers Clark* (Philadelphia and New York, J. B. Lippincott Co., 1957), pp. 123-124.

16. Temple Bodley, *George Rogers Clark: His Life and Public Services* (Boston and New York, Houghton Miflin Co, 1926), p. 107. Bodley clears up the confusion among historians regarding how many men made the march with Clark. Though Bowman's journal (p. 111) reported 170 men, he was counting the artillerists on board the boat.

17. Robert J. Holden, "Clark's Attack on Vincennes, February 5-25, 1779" in *The Life of George Rogers Clark, 1752-1816: Triumphs and Tragedies*, edited by Kenneth C. Carstens and Nancy Son Carstens (Westport CT, Praeger 2004), p. 93.

18. Holden, "Clark's Attack on Vincennes," pp. 96-97.

19. Bakeless, *Background to Glory*, pp. 200-204.

20. Bakeless, *Background to Glory*, pp. 213-218.

3. Growing Up in Albemarle

21. George Gilmer, *Sketches of Some of the First Settlers of Upper Georgia* (Baltimore, Genealogical Publishing Co.,1965 Indexed Edition, first published, 1855), p. 112-113. Dr. George Gilmer was in charge of the First Independent Company of Albemarle which William Lewis served in.

22. *Ibid*, pp. 112-13,

23. All information comes from the website of First Virginia re-enactors, based in Alexandria. (www.1va.org, accessed 12/28/13)

24. Gilmer, *Sketches*, p. 82.

25. *Ibid*, p. 70.

26. *Ibid*, p. 74.

27. Edgar Woods, *History of Albemarle County* (Albemarle, Michie Co., 1901: Google Books), pp. 31-33.

28. Peter McGhee "Meriwether Lewis" transcript, University of Virginia Special Collections, MSS 4730-a., p. 2.

29. http://valleyforgemusterroll.org

30. Zontine, "Lucy Meriwether Lewis Marks," Monticello website.

31. Linda Cooper and Alice Roker with the Town of Yorktown, *Images of America: Yorktown* (Charleston SC, Chicago IL, Portsmouth NH, San Francisco CA, Arcadia Publishing, 2003) pp. 11-12, (citing *Yorktown at War* by John Martino).

32. Woods, *History of Albemarle County*, pp. 37-38.

33. Temple Bodley: *George Rogers Clark: His Life and Public Services* (Houghton MifflinCompany, 1926) pp. 330-333.

34. Jean L. Cooper, *A Guide to Historic Charlottesville & Albemarle County, Virginia* (Charleston, History Press, 2007), p. 58. Citing *The Life of Thomas Jefferson* by B. L. Rayner, 1834, pp. 172-173. (http.//archive.org, accessed 1/7/13).

35. Virginus Dabney, "Captain Jack Jouett, Junior: American Revolutionary War Hero (1754-1822)," Jack Jouett Chapter, Daughters of the American Revolution reprint, 1966. Courtesy of the Albemarle Charlottesville Historical Society.

36. Dabney, "Captain Jack Jouett," p. 1.

37. Steven Meriwether Long, "British Lieutenant Colonel Banastre Tarleton and the American Revolution: Drama on the Plantations of Charlottesville," *Meriwether Connections,* vol 24, no. 2, Apr-June, 2005. p. 5. See http://homepages.rootsweb.ancestry.com/~meriweth/ (accessed 1/7/13).

38. *Ibid,* p. 6.

39. *Ibid,* p. 7.

40. *Ibid,* p. 5.

41. Woods, *History of Albemarle County,* pp. 46-47.

42. Zontine, "Lucy Meriwether Lewis Marks," Monticello website.

43. Thomas Jefferson, "Biographical Sketch of Lewis" circa August 18, 1813, from *Letters of the Lewis and Clark Expedition with Related Documents (1783-1854)* edited by Donald Jackson, (Urbana and Chicago, IL, University of Illinois Press, 2nd edition, 1978), vol. 2, p. 593.

44. Woods, *History of Albemarle County,* pp. 263, 379.

45. Jefferson, "Biographical Sketch," p. 593.

46. Bakeless, *Background to Glory,* p. 18.

47. McGhee, "Meriwether Lewis," pp. 2-3.

48. Wikipedia, "James Waddell (1739-1805)."

49. Bakeless, *Background to Glory,* pp. 19-21.

50. Jefferson, "Biographical Sketch," p. 593.

4. The Whiskey Rebellion

51. Leland D. Baldwin, *Whiskey Rebels: The Story of a Frontier Uprising* (Pittsburgh, University of Pittsburgh Press, 1939), p. 70.

52. *Ibid*, pp. 107-108. Citing Coxe in the *Pittsburgh Gazette,* September 20, 1794.

53. *Ibid,* p. 69.

54. Thomas P. Slaughter, *The Whiskey Rebellion: Frontier Epilogue to the American Revolution* (New York and Oxford, Oxford University Press, 1986), pp. 3, 5.

55. Baldwin, *Whiskey Rebels,* pp. 110-112.

56. Slaughter, *The Whiskey Rebellion,* pp. 223-24.

57. Baldwin, *Whiskey Rebels,* p. 255.

58. Meriwether Lewis to Lucy Marks, Nov. 24, 1794, Grace Lewis Miller Papers,

Meriwether Lewis letters, 1790's (St. Louis, Jefferson National Expansion Memorial). The originals are in the Lewis Archives at Missouri History Museum.

59. Lewis to Marks, December 7, 1794.

60. Captain Thomas Walker (1774-94) probably was Thomas Hudson Walker from Jump Creek Mountain, Rockbridge County, Virginia, about 70 miles west of Charlottesville. The location is near Waynesboro on I-64, the route of Three Notched road. Since Lewis calls him an "old friend," they must have been schoolmates. A search of a genealogy website for Thomas Walker, born 1774, turned up this name. Genealogy records for Dr. Thomas Walker of Albemarle reveal no connection to a 20 year old, although there are three generations of Thomas Walkers. Dr. Walker became the guardian of Thomas Jefferson after the death of his father, Peter Jefferson. Walker, a partner of Jefferson's in the Loyal Land Company, explored Kentucky and named the Cumberland Gap in 1750. He was related to Meriwether Lewis through his marriage to Lewis's great aunt, Mildred Thorton, the widow of Nicholas Meriwether, by which he came into possession of Castle Hill.

61. Lewis to Marks, December 24, 1794.

62. Lewis to Marks, April 6, 1795.

63. Lewis to Marks, May 22, 1795.

64. Lewis to Walker, May 23, 1795. See also Ruth Colter Frick's, *Courageous Colter and Companions* (Washington MO, L. R. Colter-Frick, 1997), pp. 304-305. Frick created the finding aid to the Grace Lewis Miller Papers, which is available on the internet. Her own book is a treasure trove of documents and research.

5. James Wilkinson

65. James Wilkinson, *Memoirs of My Own Times* (Philadelphia, 1816, reprint, three volumes), Volume 1, p. 324. Letter of Horatio Gates to John Hancock, October 18, 1777.

66. Jacobs, *Tarnished Warrior*, p. 43-45; Wilkinson, *Memoirs,* I: 330-31.

67. Wilkinson, *Memoirs*, I: 386-87.

68. William McWilliams was treasurer of George Washington's Masonic Lodge #4, Fredericksburg, Virginia See: *The Craftsman, and Freemason's Guide: Containing a Delineation of the Rituals of Freemasonry,* 1866; Google Books and Jacobs, *Tarnished Warrior,* pp. 44, 47, 51.

69. Harlow Giles Unger, *The French War Against America: How a Trusted Ally Betrayed Washington and the Founding Fathers* (Hoboken NY, John Wiley & Sons, Inc., 2005), pp. 76-77, 83-86, 88.

70. Jacobs, *Tarnished Warrior,* p. 59-61.

71. Wilkinson, *Memoirs*, I: 410. Jacobs, *Tarnished Warrior*, pp. 62-69

72. Jacobs, *Tarnished Warrior,* p. 72.

73. James, *The Life of George Rogers Clark,* pp. 268-275.

74. *Ibid,* pp. 305-606, citing Jefferson to Clark, March 4, 1784.

75. *Ibid,* pp. 298-300.

76. *Ibid*, p. 325, citing a report to the Secretary of War in 1790.

77. Bodley, *George Rogers Clark*, pp. 276-286.

78. James, *The Life of George Rogers Clark*, pp. 356-357.

79. Jay Gitlin, *Bourgeois Frontier: French Towns, French Traders & American Expansion* (New Haven & London, Yale University Press, 2010), p. 43; R. Douglas Hurt, *The Ohio Frontier: Crucible of the Old Northwest, 1720-1830* (Bloomington & Indianapolis, Indiana University Press, 1996), pp. 207-208. Bodley, *George Rogers Clark,* p. 220, quoting Clark.

80. Bodley, *George Rogers Clark,* pp. 287-299.

81. *Ibid*, pp. 317-323; Appendix, pp. 379-404.

82. Arthur Preston Whitaker, "James Wilkinson's First Descent to New Orleans in 1787," *The Hispanic American Historical Review,* vol 8: I (February, 1926) pp. 86, 90-95.

83. Bodley, *George Rogers Clark*, pp 317-323.

84. Alcée Fortier, *A History of Louisiana, The Spanish Domination and the Cession to the United States, 1769-1803* (New York, Manzi, Joyant & Co., Successors to Goupil & Co of Paris, 1904; Google Books) V 2, p. 136

85. Wilkinson, *Memoirs,* vol. 2, pp. 114-115.

86. William R. Shepherd, "Wilkinson and the Beginnings of the Spanish Conspiracy," *The American Historical Review,* vol. 9: 3 (April, 1904), pp. 502-503.

87. Fortier, *A History of Louisiana,* vol. 2, pp. 141-42. This quote (partially quoted) is at the front of Jacobs' *Tarnished Warrior.*

88. John Mason Brown, *The Political Beginnings of Kentucky* (Google eBook, 1890) pp. 98-99.

89. Whitaker, "James Wilkinson's First Descent to New Orleans in 1787," p. 90. The letter from Clark to Gardoquui was written at the Falls of the Ohio, March 15, 1788. Bodley, *George Rogers Clark*, pp. 329-333.

90. Jacobs, *Tarnished Warrior,* pp. 89-90, citing Wilkinson's letter to Miro, February 12, 1789.

91. Jacobs, *Tarnished Warrior,* pp. 100-101.

92. Daniel Clark, *Proofs of the Corruption of Gen. James Wilkinson, and of His Connexion with Aaron Burr: A Full Refutation of His Slanderous Allegations in Relation to the Character of the Principal Witness Against Him* (Honolulu, University Press of the Pacific, 2005/1809); Google Books, p. 21, Note #5, deposition of Joseph Ballinger, January 12, 1809.

93. Thomas Jefferson letter to Archibald Stuart, January 25, 1786. (*Thomas Jefferson Papers*, Library of Congress, American Memory website).

94. Richard M. Lytle, *The Soldiers of America's First Army, 1791* (Lanham MD, Toronto, Oxford, The Scarecrow Press, 2004), p. 84.

95. Jacobs, *Tarnished Warrior,* p. 115.

96. Thomas Robson Hay and M. R. Werner, *The Admirable Trumpeteer: A Biography of General James Wilkinson* (Garden City NY, Doubleday, Doran & Company, 1941), p. 109.

97. Hay and Werner, *The Admirable Trumpeteer*, p. 197. See also: John Thomas Scharf's *History of St. Louis City, From the Earliest Periods to the Present Day* (1883, Google eBook), vol. 1, p. 241.

6. The Legion of the United States

98. "Narrative of John Heckewelder's Journey to the Wabash in 1792," *Pennsylvania Magazine of History and Biography,* vol. 12, 1888, p. 42.

99. "Narrative of John Heckewelder," p. 169.

100. Charles Theodore Greve, *Centennial History of Cincinnati and Representative Citizens* (Google eBook, vol. 1, 1904), p. 252.

101. Humphrey Marshall, *The History of Kentucky Exhibiting an Account of the Modern Discovery; Settlement; Progressive Improvement; Civil and Military Transactions; and the Present State of the Country*. (Google eBook, 1812), vol. 2, p. 41.

102. Royal Ornan Shreve, *The Finished Scoundrel: General James Wilkinson Commander-in-Chief of the Army of the United States, who made intrigue a trade and treason a profession* (Indianapolis, Bobbs-Merrill Company, 1933), pp. 34-35; Wilkinson's *Memoirs,* vol. 1, pp. 233-235.

103. Humphrey Marshall, *The History of Kentucky* (Google eBook, 1824), vol. 2, pp. 41-43.

104. A. C. Quisenberry, *The Life and Times of Hon. Humphrey Marshall* (Google eBook, 1892), pp. 31-35. See also: p. 66. Green says that a protege of Wilkinson was paid by Wilkinson to write letters under the name of Jordan Harris, who tried to assassinate Marshall. The letters attacking Marshall were published in the *Lexington Gazette.*

105. James Ripley Jacobs, *The Beginning of the U. S. Army, 1783–1812,* Princeton, NJ, Princeton University Press, 1947), pp. 171. See also: *The Forts of Anthony Wayne* by David A. Simmons (Historic Fort Wayne pamphlet, 1977).

106. Alan D. Graf, *Bayonets in the Wilderness: Anthony Wayne's Legion in the Old Northwest* (Norman, University of Oklahoma Press, 2004), pp. 184-88.

107. *Ibid*, p. 203.

108. *Ibid*, p. 235-236.

109. *Ibid,* p. 238-239.

110. *Ibid,* p. 242.

111. Richard C. Knopf, *Anthony Wayne: A Name in Arms* (University of Pittsburgh Press, 1960), Wayne's letter of July 7, 1794 to Secretary of War Henry Knox.

112. Graf, *Bayonets in the Wilderness*, p. 247.

113. Landon B. Jones, *William Clark and the Shaping of the West* (New York, Hill and Wang, A Division of Farrar, Straus and Giroux, 2004) pp. 70-71.

114. Knopf, *Anthony Wayne: A Name in Arms*, pp. 383-384, letter dated January 29, 1795.

115. William Clark, "Journal of General Anthony Wayne's Campaign Against the Shawnee Indians in Ohio, 1794-1795," edited by R. C. McGrane (*Mississippi Valley Historical Review,* vol. 1, Dec. 1914; Google eBook)

116. Dwight L. Smith, editor, *From Greene Ville to Fallen Timbers: A Journal of the Wayne Campaign: July 28–September 14, 1794* (Indianapolis, Indiana Historical Society, 1952), p. 262.

117. Paul David Nelson, *Anthony Wayne: Soldier of the Early Republic* (Bloomington, Indiana University Press, 1985), p. 261.

118. Graf, *Bayonets in the Wilderness*, p. 274.

119. Nelson, *Anthony Wayne,* p. 262.

120. Timothy M. Rusche, "Treachery Within The United States Army," *Pennsylvania History*, vol. 65, Autumn, 1998, pp. 478-491. See p. 481 in which Rusche concludes: "The available evidence suggests that Newman was probably employed by Wayne."

121. Graf, *Bayonets in the Wilderness,* p. 279-280. See also John F. Winkler's, *Fallen Timbers 1794: The US Army's First Victory* (Oxford and Long Island City NY, Osprey Publishing, 2013), p. 56.

122. Clark, "Journal," p. 423 (August 6, 1794).

123. Graf. *Bayonets in the Wilderness*, p. 288.

124. *Ibid*, p. 290. 60 or more Chickasaw and Chocktaw warriors must have joined the original group which Clark brought up from Chickasaw Bluffs.

125. Simmons, *Forts of Anthony Wayne*, pp. 18-19.

126. Graf, *Bayonets in the Wilderness*, p. 297.

127. *Ibid,* p. 312. The battlefield site and the site of Fort Miamis are located in Maumee, Ohio, on the southern edge of Toledo, Ohio.

128. Jacobs, *The Beginnings of the U. S. Army*, pp. 174-175.

129. Graf, *Bayonets in the Wilderness,* pp. 317, 327.

130. *Ibid*, pp. 299-300.

131. *Ibid,* p. 320.

132. *Ibid,* p. 324-327.

133. Jacobs, *Tarnished Warrior,* p. 143.

134. Jacobs, *Tarnished Warrior,* p. 143-145.

135. Rusche, "Treachery Within the United States Army," See pages 480-484. The Simcoe letter to R. G. England was written August 19, 1794 (*Correspondence of Simcoe,* II), p. 392. This analysis contradicts that of some historians (Graf and Nelson), but is supported by Richard Knopf. The evidence in favor of this view is Simcoe's letter and the testimony of the man who received payment of his debt from Newman.

136. Knopf, *Anthony Wayne: A Name in Arms*, p. 383.

137. Harry Ammon, *The Genet Mission* (New York, Norton & Co, 1973), p. 82. See also, Frederick Jackson Turner's *The Policy of France Toward the Mississippi Valley in the Period of Washington and Adams* (Google eBook, 1905), p. 969, letter from Clark to Genet, dated Louisville, Febry 5, 1793 in the Draper Collection. An earlier letter from Clark had been received in France before Genet departed. The existence of the letter is noted in French records, but the letter has not been found.

138 *Ibid,* p. 44. Privateers were legalized pirate ships, serving under government license during times of war. The proceeds from the sale of captured ships and cargo were split between investors, officers and crew, and the government which sponsored them. See also: James, *George Rogers Clark*, p. 419.

139. Jackson, *Letters of the Lewis and Clark Expedition*, vol. 2, p. 589. Jefferson to Paul Allen, August 18, 1813.

140. Andre Michaux, *Journal of Andre Michaux, 1793-1796* (Nabu Public Domain Reprints), pp. 41-42.

141. Graf, *Bayonets in the Wilderness*, pp. 212-213 has the best account. See also Nancy Son Carstens, "George Rogers Clark and the French Conspiracy, 1793–1801," *The Life of George Rogers Clark, 1752-–1816: Triumph and Tragedies* edited by Kenneth C. Carstens and Nancy Son Carstens (Westport, CT and London, Praeger, 2004), pp. 230-256. See also, James, *The Life of George Rogers Clark*, pp. 417-431.

142. Graf, *Bayonets in the Wilderness*, p. 215-216.

143. James, *The Life of George Rogers Clark*, p. 427.

144. James, *The Life of George Rogers Clark*, pp. 417, 418, 425.

145. This is my own speculation. For a discussion of the Genêt mission, Jefferson's role in it, and Washington's cabinet, see François Furstenberg, *When the United States Spoke French: Five Refuges Who Shaped a Nation* (New York, The Penguin Press, 2014), pp. 301-317.

146. Jacobs, *Tarnished Warrior*, pp. 135-138. See also: Andro Linklater, *Artist in Treason: The Extraordinary Double Life of General James Wilkinson* (New York,

Walker Publishing Company, 2009), pp. 144-146. See also: Abernethy, *The South in the New Nation 1789–1819* (Baton Rouge, Louisiana State University, 1961, 1989) pp. 204-206 Wayne's reference to tobacco is in a private letter to McHenry, dated October 28, 1796; see Knopf, Anthony Wayne, pp. 536-538.

147. Jacobs, *Tarnished Warrior*, pp. 148, 152.

148. Graf, *Bayonets in the Wilderness*, pp. 365-366. See also: Landon Y. Jones, *William Clark and the Shaping of the West* (New York, Hill and Wang, Division of Farrar, Straus and Giroux, 2004), p. 85.

149, Dick Steward, *Duels and the Roots of Violence in Missouri* (Columbia and London, University of Missouri Press, 2000), p. 6.

150. Thomas Danisi, *Uncovering the Truth About Meriwether Lewis* (Amherst, NY, Prometheus Books, 2012), pp. 251-269 for transcript of court martial. See also Grace Lewis Miller Papers (St. Louis, Jefferson National Expansion Memorial, 1999) Series 1:1 Lewis Papers, Anthony Wayne document, dated Nov. 17, 1795.

151. Knopf, *Anthony Wayne,* letter from Henry Knox, Secretary of War to General Wayne, dated March 31, 1794. pp. 313-320. (Excerpt pp. 317-18.)

> The late intention of some restless people of the frontier settlements to make hostile inroads into the dominions of Spain, renders it indispensible that you should immediately order as respectable a detachment as you can to take post at Fort Massac and to erect a strong redoubt and block house with some suitable cannon from Fort Washington.

> Wayne's secret instructions may be read in the letter from Henry Knox. The quotation from Baron de Carondelet is found in Frederick Jackson Turner's "Correspondence of Clark and Genet: Selections from the Draper Manuscripts" in the *Annual Report of the American Historical Association, 1896.* (https://archive.org, Internet Archive). Carondelet's secret letter to Alcudia is #67 in the collection found on pp. 1089-1091.

152. James Holmberg, editor, *Dear Brother: Letters of William Clark to Jonathan Clark* (New Haven & London, Yale University Press, and the Filson Historical Society, 2002), letter from William Clark to Fanny Clark O'Fallon, May 9, 1795. See also Knopf, *Anthony Wayne*, pp. 486-487, letter dated June 24, 1796 at Pittsburgh, stating Wayne will arrive at Fort Washington in ten days.

153. Jackson, *Letters of the Lewis and Clark Expedition*, Clark to Nicholas Biddle dated August 15, 1811.(#345).

154. Jacobs, *Beginning of the U. S. Army*, p. 183.

155. Knopf, *Anthony Wayne*, pp. 481-484, 489-491.

156. *Ibid,* pp. 495-497. Some months later—in what may have been another assassination engineered by Wilkinson—Collot's map maker, Joseph Warin, was attacked by two Chickasaw Indians who followed them from Illinois country. General Collot believed the attempted assassination was meant for him and was a case of mistaken identity. Warin died of his injuries, and Collot was arrested and imprisoned in New Orleans by Carondelet. See "General Collot's Reconnoitering Trip Down

the Mississippi and his Arrest in New Orleans by order of the Baron de Carondelet, Governor of Louisiana," *Louisiana Historical Quarterly,* vol. 1, no. 4, (1918) 303-329; or Google eBook).

157. Knopf, *Anthony Wayne,* p. 498.

158. Rusche, "Treachery Within the U. S. Army," p. 483.

159. Knopf, *Anthony Wayne,* letter of October 28, 1796, Wayne to McHenry, pp. 536-538.

160. Daniel Clark, *Proofs of the Corruption of Gen. James Wilkinson, and of His Connexion with Aaron Burr: A Full Refutation of His Slanderous Allegations in Relation to the Character of the Principal Witness Against Him* (Honolulu, University Press of the Pacific, 2005, reprint from 1809; see also Google eBook), pp. 33-49.

161. Richard H. Kohn, *Eagle and Sword: The Beginnings of the Military Establishment in America* (New York & London, The Free Press, A Division of MacMillan Publishing Co., 1975), pp. 184-185.

162. *Ibid,* p. 186.

163. *Ibid,* p. 187.

164. Knopf, *Anthony Wayne: A Name in Letters,* pp. 525-529.

165. *Ibid,* pp. 536-538.

166. *Ibid,* pp. 543-546. See also *Papers of the War Department,* Nov.21, 1796 the issuance of a government horse, pistol and holster to Lewis at Pittsburgh, to be returned at Philadelphia. (wardeptpapers.org). "Letters Relating to the Death of Maj.-Gen. Anthony Wayne, *The Pennsylvania Magazine of History and Biography,* 19:1 (1895) pp. 112-115.

167. Grace Lewis Miller Papers, Jefferson National Expansion Memorial, "Papers of Meriwether Lewis," Sub-Series 1, Record Unit 116: Box 2, Folder 2. See also "Detroit, 1796," Record Unit 116; Box 37, Folder 7.

168. "Letters Relating to the Death of Maj.-Gen. Anthony Wayne," pp. 112-115.

169. Letter, George Balfour to Samuel Hodgdon, Dec. 14, 1796, RG 94, Collection National Archives and Records Adminstration, Papers of the War Department. (wardepartmentpapers.org).

170. Letter, Thomas Truxton to James McHenry, November 6, 1797, Papers of the War Department (wardepartmentpapers.org).

171. *History of Erie County,* Pennsylvania, vol. 1 (Google eBook, 1884), pp,.210-212.

172. R. A. Witthaus and Tracy Becker, *Medical Jurisprudence Forensic Medicine and Toxicology,* vol. 4 (Google eBook, 1896), pp. 436-441.

7. Mississippi Valley Conspiracies: 1797-98

173. Jacobs, *Tarnished Warrior*, p. 161.

174. Alexander De Conde, *The Quasi-War: The Politics and Diplomacy of the Undeclared War with France, 1797-1801* (New York, Charles Scribner's Sons, 1966), pp. 8-11.

175. Arthur Preston Whitaker, *The Mississippi Question, 1795–1803: A Study in Trade, Politics and Diplomacy* (New York and London, D. Appleton–Century Company and the American Historical Association, 1934), p. 180.

176. *Ibid*, pp. 102-103.

177. Jacobs, *Tarnished Warrior*, pp. 162-166 and p. 152 (finances).

178. Isaac Joslin Cox, editor, "Documents Relating to Zachariah Cox," *Quarterly Publication of the Historical and Philosophical Society of Ohio*, vol. VIII, 1913, Nos. 2 &3, April-June, July-September, pp. 39-40. See also: Abernethy, *The South in the New Nation*, pp. 155-160; Robert V. Haynes, *The Mississippi Territory and the Southwest Frontier, 1795–1817,* (Lexington, University Press of Kentucky, 2010), pp. 32-33.

179. Abernethy, *The South in the New Nation*, pp. 159-160.

180. Cox, "Documents Relating to Zachariah Cox," p. 56, citing a letter from McHenry to President Washington, dated Nov. 6, 1799.

181. Jack D. Holmes, *Gayoso: The Life of a Spanish Governor in the Mississippi Valley, 1789–1799* (Baton Rouge, Lousiana State University Press, 1965), pp. 257-258.

182. Clarence Burton, editor-in-chief, *The City of Detroit, Michigan, 1701–1922* vol. 1, (Google eBook, 1922) pp. 247-248.

183. Mary-Jo Kline, *Political Correspondence and Public Papers of Aaron Burr*, 2 vol (Princeton University Press), vol 1, pp. 257, 268-269.

184. Frederick Jackson Turner, "Documents on the Blount Conspiracy," *American Historical Review,* vol. X, No. 3, April, 1905, pp. 574, 582-583.

185. Abernethy, *The South in the New Nation,* p. 176.

186. *Ibid,* p. 177.

187. Quotes from Turner, "Documents Relating to the Blount Conspiracy" pp. 600-601. See also: *Annals of the Fifth Congress,* II, 2356-2365.

188. Bucker F. Melton, Jr. *The First Impeachment: The Constitution's Framers and the Case of Senator William Blount* (Macon, GA, Mercer University Press, 1998). pp. 99-103.

189. Allan Potopsky, "The 'Non-Aligned Status' of French Emigres and Refugees in Philadelphia, 1793-1798," *Transatlantica, American Studies Journal,* vol. 2, 2006 (www.transatlantica.org). Potopsky uses the official 1790 census for his Philadelphia population. Wikipedia says there were about 50,000 in residence in 1793. The Philadelphia Encyclopedia on the internet says the 1790 census reported 44,096

residents in Philadelphia and the adjacent communities. For the French refugees in Philadelphia see Furstenberg's *When the United States Spoke French.*

190. Stanley Elkins and Eric McKitrick, *The Age of Federalism: The Early American Republic, 1788-1800* (New York &Oxford, Oxford Univ. Press, 1993), p. 365.

191. "Yellow Fever Epidemic of 1793" (en.wikipedia.org).

192. Frances Sergeant Childs, *French Refugee Life in the United States, 1790–1800: An American Chapter of the French Revolution* (Baltimore, Johns Hopkins Press, 1940, Google Book reprint), pp. 65, 85.

193. John L Earl III, "Talleyrand in Philadelphia, 1794–1796," *Pennsylvania Magazine of History and Biography,* 91:3 (July, 1967), pp. 294-296.

194. Ron Chernow, *Alexander Hamilton* (NY, The Penguin Press, 2004), p. 466.

195. *Ibid,* p. 549.

196. "Talleyrand and Jaudenes, 1795," *The American Historical Review,* 30:4 (July, 1925), pp. 778-787.

197. Don Hickey, "America's Response to the Slave Revolt in Haiti, 1791–1806," *Journal of the Early Republic,* vol. 2, no. 4 (Winter, 1982), pp. 361-379, see 363-366,

198. *Ibid.* p. 371.

199. Andrew Ellicott, *The Journal of Andrew Ellicott* (Google eBook, 1803), pp. 43-44.

200. Catharine Van Cortlandt Matthews, *Andrew Ellicott: His Life and Letters* (New York, The Grafton Press, 1908; Google eBook), pp. 161-163.Elicott named General Wilkinson, U. S. Senator John Brown, Judge Sebastian, Mr. Lackasang, Mr. Murry, as being in the pay of Spain, and several others as receiving occasional payments. The translated cipher code is in the Pickering Papers, Massachusetts Historical Society, 21:341.

201. Ellicott, *Journal,* p. 160. Letter to his wife dated November 8, 1798. See also Jacobs, *Tarnished Warrior*, pp. 180-181.

202. Dan L. Flores, *Southern Counterpart to Lewis and Clark: The Freeman and Custis Expedition of 1806* (Norman, Univ. of Oklahoma Press, 1984), pp. 51, 99.

203. In 1799, Fort Adams was built on the river, near the site of the old Spanish Fort Ferdinand (Fernando de las Barrancas). It was the third fort named for President Adams. Floods and malaria caused the fort to be rebuilt on the bluff at the location of the old Spanish fort in 1801, when it was renamed Fort Pickering. Fort Pickering was in existence until about 1812. and served as the Chickasaw Indian Agency from 1814-1818. It was later rebuilt as much larger Civil War Fort. (www.historic-memphis.com, www.northamericanforts.com, www.fortwiki.com)

204. Landon B. Jones, *William Clark and the Shaping of the West* (New York, Hill and Wang, Division of Farrar, Straus and Giroux, 2004). p. 98-106. Jones notes that the "Captain Lewis" that Clark met at Chickasaw Bluffs was not Meriwether Lewis, but rather an army captain named Thomas Lewis. There has been some confusion

regarding this report of Lewis at Chickasaw Bluffs. There is no evidence that Meriwether Lewis visited Fort Pickering prior to 1809. War Department Papers confirm the presence of Captain Thomas Lewis in the area.

205. The original of these drawings are in the Missouri History Museum William Clark Papers, Box 11, Folder 6.

8. Conflicts and Secrets: 1798–1801

206. Richard Dillon, *Meriwether Lewis: A Biography* (Lafayette CA, Great Western Books, 2003), p. 22. See also Widow's Sons' Masonic Lodge No. 60 http://wsl.avenue.org) The lodge was called Widow's Son in the singular until February 14, 1801 when it changed its name to Widows Sons. See www.lewis-clark.org for more information about his masonic connections, and to view his masonic apron. *The Freemason's Monthly Magazine,* vol. 31, no. 9, July 1, 1872 (available on the internet) says he attended the Door to Virtue Lodge in Charlottesville in June, 1798, joined the lodge in 1799, and attended meetings regularly and held office in the lodge until March, 1799. The Door to Virtue Lodge closed in 1801 and its members transferred to the Widow's Son Lodge. See pp. 265-66. This account of his degrees agrees with Dillon's account and Joseph Musselman's account on lewis-clark.org.

207. Dillon, *Meriwether Lewis*, p. 24.

208. Miller, Grace Lewis, Miller Archives, JNEM Archives: Record Unit #16, NPS Catalog # 9359, Box 37, Folder 11, "Recruiting II: C. O. Charlottesville; Adjutant, Staunton." See also War Department Papers on the internet for letter dated May 20, 1799 by William Simmons, accountant, authorizing the pay of $1403.50 to Lewis and his detachment of recruits at Charlottesville, for the period of January 1–April 30, 1799 and letter of June 19, 1799 authorizing payment of $188 for the month of May at the camp near Staunton.

NOTE: Grace Lewis Miller developed a theory based on the financial records of the War Department that Meriwether Lewis was secretly buying 1 500 guns for Jefferson. This was based on the location of two payments to gun manufacturers on the same page as the end of 3 1/2 pages of Meriwether Lewis's account. The date of the entries is March 11, 1802. I went to the National Archives in Washington DC in August, 2013 to locate the original volume. It was found in Record Group 217, Entry 366, *Second Auditors Journals of the Accountants for the War Dept*, Volume No. 8 ("H") pp. 4302-4305. Subsequent research, based on internet search of War Department Papers and American State Papers established it was payment for guns ordered directly by the federal government.

Ms. Miller devoted many years to research in the 1950's, when it was extremely difficult and expensive to do research. Her extensive archives are well worth investigating. Because she located these accountant records, it is possible to know much more about Lewis's army service. It is on these records that the dates of Lewis's service and thirty nine recruits who received bounty payments is based.

209. Malone, *Jefferson and the Ordeal of Liberty* (Boston, Little Brown and Company, 1962), p. 380.

210. *Thomas Jefferson Papers*, American Memory, Library of Congress, letter to Ar-

chibald Stuart, February 13, 1799. (www.memory.loc.gov)

211. Adrienne Koch, *Jefferson and Madison: The Great Collaboration* (Old Saybrooke, CT, Konecky & Konecky, no date), pp. 186-187.

212. Richard Beale Davis, editor, *Jeffersonian America: Notes on the United States of America Collected in the Years 1805-6-7 and 11-12 by Sir Augustus John Foster, Bart.* (Westport CT, Greenwood Press, 1954), p. 143.

213. *Second Auditors Journals for the Accountants of the War Dept Journals*, Volume 8 ("H"), p. 4302.

214. Adrienne Koch and Harry Ammons, "The Virginia and Kentucky Resolutions: An Episode in Jefferson's and Madison's Defense of Civil Liberties," *The William and Mary Quarterly*, Third Series, Vol. 5, No. 2 (April, 1948) p. 164.

215. James Iredell, *Life and Correspondence of James Iredell: One of the Associate Justices of the Supreme Court* (Google ebooks, 1858), Vol. 2, p. 577.

216. Philip G. Davidson, "Virginia and the Alien and Sedition Laws," *The American Historical Review*, Vol. 36, No. 2 (Jan, 1931), pp. 336-342. See pp. 339-340. The bill authorizing the purchase of 20,000 guns, and the building of two arsenals and an armory was passed on January 23, 1798 by the Virginia Legislature.

217. "Clark and the French Conspiracy, 1793-1801," by Nancy Son Carstens in *The Life of George Rogers Clark, 1752--1816: Triumph and Tragedies*, edited by Kenneth C. Carstens and Nancy Son Carstens (Westport, CT and London, Praeger, 2004), pp. 243-244; citing Thomas, "Unpublished Papers of George Rogers Clark" (Historic Locust Grove), letter from Clark in Philadelphia to Samuel Fulton in Paris, June 3, 1798.

218. *Ibid*, p. 244, citing June 3, 1798 letter.

219. James, *The Life of George Rogers Clark*, pp. 445, 449, 457.

220. See note 205.

221. Jones, *William Clark and the Shaping of the West*, p. 97.

222. *Ibid*, p. 106.

223. *Alexander Hamilton, Portrait in Paradox* by John C. Miller (Konecky & Koneky, by special arrangement with HarperCollins Publishers, 1959), p.452.

224. *Ibid*, p. 468.

225. Kohn, *Eagle and Sword*, p. 229.

226. Miller, *Alexander Hamilton*, p. 458.

227. Kohn, *Eagle and Sword*, p. 216-217.

228. *Ibid*, p. 230. January 7, 1797, Adams Papers, Massachusetts Historical Society.

229. Chernow, *Alexander Hamilton*, pp. 566-567.

230. *Ibid*, pp. 567-568. Chernow cites *The Papers of Alexander Hamilton*, Vol. 22, letter to Harrison Gray Otis, January 26, 1799.

231. *Ibid*, pp. 577. Chernow cites the *Papers of Alexander Hamilton,* volume 22, letter to Harrison Gray Otis, December 27, 1798.

232. David McCullough, *John Adams* (NY, Simon & Schuster, 2001), p. 522-523.

233. *Ibid*, p. 531.

234. Miller, *Alexander Hamilton,* p. 503-504, citing *John Adams, Works,* X, p. 113.

235. James Wilkinson, *Memoirs of My Own Times* (1816) Volume 1, pp. 435-458. See p. 439 for comment about Burr. I am indebted to David P. Mayer, whose unpublished paper, "Cantonment Wilkinsonville 1801: Burr, Wilkinson, the U. S. Army and Treason," first cited this reference. The paper, dated March 7, 1985, Erie, Pennsylvania is available at the Cairo Public Library, Cairo, Illinois. Mayer also noted that the *Memoirs* contain no reference to Wilkinsonville.

236. Jacobs, *Tarnished Warrior*, pp. 190-195. See also, De Conde, *The Quasi-War,* p. 122. See also, Norman Caldwell, "Cantonment Wilkinsonville," *Mid-America: An Historical Review*, 31:1, Jan, 1949, pp. 7-13.

237. James Roger Sharp, *American Politics in the Early Republic*: *The New Nation in Crisis* (New Haven, Yale University Press, 1993), pp. 250-252. Despite the destruction of records in the War Department fire, some 55,000 documents have been located in other collections and repositories. They are available online at www.wardepartmentpapers.org. The Roy Rosenzweig Center for History and New Media at George Mason Univerity is managing the project.

238. Caldwell, "Cantonment Wilkinsonville," pp. 3-4 and 13.

239. James, *Tarnished Warrior,* pp. 193-194.

240. Reuben Gold Thwaites, *On the Storied Ohio* (Chicago, A. C. McClurg and Co., 1903), p. 291. Thwaites made a lesiurely river trip in 1894 in a small boat with three companions, and reached the site of "Fort Wilkinson,"which he dates to the War of 1812-15, at 942 miles, in about five weeks time.

241. Mark Wagner, "Officer Country: Clothing and Personal Items from Cantonment Wilkinson (11PU282) in Southern Illinois" *Midcontinental Journal of Archeology*, 37:2, Fall, 2012, pp. 178-180. See also Thomas Rodney, *A Journey Through the West: Thomas Rodney's 1803 Journal from Delaware to the Mississippi Territory*, edited by Dwight L. Smith and Ray Swick (Ohio University Press, 1997), pp. 165-168. Rodney visited Wilkinsonville on November 7-8, 1803 and found it to be no longer a military post, but instead was inhabited by about 200 Cherokee Indians and other squatters. See also, Zebulon Montgomery Pike, *The Expeditions of Zebulon Montgomery Pike*, edited by Elliott Coues (1895), letter of Jonathan R. Williams, August 26, 1846, pp. xxiii-xxx, regarding establishing the cantonment. This Williams is not to be confused with Wilkinson's friend, Jonathan B. Williams, the first Superintendent of West Point. Jonathan R. Williams was the commissary agent at the post, and later became the first elected mayor of Detroit. The actual number of men in the U. S. Army when Jefferson took office was less than 3600.

242. Donald Jackson, editor, *Letters of the Lewis and Clark Expedition with Related Documents, 1783–1854* (Urbana and Chicago, University of Illinois Press, 1978,

two volume edition). See pp. 675-677, #430, volume 2.

243. Alexander De Conde, *The Quasi War: The Politics and Diplomacy of the Undeclared War with France, 1797-1801* (NY, Charles Scribner's Sons, 1966), pp. 122-23 and footnote 34, p. 417.

244. Maurine T. Wilson and Jack Jackson, *Philip Nolan and Texas: Expeditions to the Unknown Land, 1791-1801* (Waco, Texas, Texian Press, 1987), pp. 2-5.

245. *Ibid,* p. 40. Citing *Life, Letters and Correspondence of William Dunbar,* letter from Jefferson to Dunbar, January 16, 1800.

246. *Ibid,* p. 40-41. Jack Jackson provides an excerpt from Wilkinson's letter to Jefferson and goes on to state:

> Of the contemplated interview we have no information, but most writers suggest that it occurred and that it had some influence in determining Nolan to undertake a new expedition into the forbidden territory. . . . past historians like Isaac Joslin Cox have perhaps been too hasty in attributing Nolan's final entry to the influence of Jefferson and his expansionist national policy. . . .

Dan L. Flores, after examining the manuscript in the Thomas Jefferson Papers, presented evidence that this meeting never took place. He has found that Jefferson entered his receipt of Wilkinson's letter of introduction on November 3rd (Jefferson noted the date of receipt on all incoming mail). By then Nolan was leading his men across the Mississippi, headed to Texas. Flores notes Nolan had started to Monticello in May of 1800, "taking with him a fine paint stallion for Jefferson." He believes Nolan sold this handsome animal in Lexington, Kentucky, after which he then decided against continuing to Virginia.

A check of the Flores account, which is a footnote in his book, *Southern Counterpart to Lewis and Clark: The Freeman and Custis Expedition of 1806* (University of Oklahoma Press, 1984) p. 33, reveals Flores says: "I believe that while in Lexington Nolan sold the stallion he was taking to Jefferson." He gives no evidence for his belief. Where is the evidence for the sale of a paint stallion destined for Jefferson in Lexington, and for Nolan turning back to Natchez after selling that particular horse? Jefferson's original letter to Dunbar says that he wants to purchase "one of his fine horses." Nolan almost certainly sold horses in Kentucky while traveling east.

Jefferson's possession of Nolan's letter, dated as being received by him on November 3, 1800 is Flores's other evidence. Jefferson was perfectly capable of postdating the receipt of a document indicating a meeting with a "Mexican traveller." He *had possession* of the letter introducing the "Bearer of this letter—Mr. P. Nolan." indicating that he did meet with Nolan.

Nolan received the letter from Wilkinson on May 22nd at Fort Adams, on the eve of Wilkinson's departure for the east coast. He had three months to bring a horse, or horses, to Monticello and return to Natchez. There is no problem with this timeline. The only other evidence Jack Jackson cites is a letter Wilkinson writes to Jefferson on September 1st that he was certain Nolan would have visited Monticello by this time, but he has received a letter from Kentucky, causing him to worry it hasn't happened. It is my contention that Nolan not only met with Jefferson, but met with supporters of Aaron Burr before going by ship back to New Orleans. If he, in fact, did only travel as far as Kentucky, then he met with Burrites there who were able to promise him British support. It is as likely that he went to the east coast and returned by ship to the Mississippi.

247. Grace King, "The Real Philip Nolan," *Publication of the Louisiana Historical Society, Proceedings & Reports,* vol. X (1917). Letters of Fanny Nolan, dated August and August 22nd, 1800. pp. 107-108.

248. Noel M. Loomis and Abraham P. Nasatir, *Pedro Vial and the Roads to Sante Fe* (Norman, University of Oklahoma Press, 1967), pp. 217-219. Original letter, dated October 21, 1800, in Papeles de Cuba, leg. 71-A. See also Wilson & Jackson, *Philip Nolan and Texas*, pp. 47-48.

249. Wilson and Jackson, *Philip Nolan and Texas*, pp. 51-52, 61.

250. *Ibid*, p. 43.

251. *Ibid*, p. 73. The footnote cites Músquiz's "Diary" in Nolan Documents, 795:29 for Nolan being killed by a cannonball. Loomis and Nasatir in *Pedro Vial*, p. 225, say he was killed by a "chance shot to the forehead." Roger Kennedy in his book, *Cotton and Conquest: How the Plantation System Acquired Texas* (Norman, University of Oklahoma Press, 2013) speculates he was killed in a "mafia-style" assassination on the orders of General Wilkinson, pp. 202-203.

251. Noel M. Loomis, "Philip Nolan's Entry into Texas in 1800," *The Spanish in the Mississippi Valley*, edited by Francis McDermott (Urbana, University of Illinois Press, 1974), p. 129.

252. J. H. Daviess, "View of the President's Conduct Concerning the Conspiracy of 1806," edited by Isaac J. Cox and Helen Swineford, *Quarterly Publication of the Historical and Philosophical Society of Ohio*, vol XII: nos. 2 & 3, April-June, July -September, 1917. p. 89.

253. Mary Jo Kline et al, *The Political Correspondence and Public Papers of Aaron Burr* (Princeton, Princeton University Press, 2 volumes, 1983) I: 457-458.

254. James Wilkinson, *Memoirs of My Own Times*, Vol. 2, pp. 270-273. Letter from Burr to Wilkinson, dated October 10, 1800, Appendix, No. LXIV, Vol. 2.

255. Carlos Marichal, *Bankruptcy of Empire: Mexican Silver and the Wars Between Spain, Britain and France, 1760-1810* (Cambridge, Cambridge University Press, 2007), p. 61.

256. Kennedy, *Cotton and Conquest,* p. 32.

257. Robin Fabel, "Bernard Lintot: A Connecticut Yankee on the Mississippi, 1775-1805," *The Florida Historical Quarterly*, Vol. 60, No. 1 (July, 1981), pp. 88-102.

258. De Conde, *The Quasi-War*, p. 200. See also, Peggy Liss, *Atlantic Empires: The Network of Trade and Revolution, 1713-1826* (Baltimore, Johns Hopkins University Press, 1983), p. 178.

259. See footnote 252.

260. Harold C. Syrett, *The Papers of Alexander Hamilton* (NY, Columbia University Press, 1977) vol. 25, p. 257. Letter to Oliver Wolcott, Jr., December 16, 1800.

261. Michael A. Bellesiles, "The Soil Will Be Soaked with Blood," *The Revolution of 1800: Democracy, Race & the New Republic,* edited by James Horn, Jan Ellen Lewis, & Peter S. Onuf (Charlottesville, University of Virginia Press, 2002), p. 65.

262. Douglas R. Egerton, *Gabriel's Rebellion: The Virginia Slave Conspiracies of 1800 & 1802* (Chapel Hill, University of North Carolina Press, 1993), pp. 50, 68.

263. *Ibid,* p. 115.

264. "Gabriel's Rebellion" or "Gabriel Prosser" on Wikipedia. (accessed 3/10/2014).

265. Sharp, *American Politics in the Early Republic,* pp. 268-271.

266. James E. Lewis, "What Is to Become of Our Government?" in *The Revolution of 1800,* p. 20 and footnote 62.

267. Richard N. Côte, *Theodosia: Theodosia Burr Alston: Portrait of a Prodigy* (Mt. Pleasant, SC, Corinthian Books, 2003), pp. 120 , 126, 141.

268. Edward J. Larson, *A Magnficent Catastrophe: The Tumultuous Election of 1800, America's First Presidential Campaign* (NY, Free Press, A Division of Simon & Schuster, 2007), p.268.

269. Grace Lewis Miller Archives, Jefferson National Expansion Memorial, #116, NPS Catalog No. 9359, Box 37, Folder 13, "Captain Lewis's Company." She notes the record of his paymaster travels is found in GAO, Journal N, 4302, Item D. His original notes are found in the journal he later used during the expedition labelled Lewis & Clark Diary # 6, Thermometrical Observations beginning 25 Feb. 1805.

9. The President's House: 1801-1803

270. Dumas Malone, *Jefferson The President: First Term, 1801-1805* (Little Brown and Company, 1970), pp 43-46. See also"Residence Overview: Jefferson's Enhancements: 1801-1809" at www.whitehousemuseum.org; and FAQ at www.whitehousehistory.org.

271. Margaret Bayard Smith, *First Forty Years of Washington Society: Portrayed by the Family Letters of Mrs. Samuel Harrison Smith (Margaret Bayard) From the Collection of Her Grandson J. Henley Smith* (Nabu Public Domain 1906), p. 385.

272. Malone, *Jefferson The President: First Term,1801-1805,* footnote, p. 43

273. William Seale, *The President's House: A History* (White House Historical Association and National Geographic, two volumes, 1986), Vol. 1, pp. 93–94. Lynne Cheney, *James Madison* (Viking Press, 2014), pp. 296-297.

274. Thomas Jefferson, *The Family Letters of Thomas Jefferson* (Charlottesville, University of Virginia Press, 1986), p. 202.

275. Malone, *Jefferson The President: First Term, 1801-1805,* p. 44 , and "dumbwaiter" at www.monticello.org.

276. Cynthia D. Earman, "Remembering the Ladies: Women, Etiquette and Diversions In Washington City, 1800-1814," *Washington History*, Vol. 12, No. 1 (Spring/Summer, 2000), pp. 102-117. See p. 110.

277. Merry Ellen Scofield, "The Fatigues of His Table: The Politics of Presidential Dining During the Jefferson Administration," *Journal of the Early Republic* (University of Pennsylvania Press), Vol. 26, No. 3 (Fall, 2006), pp. 449-469. Scofield writes: "All available evidence indicates that the only people omitted from the[dining] records were family members, staff, and house guests." p. 464. See also Seale, *The President's House,* p. 94.

278. Smith, *First Forty Years of Washington Society*, pp. 391-392.

279. Scofield, "The Fatigues of His Table," p. 465.

280. Donald Jackson, *Letters of the Lewis and Clark Expedition with Related Documents, 1783–1854* (Urbana and Chicago, University of Illinois Press, 1978, 2 volume edition), Jefferson to Lewis, February 23m 1801, pp. 2-3.

281. Theodore Crackel, *Mr. Jefferson's Army: Political and Social Reform of the Military Establishment, 1801-1809* (NY & London, New York University Press, 1987) pp. 45-53, 58-62 and footnote 3, p. 193; and Donald Jackson, "Jefferson, Meriwether Lewis, and the Reduction of the United States Army, *Proceedings of the American Philosophical Society,* Vol. 124, No. 2 (April 29, 1980), pp. 91-96, see pp. 95-96.

282. For the coded list of army officers see Thomas Jefferson Papers, American Memory, Library of Congress. Do a search for July 24, 1801, "War Department, List of Army Officers." (www.memory.loc.gov).

283. Crackel, *Mr. Jefferson's Army,* pp. 45-47, 50-51.

284. Stephen Witte, "Republicanizing the Army: The Military Peace Establishment Act of 1802" in *Discovering Lewis & Clark* at www.lewis-clark.org.

285. Jonathan Williams Archives (The Lilly Library, Indiana University, Bloomington), letter from Wilkinson to Williams, Pittsburgh, dated July 2, 1801.

286. "David Strong, Sr. (1744-1801)" on www.geni.com. Family historian Jerry Templeton's biography of Colonel Strong lists a source, Mayer:1985, for his information. "Various records indicate that all of the 1st, 2nd, 3rd, and 4th Infantry Regiment companies . . . . were present at this time." Templeton also writes:

> A large amount of artillery ordinance was dispatched to Cantonment Wilkinsonville on July 1-8, 1801 in the anticipation of the arrival of General Wilkinson and Major Williams of the artillery. These included one 9 pound cannon, one 12 pound cannon, two 6 pound brass cannons, 40 barrels (4,000 pounds) of gunpowder, gun carriages, and hundreds of different types of round, case, grape—and possibly shrapnel—shot (Issac Craig Papers, 217, 219).

David P. Mayer's unpublished paper, "Cantonment Wilkinsonville 1801: Burr, Wilkinson, the U. S. Army and Treason," (March 7, 1985, Erie, PA) is available at the Cairo Public Library, Cairo, IL. It contains a great deal of well researched information and supports the idea that the cantonment was a staging area for an invasion of Spanish territory if Burr had been elected president.

287. *Ibid.*

288. Descendant Amy Schneider contacted me in January, 2011 after reading an internet blog posting of mine regarding my speculation concerning Wilkinson's role in the assassination of Lewis, saying "I have always secretly believed that Gen. Wilkinson murdered my ancestor, Lt. Col. David Strong."

289. Norman Caldwell, "Cantonment Wilkinsonville," *Mid-America: An Historical Review,* 31:1, Jan, 1949, p. 24.

290. Mark J. Wagner, "Searching for Cantonment Wilkinson: An 1801-1802 U.S. Army Post in Pulaski County, Illinois, 2003 [November], Paper Presented At the Illinois Archaeological Survey Annual Meetings, Southern Illinois University, Carbondale, IL, p. 5. Original excerpt of Wilkinson letter dated September 8, 1801 found in Strong geneaology records, managed by Jerry Templeton, at www.geni.com. The letter is noted as received in War Department records, but research at the National Archives has failed to locate the letter itself.

291. *The Western Spy* newspaper, October 10, 1801. Courtesy of the Cincinnati Museum Center.

292. Strong Family Papers at the Cincinnati Museum Center. Handwritten notes taken from "Connecticut in the War" and Gen. Wayne's Log Book by Mrs. R. C. Belt, Milford, Ohio (no date).

293. Letter from General Wilkinson to Major Williams, September 6, 1801, Jonathan William Archives, The Lilly Library, Indiana University, Bloomington.

294. *The Western Spy* newspaper, November 7, 1801, Cincinnati Museum Center.

295. The information on the location of camps is taken from Mark J. Wagner's article, "Searching for Cantonment Wilkinsonville," which was on the Southernmost Illinois History website in 2007. It was described as an "still in progress section of a report (chapter 3) on the archeaology and history of Cantonment Wilkinson." The reference to Wilkinsonville as a customs house and port of entry in April, 1802 comes from a French traveler's account by Francois Marie Perrin du Lac, *Travels Through the Two Louisianas, and Among the Savage Nations of the Missouri; Also in the United States, Along the Ohio and the Adjacent Provinces, in 1801, 1802 & 1803. With a Sketch of the Manners, Customs, Character, and the Civil and Religious Ceremonies of the People of Those Countries. Translated from the French.*

296. American State Papers, Indian Affairs, Volume 1, (7th Congress, 1st session), pp. 684-653.

297. Jacobs, *Tarnished Warrior*, p. 197.

298. Wilkinson, *Memoirs of My Own Times*, Volume 1, p. vii.

299. Thomas Robson Hay and M.R. Werner, *The Admirable Trumpeteer: A Biography of General James Wilkinson* (Garden City NY, Doubleday, Doran & Company, 1941), p. 192.

300. Barbara Oberg, editor, *The Papers of Thomas Jefferson* (Princeton and Oxford,

Princeton University Press, 2009) Volume 36, 1 Dec. 1801 to 3 March 1802, pp. vii.

301. John Whitcomb and Claire Whitcomb, *Real Life at the White House: Two Hundred Years of Daily Life at America's Most Famous Residence* (New York and London, Routledge, 2000). pp. 16-17.

302. Merrill D. Peterson, *Thomas Jefferson and the New Nation,* pp. 722-723, and Wikipedia, "Chesire Mammoth Cheese."

303. Peterson, *Thomas Jefferson and the New Nation,* pp. 725-727.

304. Barbara Oberg, editor, *The Papers of Thomas Jefferson,* Volume 36, *xlvi-xlvii* and Merrill Peterson, *Thomas Jefferson and the New Nation,*, pp. 720-721.

305. Jackson, *Letters of the Lewis & Clark Expedition,* Vol 2, pp. 677-684, Excerpts from the Diary of Mahlon Dickerson. Quote from a letter to Silas Dickerson, April 21, 1803 (New Jersey Historical Society).

306. *Ibid,* pp. 678-679. See also, Oberg, *Papers of Thomas Jefferson,* Volume 36, *xlvi-xlvii,* and p. 406.

307. Charles Coleman Sellers, *Charles Willson Peale: A Biography,* (New York, Charles Scribner's Sons, 1969), pp. 306-308.

308. *Ibid,* pp. 299-300. Oberg, *Papers of Thomas Jefferson,* Vol. 37, letter from Peale to TJ, June 6, 1802.

309. Paul Semonin, *American Monster: How the Nation's First Prehistoric Creature Became a Symbol of National Identity* (New York & Londoln, New York University Press, 2000), pp. 4-6, 362-364.

310. *Ibid,* pp. 124-126.

311. Thomas Jefferson, *Notes on the State of Virginia* (original publication date of 1787, Google books), p. 83.

312. Oberg, *Papers of Thomas Jefferson,* Vol. 37, pp. 431-437, John Vaughan to T J, May 8, 1802.

313. Donald Jackson, *Thomas Jefferson & the Stony Mountains: Exploring the West from Monticello* (Urbana, University of Illinois Press, 1981 and Norman, University of Oklahoma Press, 1993), pp. 122-124, and note 12, p. 158.

314. *Ibid,* p. 95, citing p. 411 in MacKenzie's book.

315. Robert J. Miller, *Native America, Discovered and Conquired: Thomas Jefferson, Lewis and Clark, and Manifest Destiny* (Westport CT and London, Praeger, 2006), pp. 19-20, 25.

316. Warren L. Cook, *Flood Tide of Empire: Spain and the Pacific Northwest, 1543-1819* (New Haven and London, Yale University Press, 1973), p. 434.

317. *Ibid,* pp. 435-39

318. A. P. Nasatir, editor, *Before Lewis and Clark: Documents Illustrating the History of the Missouri, 1785-1804* (Lincoln and London, University of Nebraska Press, 1990, 2 volumes), letter from Trudeau to Carondelet, dated St. Louis, Octber 20,

1792, Vol. 1, pp. 160-161.

319. David Williams, "John Evans' Strange Journey: Part I. The Welsh Indians and Part II. Following the Trail." *The American Historical Review,* Vol. 54, No. 2 & 3 (Jan. & Apr., 1949) (Oxford University Press on behalf of the American Historical Association), pp. 277-295 and 508-529. See p. 294.

320. Gwyn A. Williams, *Madoc: The Legend of the Welsh Discovery of America* (Oxford & New York, Oxford University Press, 1987), p. 149. See also W. Raymond Wood, *Prologue to Lewis and Clark: The Mackay and Evans Expedition* (Norman, University of Oklahoma Press, 2003), pp. 190-193, letter of John Evans to Samuel Jones, St. Louis, July 15, 1797.

321. Wood, *Prologue to Lewis and Clark,* p. 192.

322. Gwynn A. Williams, "Welsh Indians:The Madoc Legend and the First Welsh Radicalism," *History Workshop*, No. 1 (Spring, 1976) (Oxford, *Oxford Journals*, Oxford University Press), pp.136-154, see p. 143.

323. Nasatir, *Before Lewis and Clark*, Vol. 1, p. 99, note.

324. Recommended reading: *Footprints of the Welsh Indians: Settlers in North American Before 1492* by William L. Traxel (Algora Publishing, 2004); *The Legend of Price Madoc and the White Indians* by Dana Olson (Jeffersonville IN, Olson Enterprises,1987); *Madoc and the Discovery of America Some New Light on an Old Controversy* by Richard Deacon (New York, George Braziller, 1966); and *Madoc: The Legend of Welsh Discovery of America* by Gwynn A. Williams (Oxford & New York, Oxford University Press, 1987).

325. Paul Jarman, "Madoc, 1795: Robert Southey's Misdated Manuscript," *The Review of English Studies,* New Series, Vod. 55, No. 220 (June, 2004), pp. 355-373 (Oxford University Press).

326. Williams, "John Evans Strange Journey," Part I, p. 291.

327. 327. W. Raymond Wood, *Prologue to Lewis and Clark: The Mackay and Evans Expedition* (Norman, University of Oklahoma Press, 2003), pp. 65-80.

328. *Ibid*, p. 192.

329. Abraham P. Nasatir, "John Evans, Explorer and Surveyor," *Missouri Historical Review*, Vol. 25, Issue 3, Apr., 1931. Part II. Mackay's letter of February 19, 1796 to Evans, pp. 447-449. See also Louis Houck's *The Spanish Regime in Missouri*, Vol. 2, p. 254 at www.archives.org.

330.

331. Nasatir, "John Evans," Part II, Document No. 5, pp. 441-447. Instructions to John Evans.

332. *Ibid*, p. 448, letter of February 19, 1796.

333. *Ibid,* pp. 449-454, "Extracts of Mr. Evans Journal," also found in Wood, *Prologue to Lewis and Clark,* pp.180-183. The post is identified as Jusseaume's Post in Wood, *Prologue,* pp. 126-128.

334. Abraham P. Nasatir, "Anglo-Spanish Rivalry on the Upper Missouri," *The Mississippi Valley Historical Review*, Vol. 16, No. 4 (Mar., 1930), pp. 507-528, see pp. 519-520. Carlos(?) Howard reported Makay told him he had "definite information" thatEvans had started on his way to the Pacific. In any event, Evans did not go west, but returned to St. Louis. (Howard to Carondelet, May 13, 1797).

335. Wood, *Prologue to Lewis and Clark*, p. 132.

336. Thomas C. Danisi and W. Raymond Wood, "Lewis and Clark's Route Map: James Mackay's Map of the Missouri River," *The Western Historical Quarterly*, Vol. 35, No. 1 (Spring, 2004) (The Western History Association), pp. 53-72.

337. Wood, *Prologue to Lewis and Clark*, pp. 152-154.

338. Nasatir, "Anglo-Spanish Rivalry," pp. 520-521. Mackay to Gayoso, letter dated June 8, 1797.

339. Wood, *Prologue to Lewis and Clark*, p. 193.

340. Nasatir, "John Evans," Part III, pp. 595-597.

341. *Ibid*, p. 603-604

342. Gwyn A. Williams, *Madoc: The Legend of Welsh Discovery of America* (Oxford & New York, Oxford University Press, 1987), p. 190.

343. Richard Deacon, *Madoc and the Discovery of America* ((New York, George Braziller, 1966), p. 149.

344. *bid*, pp. 149-150. Deacon cites the *Reglamentos de Don Manuel Gayoso de Lemos, New Orleans, 1797.*

345. Jacobs, *Tarnished Warrior*, 187-190.

346. Williams, "John Evans Strange Journey," Part II, p. 528.

347. Deacon, *Madoc*, p.149.

348. Oberg, *The Papers of Thomas Jefferson*, Vol. 36, pp. 206-207, Livingston to Jefferson, December 26, 1801.

349. Jon Kukla, *A Wilderness So Immense: The Louisiana Purchase and the Destiny of America* (New York, Alfred A. Knopf, 2003), pp. 216-25.

350. Mary P. Adams, "Jefferson's Reaction to the Treaty of San Ildefonso," *The Journal of Southern History*, Vol. 21, No. 2 , May, 1955, p. 175.

351. *Ibid*, p. 174. Letter from Jefferson to Horatio Gates, July 11, 1803.

352. Oberg, *Papers of Thomas Jefferson*, Vol. 37, pp. 263-66, letter from Jefferson to Robert R. Livingston, April 18, 1802.

353. *Ibid*, p. 266.

354. Ronald D. Smith, "Napoleon and Louisiana: Failure of the Proposed Expedition to Occupy and Defend Louisiana," *Louisiana History: The Journal of the Louisiana Historical Association*, 12: 1, Winter, 1971, p. 26.

355. Kukla, *A Wilderness So Immese*, pp. 248-49.

356. *Ibid*, p. 249-50.

357. *Ibid*, pp. 266-268. Smith, "Napoleon and Louisiana," pp. 31-34.

358. Smith, "Napoleon and Louisiana," p. 35.

359. Junius P. Rodriguez, editor, *The Louisiana Purchase: A Historical and Geographical Encyclopedia* (Santa Barbara, ABC CLIO, 2002), p. 408.

360. Kukla, *A Wilderness So Immense,* p. 327.

361. Donald Jackson, editor, *Letters of the Lewis and Clark Expedition with Related Documents, 1783-1854* (Urbana and Chicago, University of Illinois Press, 1978, second edition, two volumes), Vol. 1, p. 18, Jefferson to Benjamin Rush, Feb. 28, 1803 (Document 13).

362. Harpers Ferry National Historical Park, www.nps.gov/hafe/meriwether-lewis.htm.

363. Paul Russell Cutright, *Contributions of Philadelphia to Lewis and Clark History*, with site maps by Frank Muhly, a booklet edited by the Philadelphia Chapter of the Lewis and Clark Trail Heritage Foundation (2001).

364. Jackson, *Letters of the Lewis and Clark Expedition,* letter of Lewis to Clark, June 19, 1803 (Document # 46), pp. 57-60.

365. *Ibid,* Preface, vii.

366. *Ibid,* Lewis to Lucy Marks, July 2, 1803 (Document 58), pp. 100-101.

10. JOURNEY TO THE PACIFIC

367. *Ibid,* Clark to Lewis, July 18, 1803 (Document 74), pp. 110-111.

368. *Ibid*, Lewis to Jefferson, July 22, 1803 (Document 75), pp. 11-112.

369. *Ibid*, Lewis to Clark, August 3, 1803 (Document 80), pp. 115--117.

370. *Ibid,* Purveyors Summary (Document 57), pp. 93-99.

371. Arlen Large, "Additions to the Party: How An Expedition Grew and Grew," *We Proceeded On* (Lewis and Clark Trail Heritage Foundation) Vol. 16, No. 1, February, 1990. pp. 4-10. PDF at www.lewisandclark.org.

372. Jackson, *Letters,* Dearborn to Russell Bissell and Amos Stoddard, July 2, 1803 (Document 64).

373. *Ibid,* Dearborn to Lewis, July 2, 1803 (Document 62).

374. Illustration from *We Proceeded On,* May, 2002, p. 2.

375. Jackson, *Letters*, Lewis to Jefferson, September 8, 1803 (Document 87), pp. 121-23.

376. Gary E. Moulton, editor, *The Definitive Journals of Lewis & Clark* (Lincoln and London, University of Nebraska Press, 2002), Volume 2, *From the Ohio to the Vermillion*, pp. 65-66, August 30, 1803 misdated entry in Lewis's journal. It was ac-

tually August 31st. The Masonic farewell at Brunot's Island is discussed in the article on the Discovering Lewis and Clark website, www.lewis-clark.org.

377. Rodney, *A Journey Through the West,* pp. 50-51.

378. Guns of the Lewis and Clark Expedition are discussed in the May, 2006 issue (Vol. 32:2) of *We Proceeded On,* the quarterly journal of the Lewis and Clark Trail Heritage Foundation (available online as a downloadable PDF at www.lewisand-clark.org). See Jim Merritt's article, "Shooting the Model 1800 short rifle and Girandoni repeating airgun," p. 25, and Robert Beeman's article, "Meriwether Lewis's Wonder Weapon," pp. 29-34. Beeman has Lewis's original gun in his collection. The rifle's mainspring has been replaced with an improvised piece of a ferrier's file, the kind of file used to trim a horses's hoof. On June 10, 1805, the expedition's blacksmith, John Shields, replaced the mainspring of Lewis's air gun. Lewis wrote on that date, "we have been much indebted to the ingenuity of this man on many occasions."

379. Ephraim G. Scuier and Edwin H. Davis, edited and with an introduction by David J. Meltzer, *Ancient Monuments of the Mississippi Valley* (Washington and London, Smithsonian Press, 1998; original date of publication, 1848), pp. 168-69.

380. Moulton, *Definitive Journals of Lewis & Clark,* Vol. 2, pp. 77-79, Lewis journal entry, Septmber 10, 1803.

381. Robert E. Lange, "Private George Shannon: The Expedition's Youngest Member—1785 or 1787–1836," *We Proceeded On,* Vol 8:3, July, 1982, pp. 10-15. Available on the WPO archives at www.lewisandclark.org. Colter's date of enlistment was October 15, 1803, and Shannon's date of enlistment was October 19, 1803, which establishes their identities as the "two young men on trial." Shannon probably joined Lewis at Pittsburgh where he was attending school. His family lived in Belmont, Ohio, however, so he may have joined Lewis at Cincinnati. Colter lived in Maysville (Limestone), Ohio.

382. Jackson, *Letters,* Jefferson to Lewis, November 16, 1803 (Document 94), pp. 136-140.

383. Potts, *George Rogers Clark,* pp 78-85. The drawing of a larger home seen in my *Lewis and Clark Road Trips* book (2006) is George Rogers Clark's home after it was enlarged in 1808. He went to live with family members in 1809 after his leg was amputated. Private communication with Jim Holmberg, Filson Historical Society, Louisville, Kentucky.

384. Jackson, *Letters,* Lewis to Henry Dearborn, January 15, 1807 (Document 236), pp. 364-73. This report by Lewis contains his "remarks on their rispective merits and services."

385. Large, "Additions to the Party," p. 8. Available as a PDF download, WPO archives, www.lewisandclark.org.

386. Jackson, *Letters,* Delassus to de Casa Calvo and Salcedo, December 9, 1803 (Document 213), pp.142-43.

387. *Ibid,* Casa de Cavo and Manuel Salcedo to Nemesio Salcedo, March 5, 1804 (Document 119), pp. 185-186.

388. Jacobs, *Tarnished Warrior*, pp. 201-11 and Isaac J. Cox, "General Wilkinson and His Later Intrigues with the Spaniards," *American Historical Review*, Vol XIX, No. 4 (July, 1914), pp. 794-812. See p. 800.

389. James Alexander Robinson and Paul Alliot, *Louisiana Under the Rule of Spain, France, and the United States, 1785-1807; Social, Economic, and Political Conditions of the Territory Represented in the Louisiana Purchase* (ULAN Press sold through Amazon, 2011, original publication date, 1923). "Reflections of Vincent Folch," pp. 323-48. See page 338 for quote.

390. Jacobs, *Tarnished Warrior*, pp. 207-11, Cox, "General Wilkinson," p. 800.

391. Cook, *Flood Tide of Empire*, pp. 446-83. For the fourth attempt, pp. 477-83.

392. Robert Hunt, "Crime and Punishment on the Lewis and Clark Expedition," *We Proceeded On*, Vol. 15, Nos. 2&3, (May & August, 1989), Part 1, p. 4. PDFs available, *WPO*, www.lewisandclark.org.

393. Moulton, *Definitive Journals,* Vol. 2, p. 206.

394. Jackson, *Letters*, Lewis to Jefferson, December 28, 1803 (Document100), Vol. 1, pp. 148-157.

395. Robert E. Hartley, *Lewis and Clark in the Illinois Country: The Little Told Story* (Westminster CO, Sniktau Publications, 2002), pp. 124-42.

396. Thomas Danisi, *Uncovering the Truth About Meriwether Lewis* (Amherst, NY, Prometheus Books, 2012), Chapter 9, "Was Governor Meriwether Lewis's Correspondence Intentionally Delayed?," pp. 108-23.

397. Jackson, *Letters,* Henry Dearborn to Lewis, 26th March, 1804 (Document 110)Vol. 1, pp.172-73; The Nicholas Biddle Notes [c. April, 1810], (Document 326), Vol 2, pp. 497-545. see p. 533; Clark to Nicholas Biddle, 15th August 1811 (Document 345) Vol. 2, pp. 571-72.

398. Moulton, *Definitive Journals,* Vol. 2, pp. 347, 518-519

399. *Ibid,* Nicholas Biddle notes, p. 534.

400. Moulton, *Definitive Journals*, Vol. 2, pp. 254-59, Lewis's detachment orders.

401. Ernest Staple Osgood, editor, *The Field Notes of Captain William Clark, 1803-1805* (New Haven and London, Yale University Press, 1964), p.64. For a good discussion of the techniques used in recording information for map making, see John Logan Allen's discussion on "Geography" at www.lewis-clark.org.

401. Jackson, *Letters,* Lewis to Lucy Marks, Fort Mandan (Document 143), Vol. 1, pp. 222-25. Moulton, *Definitive Journals,* Vol. 2, pp. 289, 302.

402. Moulton, *Definitive Journals,* Vol. 2, pp. 305-07.

403. Paul Russell Cutright, *A History of the Lewis and Clark Journals* (Norman, University of Oklahoma Press, 1976), pp. 8-11.

404. Jackson, *Letters,* pp. 234-42, Lewis to Jefferson, April 7, 1805 (Document 149).

405. Jackson, *Thomas Jefferson and the Stony Mountains*, pp. 192-195. See also Gary

E. Moulton, "The Missing Journals of Meriwether Lewis," *The Montana Magazine of Western History,* Vol. 35, No. 3 (Summer, 1985), pp. 28-39.

406. Horace W. Sellers, editor, "Letters of Thomas Jefferson to Charles Willson Peale, 1795-1825," *The Pennsylvania Magazine of History and Biography,* Vol. 28, No. 3 (1904), pp. 295-319. Letters dated October 6 & 9, 1805.

407. Jackson, *Letters,* pp. 234-42, Lewis to Jefferson, April 7, 1805 (Document 149).

408. Bob Saindon, "The Lost Vocabularies of the Lewis & Clark Expedition," *We Proceeded On,* Vol.3, No.3, July, 1977. See archives at www.lewisandclark.org . James P. Ronda, *Lewis and Clark Among the Indians* (Lincoln and London, University of Nebraska Press, 1984), "Lewis and Clark as Plains Ethnographers," pp 113-32.

409. For a further discussion, see Robert R. Hunt, "Crime and punishment on the Lewis and Clark Expedition," *We Proceeded On,* Vol. 15: 2&3, May and August, 1989. www.lewisandclark.org

410. Moulton, *Definitive Journals*, Vol. 2, p. 492.

411. James Holmberg editor, *Exploring with Lewis and Clark: The 1804 Journal of Charles Floyd* (Norman, University of Oklahoma Press, 2004) pp. 70-71. This facsimile edition contains the handwritten manuscript, transcript and extensive notes.

412. David J. Peck, D.O.,*Or Perish in the Attempt: Wilderness Medicine in the Lewis & Clark Expedition* (Helena, MT, Far Country Press, 2002), pp. 42-43 102-07.

413. Moulton, *Definitive Journals*, Vol. 2, p. 495.

414. Moulton, *Definitive Journals*, Vol. 3, p. 77.

415. Paul Russell Cutright, *Lewis & Clark Pioneering Naturalists* (Lincoln and London, University of Nebraska Press, 1969), pp. 115-17.

416. *Ibid,* pp.79-89.

417. Jackson, *Letters,* Jefferson to Lewis, November 16, 1803 (Document 94), pp. 136-40. Quote, pp. 138-39.

418. Richard E. Oglesby, *Manuel Lisa and the Opening of the Missouri Fur Trade* (Norman and London, University of Oklahoma Press, 1963, pp. 151-52. John Sugden, *Tecumseh's Last Stand* (Norman and London, University of Oklahoma Press, 1985), p. 28. Doane Robinson, editor, "A Sioux Indians View of the Last War with England," *South Dakota State Historical Collections,* Vol. 5 (1910). Kira Gale, *Defending the Western Frontier: Manuel Lisa and the War of 1812 in the Council Bluffs Area* (Omaha, River Junction Press LLC, 2000).

419. Oglesby, *Manuel Lisa*, pp. 156-57. Walter B. Douglas,annotated and edited by Abraham Nasatir, *Manuel Lisa* (New York, Argosy-Antiquarian, 1964), pp. 374-386. John Bradbury, *Travels in the Interior of American in the Years 1809, 1810, and 1811* (Google Books). Henry Marie Brackenridge, *Journal of a Voyage Up the Missouri River in 1811* (Massachusetts, Applewood Press, reprint of second edition, 1816). The second edition of Brackenridge's Journal contained Big Elk's oration, but this reprint omits it and refers to Bradbury's *Travels.*

420. Ronda, *Lewis and Clark Among the Indians,* "The Arikara Interlude," pp.42-66. Roger Nichols, "The Arikara Indians and the Missouri River Trade," *Great Plains Quarterly,* Paper 1656 (http://digitalcommons.unl.edu/greatplainsquarterly/1656).

421. Moulton, *Definitive Journals,* Vol. 3, pp. 163-64, and Vol. 10 (Gass), pp. 54-55.

422. Jackson, *Letters,* Vol. 2, p. 537-38.

423. Moulton, *Definitive Journals,* Vol. 3, pp. 167-173. Jackson, *Letters,* Vol. 1, Lewis to Henry Dearborn, January 15, 1807, pp. 364-373.

424. Cutright, *History of the Lewis and Clark Journals,* pp. 263-64.

425. Hunt, "Crime and Punishment," *WPO,* May and August, 1989.

426. Moulton, *Definitive Journals,* Vol. 3, pp. 186-88 (October 20, 1804).

427. *Ibid,* pp. 401-405 ("Estimate of Eastern Indians").

428. Ronda, *Lewis and Clark Among the Indians,* p.75.

429. Moulton, *Definitive Journals,* Vol. 3, p. 218 (October 31, 1804).

430. Tracey Potter, *Sheheke, Mandan Indian Diplomat: The Story of White Coyote, Thomas Jefferson and Lewis and Clark* (Far Country Press and Fort Mandan Press, 2003), p. 91.

431. Moulton, *Definitive Journals,* Vol. 3, pp. 226-27 (November 3, 1804).

432. *Ibid,* pp.233-34 (November 12, 1804).

433. *Ibid,* p. 261 (December 25, 1804).

444. *Ibid,* p. 267 (January 1, 1805).

445. David Wishart, editor, *Encyclopedia of the Great Plains,* "Gros Ventres," at www.plainshumanities.unl.edu/encyclopedia/doc/egp.na.031

446. Thomas P. Lowry, *Veneral Disease and the Lewis and Clark Expedition* (Lincoln and London, University of Nebraska Press, 2004), pp. 44-45 and Moulton, *Definitive Journals,* Vol. 3, p. 273 (January 14, 2014), pp. 323-324 (March 31, 1805).

447. Moulton, *Definitive Journals,* Vol. 3, p. 308.

448. Jackson, *Letters,* Vol. 1, Lewis to Jefferson, April 7, 1805 (Document 149), pp. 231-242.

449. Jackson, *Letters,* Vol. 1, Etienne Lemaire to Jefferson August 20, 1805 (Document 166), p. 256.

450. *Ibid,* James Wilkinson to Henry Dearborn, Sept. 22, 1805 (#170), Oct. 8, 1805 (#172), Oct. 22, 1805 (#176);Wilkinson to Jefferson, Dec. 23, 1805 (#184); Dearborn to Wilkinson April 9, 1806 (#200), pp. 259-267, 272-75, 303-05.

451. Moulton, *Definitive Journals,* Vol. 4, p. 9 (April 7, 1805).

452. "Assiniboine Chief Rosebud Remembers Lewis & Clark," You Tube video. The Indian history states that because they had a woman and baby with them, they were

not a war party and would not be attacked.

453. Eldon G. Chuinard, M.D., *Only One Man Died: The Medical Aspects of the Lewis and Clark Expedition* (Fairfield, WA, Ye Galleon Press,1979), p. 24.

454. Moulton, *Definitive Journals,* Vol. 4, p. 136 (May 10, 1805).

455. See 452, You Tube video.

456. Moulton, *Definitive Journals,* Vol. 4, p. 141 (May 11, 1805).

457. Moulton, *Definitive Journals,* Vol. 4, p. 225 (May 31, 1805).

458. Moulton, *Definitive Journals,* Vol. 4, pp. 248-49 (June 15, 1805).

459. Jackson, *Letters*, Israel Weeling Dr. to George Ludlam, May 25, 1803 (Document 55), Vol. 1, pp. 80-81.

460. See note 462.

461. Moulton, *Definitive Journals,* Vol. 4, pp. 297-98.

462. Peck, *Or Perish in the Attempt,* pp. 159-62. Dr. David Peck speculates that it was venereal disease, but that it might also have been a parasite roundworm infection caused by *Trichinella,* which could have happened when they ate grizzly meat in early June. The parasite symptoms are nausea, abdominal pain, vomiting and fever—which sound like Lewis's symptoms.

463. Moulton, *Definitive Journals,* Vol. 4, p. 341 (June 29, 1805).

464. *Ibid,* p. 342 (June 29, 1805).

465. *Ibid,* p. 330 (June 24, 1805).

466. *Ibid,* p. 361 (July 4, 1805).

467. *Ibid,* p. 382 (July 15, 1805).

468. *Ibid,* p. 387 (July 16, 1805) and p. 412 (July 21, 1805).

469. *Ibid,* p. 405 (July 19, 1805).

470. *Ibid,* p. 423 (July 24, 1805).

471. Moulton, *Definitive Journals*, Vol. 5, p. 9 (July 28, 1805).

472. *Ibid,* p. 54 (August 6, 1805).

473. *Ibid,* p. 59 (August 8, 1805) and p. 63 (August 9, 1805).

474. *Ibid,* p. 74 (August 12, 1805).

475. *Ibid*, p. 79 (August 13, 1805).

476. *Ibid*, p. 97 (August 15, 1805).

477. *Ibid*, pp. 104-05 (August 16, 1805).

478. *Ibid*, p. 106 (August 16, 1805).

479. *Ibid*, p. 109 (August 17, 1805).

480. *Ibid*, p. 11 (August 18, 1805).

481. Moulton, *Definitive Journals*, Joseph Whitehouse, Volume 11, p.303 (September 5, 1805).

482. Salish-Pend d'Oreille Culture Committee and Elders Cultural Advisory Council Confederated Salish and Kootenai Tribes, *The Salish People and the Lewis and Clark Expedition* (Lincoln & London, University of Nebraska Press, 2005), pp. 12, 91-108. The oral histories contain numerous variations on the same themes, and are presented equally. The book is one of the best books to come out of the Lewis and Clark Bicentennial.

483. Ted S. Hall, *The Trail Between The Rivers: The Travels of Meriwether Lewis and William Clark; August 24 thru September 26, 1805,* (Stevensville, MT, Stoneydale Press Publishing Company, 2000), Volume 1, pp. 122-29 (September 15, 1805). My map markers indicate 6,692 ft for the campsite on Sept. 15, and 3,439 ft. for the campsite of the 14th, near Powell Ranger Station at White Sand (Killed Colt) Creek.

484. Moulton, *Definitive Journals*, Vol. 5, p. 209 (September 16, 1805).

485. Leandra Zim Holland, *Feasting and Fasting with Lewis & Clark: A Food and Social History of the Early 1800s* (Emigrant, Montana, Old Yellowstone Publishing, 2003), pp. 158-160, 220-21. Robert Moore, "The Making of a Myth: Did the Corps of Discovery Actually Eat Candles," *We Proceeded On,* Vol. 25, No. 2, May, 1999, pp. 22-24. Available at www.lewisandclark.org, *WPO* archives.

486. Zoa L. Swayne, *Do Them No Harm!: Lewis and Clark Among the Nez Perce* (Caldwell, Idaho, Caxton Press, 2003), pp. 16-22. Allen V. Pinkham, and Steven R. Evans, *Lewis and Clark Among the Nez Perce: Strangers in the Land of the Nimiipuu* (Washburn, North Dakota, The Dakota Institute Press of the Lewis & Clark Fort Mandan Association, 2013), pp. ii-iii. "Discovering Lewis & Clark," www.lewis-clark.org for discussion of size and pictures of old Newfoundland dogs.

487. Peck, *Or Perish in the Attempt*, pp. 207-09.

488. Swayne, *Do Them No Harm!*, pp. 4-6, 28-42. Pinkham and Evans, *Lewis and Clark Among the Nez Perce*, pp. 35-43, "The Plot to Kill Lewis and Clark," as told by Irvin Winters. This story by Winters has different details. Here the sex of the baby is unknown, and he goes across the ocean and she waits for him a year before returning home.

489. Pinkham and Evans, *Lewis and Clark Among the Nez Perce*, p.30.

490. Pinkham and Evans, *Lewis and Clark Among the Nez Perce*, pp. 52-55.

491. Moulton, *Definitive Journals*, Vol. 5, pp. 252-53 (October 9, 1805).

492. Jackson, *Letters of the Lewis and Clark Expedition*, Vol. 1, Document 47, Jefferson's Instructions to Lewis, June 20, 1803, pp. 61-66.

493. Eric Jay Dolan, *Fur, Fortune, and Empire: The Epic History of the Fur Trade in America* (New York and London, W.W. Norton & Company, 2010), pp. 147-65.

494. Jackson, *Letters*, Vol. 1, Document 207, pp. 319-25.

495. The 17 miles was measured from the top of a Milk River tributary on a map

program. The Treaty of 1818 between the U.S. and Great Britain, swapped the Red River Basin south of the 49th parallel for the Milk River and other watersheds north of the 49th parallel.

496. Holland, *Feasting & Fasting with Lewis and Clark*, pp. 220-21. Moulton, *Definitive Journals*, Vol. 5, p. 256 (October 10, 1805).

497. Moulton, *Definitive Journals*, Vol. 5, p. 268 (October 13, 1805).

498. *Ibid*, pp. 318-19 (October 21, 1805).

499. Ronda, *Lewis & Clark Among the Indians*, pp. 170-71.

500. Moulton, *Definitive Journals*, Vol. 5, pp. 331-33 (October 24, 1805).

501. *Ibid*, p. 339. "The Oregon History Project," Fort Rock Sandals, Oregon Historical Society, www.ohs.org.

502. Moulton, *Definitive Journals*, Vol. 6, p. 41 (November 11, 1805).

503. *Ibid*, p. 416 (March 15, 1806).

504. *Ibid*, pp. 164-66 (January 4, 1806).

505. *Ibid*, pp. 167 (January 5, 1806).

506. *Ibid*, pp. 168 (January 6, 1806).

507. *Ibid*, p. 179 (January 8, 1806) and p. 187 (January 9, 1806).

508. *Ibid*, pp. 330-31 (February 20, 1806).

509. *Ibid*, p. 421 (March 16, 1806) and pp. 263-65 (February 1, 1806).

510. *Ibid*, pp. 425-26 (March 17, 1806).

511. Elin Woodger and Brandon Toropov, *Encyclopedia of the Lewis & Clark Expedition* (NY, Checkmark Books, imprint of Facts on File, 2004), p. 145.

512. *Ibid*, p. 421 (March 16, 1806) and pp. 425-26 (March 17, 1806).

513. *Ibid*, pp. 429-30 (March 18, 1806).

514. *Ibid*, p. 444 (March 22, 1806).

515. Moulton, *Definitive Journals*, Vol. 7, p. 149 (April 20, 1806).

516. *Ibid*, p. 156 (April 22, 1806).

517. *Ibid*, p. 180 (April 28, 1806). See also "Chemical Drugs" on Discovering Lewis and Clark website, www.lewis-clark.org

518. *Ibid*, pp. 196-97 (May 1, 1806).

519. William R. Swagerty, *The Indianization of Lewis and Clark* (Norman, OK, The Arthur H. Clark Company, a division of the University of Oklahoma Press, 2012, 2 volumes), pp.454-58.

520. Pinkham and Evans, *Lewis and Clark Among the Nez Perce*, pp. 165-67 and 236-37. The Nez Perce name of Clark's son was Halahtookit. They report the second man who died in St. Louis was Speaking Eagle, a Nez Perce. William Clark

Kennerly (as told to Elizabeth Russell), *Persimmon Hill: A Narrative of Old St. Louis and the Far West* (Norman, University of Oklahoma, 1948) pp. 62-64. Kennerly was the son of Harriet Kennerly Clark's brother. His cousin Jeff was raised in their household after Harriet died. They were the same age, born in 1824. Kennerly says Man of the Morning was a Flathead. Zoa Swayne in *Do Them No Harm!* says the mother of Clark's child left her tribe while she was pregnant and went to live with the Flatheads in the Bitterroot Valley, pp. 169 and 311-314. See also William E. Foley, *Wilderness Journey: The Life of William Clark* (Columbia, MO, University of Missouri Press, 2004, pp. 252-55; Jay H. Buckley, *William Clark: Indian Diplomat* (Norman and London, University of Oklahoma Press, 2008) pp. 219-221; Alvin M. Josephy, Jr., *The Nez Perce Indians and the Opening of the Northwest* (New Haven and London, Yale University Press, 1965), pp. 96-103. Josephy, an authority on the Nez Perce, says Man of the Morning (of the Dawn Light) was the son of a Nez Perce buffalo hunting leader and a Flathead mother.

521. Jackson, Donald, "Call Him a Good Old Dog, But Don't Call Him Scannon," *We Proceeded On* (Lewis and Clark Trail Heritage Foundation Quarterly, Vol. 2, No.3, August, 1985). Online archives, *WPO*, www.lewisandclark.org

522. Moulton, *Definitive Journals*, Volume 8, p. 113 (July 17, 1806).

523. The boundary matter was settled by the London Convention Treaty of 1818 in which the Red River Valley in North Dakota and Minnesota was swapped for tributaries of the Missouri extending further north of the U.S.-Canadian border (the Milk River, Poplar River and Big Muddy Creek). See Discovering Lewis & Clark at www.lewis-clark.org for more information regarding geography.

524. Moulton, *Definitive Journals*, Volume 8, p. 129 (July 26, 1806).

525. John C. Jackson, *The Piikani Blackfeet: A Culture Under Seige* (Missoula MT, Mountain Press Publishing Co, 2000), pp. 56-57, and note 114, p. 221. The quote is from Edmonton House Journal, Hudson's Bay Company Archives, B60/a/ 27, 11.

526. Olin D. Wheeler, *The Trail of Lewis and Clark* (Scituate MA, DSI Digital Scanning, Inc. 2002, reprint of 1904 edition, 2 volumes), Vol. 2, pp. 311-313. John Bradbury, *Travels in the Interior of American in the Years 1809, 1810, and 1811* (Google Books and University of Nebraska Press, 1986). Bradbury relates that John Colter told him that one Blackfeet had been killed by Lewis, see pp. 44-47.

527. *David Thompson's Narrative of His Explorations in Western America, 1784-1812* (Google Books), p. 375.

528. Thomas James, and Walter B. Douglas (editor, with notes & biographical sketches), *Three Years Among the Indians and Mexicans* (St. Louis, Missouri Historical Society, 1916), pp. 57-64. The reprint edition by BiblioLife contains the complete version of James's acount, with appendices. John Bradbury, *Travels in the Interior of American in the Years 1809, 1810, and 1811* (Google Books and University of Nebraska Press, 1986), pp. 44-47. Bradbury says it was a raft, James a two story beaver house. Both heard the story of his adventures directly from Colter.

529. M. O. Skarsten, *George Drouillard: Hunter and Interpreter for Lewis & Clark and the Fur Trade, 1807-1810* (Spokane WA, The Arthur H. Clark Com-

pany, 2003), pp. 282-311. Menard newspaper article, pp. 310-11. James, *Three Years Among the Indians and Mexicans*, notes by Douglas, pp. 283-86.

530. Moulton, *Definitive Journals*, Volume 8, pp. 154-55 (August 11, 1806).

531. *Ibid*, p. 158 (August 12, 1806).

532. *Ibid*, p. 290 (August 12, 1806).

533. Potter, *Sheheke*, pp.112-18,

534. Moulton, *Definitive Journals*, Volume 8, p. 319 (August 22, 1806).

535. *Ibid*, pp. 346-7 (September 3, 1806).

536. *Ibid*, p. 349 (September 4, 1806). Jackson, *Letters*, Nicholas Biddle notes, c. April, 1810, Document No. 326, pp. 541-42.

11. The Burr-Wilkinson Conspiracy: 1804–07

537. Thomas Fleming, *Duel: Alexander Hamilton, Aaron Burr and the Future of America* (Basic Books, A Member of the Perseus Books Group, 1999), pp.260--63.
538. Charles Raymond Brown, *The Northern Confederacy: According to the Plans of the Essex Junto, 1796-1814* (Kessinger Publishing's Rare Reprints, first published 1915), p. 32. See also Chernow, *Alexander Hamilton,* pp. 678-79.

539. Fleming, *Duel*, pp. 269-270. Original letter at National Archives. website. www.founders.archives.gov. See also *The True George Washington* by Paul Leicester Ford, 1900, cited in Wikipedia article on Morris.

540. Robert E. May, "Young American Males and Filibustering in the Age of Manifest Destiny: The United States Army as a Cultural Mirror," *The Journal of American History*, Vol. 78, No. 3 (Dec.,1991), pp. 857-886. Rafael de la Cova, "Filibusters and Freemasons: The Sworn Obligation," *Journal of the Early Republic,* Vol. 17, No. 1 (Spring, 1997), pp. 95-120.

541. Fleming, *Duel,* p. 280. Helen I. Cowan, *Charles Williamson: Genesee Promoter* (Clifton NJ, Augustus M. Kelley Publishers, 1973), pp. 270 and 278-79.

542. Fleming, *Duel*, pp. 231-33, 301-03, 314, 342. Letter to Thomas Sedgewick online at Massachusetts Historical Society, www.masshist.org. Chernow, *Alexander Hamilton*, p. 697,

543. *Ibid*, pp. 351-52.

544. Thomas Robson Hay, "Charles Williamson and the Burr Conspiracy," *The Journal of Southern History,* Vol. 2, No. 2 (May, 1936), p. 184.

545. Isaac Joslin Cox, "Opening the Sante Fe Trail," *Missouri Historical Review*, Vol. 25 (Oct. 1930), p. 37. Letter to Charles Biddle, March 18, 1805.

546. Abernethy, *The Burr Conspiracy*, pp.24-26.

547. *Ibid*, pp. 28-30. See also Stewart, *American Emperor,* pp. 104-05, 113.

548. *Ibid*, pp. 53-54.

549. Crackel, *Mr. Jefferson's Army,* p. 115.

550. Wilkinson, *Memoirs,* Vol. 2, Appendix XXXIII.

551. Wilkinson to Dearborn, September 8, 1805 marked "Private," in the Library of Congress, American Memory, *The First American West: The Ohio River Valley, 1750-1820* Collection. This collection includes typed transcripts and original handwritten texts. The letter is in the handwriting of Lewis Edwards, the War Dept. clerk who wrote the so called "Russell Statement" for Wilkinson during his court martial in 1811. The letter is found in a private collection—not entered as received by the War Department—and so it may be a copy but it is signed JW in Wilkinson's handwriting. The letter probably came from the collection of Reuben Durrett, founder of the Filson Historical Society in Louisville, Kentucky. No other writings found in this handwriting connected to Wilkinson have been located, other than Wilkinson's notes during his congressional investigations in 1808 and his court martial in 1811.

552. Milton Lomask, *Aaron Burr: The Conspiracy and Years of Exile, 1805-1836* (New York, Farrar, Straus, Giroux, 1982), pp. 75-78.

553. Kline, *Political Correspondence and Papers of Aaron Burr*, pp. 968-69.

554. *Ibid*, 943-48.

555. Stewart, *American Emperor*, pp. 118-120. Abernethy, *The Burr Conspiracy*, pp 38-40, 55-56. Stewart says $3,000, Abernethy says $2,500.

556. "William S. Smith and Samuel G. Ogeden Trials: 1806." Law Library, www.jrank.org

556. Stewart, *American Emperor*, pp. 120-21.

558. Crackel, *Mr. Jefferson's Army,* p. 124 and www.avalon.law.yale.edu.

559. Donald Jackson, *The Journals of Zebulon Montgomery Pike with Letters and Related Documents* (Norman, University of Oklahoma Press, 1966, 2 volumes), Vol. 2, pp. 122-26.

560. David E. Narrett, "Liberation and Conquest: John Hamilton Robinson and U. S. Adventurism toward Mexico, 1806-1819," Western Historical Quarterly, (Utah State University and Western History Association), Vol. 40, No. 1, (Spring 2009).

561. Jackson, *Journals of Zebulon Montgomery Pike,* p. 441.

562. *Ibid,* pp. 441-42.

563. Abernethy, *The Burr Conspiracy*, p. 163. American State Papers, House of Representatives, 11th Congress, 2nd sessions, Miscellaneous, Vol. 2, p. 113.

564. Kennedy, *Burr, Hamilton and Jefferson*, p. 246.

565. Abernethy, *The Burr Conspiracy,* pp. 73-77.

566. *Ibid*, pp. 74-75.

567. Jacobs, *Tarnished Warrior,* pp. 218, 280-82.

568. Abernethy, *The Burr Conspiracy*, pp. 38-40.

569. *Ibid*, pp 39, 57-58.

570. *Ibid*, pp. 57-58.

571. McCaleb, *The Aaron Burr Conspiracy*, pp. 74-75.

572. *Ibid*, pp. 74-75 and Abernethy, *The Burr Conspiracy*, 72-73.

573. Abernethy, *The Burr Conspiracy*, p. 103.

574. *Ibid*, pp. 192, 197-98. Kline, *Political Correspondence and Papers of Aaron Burr*, pp. 1051-56, 1059-60.

575. *Ibid*, pp. 141-46, 180-81.

576. *Ibid*, p. 146.

577. *Ibid*, pp. 148-49.

578. *Ibid*, pp. 150-51.

579. *Ibid*, pp.160-163.

580. William Claiborne, *Official Letter Books of W. C. C. Claiborne, 1801–1816* (www.rarebooksclub.com, Nabu Public Domain Reprints, 2012, first published, 1917), Volume 4 (Sept 5, 1806–November 5, 1809), pp,53-54 (Jackson to Claiborne, November 12, 1806) and pp. 114-15 (Claiborne to Madison, Jan. 29, 1807).

581. Daniel Bissell's statement concerning Burr's activities in the vicinity of Fort Massac is found at *Letters Received by the Secretary of War*, Registered Series, 1801-1870, Microfilm M221, Roll 4 (Vol.3, May 1, 1806–Dec. 31, 1808), National Archives and Records Administration. Bissell was the fort commander. A transcript is on the internet at chribbs.wordpress.com. Look under documents/Chribbs/ "1807–Daniel Bissell's Account of Aaron Burr." The other reference to Mr. Hopkins is found in Claiborne's, *Official Letter Books*, Vol. 5, letter from Claiborne to Robert Smith, December 31, 1809. Volume 5 of the *Letter Books* is available as a reprint from the University of California Libraries. Digital versions at archives.org.

582. Thomas Jefferson Papers, American Memories, Library of Congress website (Jefferson to Wilkinson, February 3, 1807). James Wilkinson. *Burr's Conspiracy Exposed; and General Wilkinson Vindicated Against the Slander of His Enemies on that Important Occasion* (Nabu Press, 2009, first published, 1811), pp. 33-36.

583. Abernethy, *The Aaron Burr Conspiracy*, pp. 199-226. Stewart, *American Emperor*, pp. 193-207.

12. Governor of Louisiana Territory: 1807-09

584. Charles vanRavenswaay, *St. Louis: An Informal History of the City and Its People, 1764-1865* (St. Louis, Missouri Historical Society Press, 1991), p. 139, and Moulton, *The Definitive Journals*, Vol. 9, John Ordway's Journal, pp. 365-66 (September 21-23, 1806).

585. Clarence Edwin Carter, *The Territorial Papers of the United States, Vol. XIV:*

The Territory of Louisiana-Missouri, 1806–1814 (Washington, U. S. Gov't Printing Office, 1949), pp. 43-46 (Easton to Jefferson, December 1, 1806).

586. *Ibid*, p. 15 (Silas Bent to Jared Mansfield, October 5, 1806)

587. William E. Foley, *The Genesis of Missouri: From Wilderness Outpost to Statehood* (Columbia and London, University of Missouri Press,1989), pp. 170-75.

588. Henry Rosecrans Burke, Papers (Missouri History Museum Archives, St. Louis). "Notes on John Smith T." (Folder 6) and "Design for Treason" (Folder 8); Dick Steward, *Frontier Swashbuckler: The Life and Legend of John Smith T* (Columbia, Missouri, University of Missouri Press, 2000), pp. 8-16. Coleman, C. W., "Geneaology of the Smith Family of Essex County, Virginia," William and Mary Quarterly Vol. 25, No. 3 (January, 1917). Burke was a descendant of Smith T's family and said another descendant showed him vases given to General Tom Smith by his "kinsman, General Wilkinson," see "Notes on John Smith T" p. 6. Coleman was not able to establish a geneaology link. Regarding the 12,000 pounds of lead for Burr, see p. 33, in manuscript "Design for Treason."

589. Steward, *Frontier Swashbuckler*, pp. 76-77.

590. John F. Darby, *Personal Recollections of John F. Darby: Mayor of St. Louis, 1835* (St. Louis, Hawthorne Publishing Company, 1880), pp. 58

591. Henry Marie Brackenridge, *Recollections of Persons and Places in the West* (Bibliolife, 2009, originally published 1868), pp. 216-222. Darby, *Personal Recollections*, pp. 57, 65.

592. Kristie C. Wolferman, *The Osage in Missouri* (Columbia and London, University of Missouri Press, 1997), p. 60; Jackson, *Letters of the Lewis and Clark Expedition*, p. 325; Robert B. Betts, with a new epilogue by James J. Holmberg, *In Search of York: The Slave Who Went to the Pacific with Lewis and Clark, Revised Edition* (Boulder, Co, Colorado Associated University Press, originally published 1985, revised edition, 2000), pp. 109-110;

593. William Clark Kennerly, as told to Elizabeth Russell, *Persimmon Hill: A Narrative of Old St. Louis and the Far West* (Norman, University of Oklahoma Press, 1948), pp. 19-20. Betts, *In Search of York,* pp. 109-110.

594. James J. Holmberg, "Lewis and Clark at the Falls of the Ohio Timeline" (Filson Historical Society, www.lewisandclarkinkentucky.org). Danisi, *Uncovering the Truth,* pp. 78-80.

595. Stewart, *American Emperor,* pp. 242, 272.

596. Holmberg, "Lewis and Clark at the Falls of Ohio Timeline."

597. Holmberg, *Epilogue, In Search of York*, pp. 156-57. Jones, *William Clark,* p. 154.

598. Danisi, *Uncovering the Truth*, pp. 81-83.

599. *Ibid*, p. 83. Jackson, *Letters,* Document 220, pp. 350-51 (Jefferson to Lewis, October 26, 1806).

600. Gene Crotty, *The Visits of Lewis & Clark to Fincastle, Virginia* (History Museum and Historical Society of Western Virginia, 2003), pp. 39-47.

601. Jackson, *Letters*, Document 236, pp. 363-64 (Dearborn to Willis Alston, January 14, 1807); Document 246, pp. 380-82 ("Messrs. Lewis & Clarke's Donation Lands," March 6, 1807).

602. For the pay amounts, see Joseph Musselman, "Discovering Lewis & Clark" website, www.lewis-clark.org, section called "Worth Their Salt." Arlen Large, "Captain Clark's Belated Bonus," *We Proceeded On* (Lewis and Clark Trail Heritage Foundation), Vol 23, No. 4, November, 1997., pp. 10-12. Available as download on the foundation website at www.lewisandclark.org.

603. Jackson, *Letters*, Document 214, p. 347 (Clark to Dearborn, Oct. 10, 1806).

604. Jay H. Buckley, "William Clark: Superintendent of Indian Affairs, 1813-38," a dissertation, University of Nebraska-Lincoln, www.digitalcommons.unl.edu. See also Buckley, *William Clark: Indian Diplomat*, p. 69.

605. Holmberg, *Dear Brother*, pp. 119-26.

606. Lewis to Clark, letter of March 13, 1807, in William Clark Papers, Missouri History Museum. This letter is a typescript, misdated March 15, 1807.

607. Jackson, *Letters*, Document 254, p. 387 (Lewis to Clark, March 15, 1807).

608. *Ibid*, Document 266, pp. 409-10 (Frazier to Jefferson, April 16, 1807).

609. Carter, "The Burr-Wilkinson Intrigue in St. Louis," p. 462. Steward, *Frontier Swashbuckler*, p. 49.

610. Thomas Maitland Marshall, *The Life and Papers of Frederick Bates* (St. Louis, Missouri Historical Society, 2 volumes, 1926). Vol 1, pp. 103-09.

611. *Ibid*, pp. 110-14 (Bates to Woodson, May 1, 1807).

612. Schroeder, *Opening the Ozarks*, p. 288.. Frederick Bates was from Goochland County, Virginia. Moses Bates was from Durham, Connecticut.

613. *Ibid*, pp. 135-142

614. Frederick A. Hodes, *Beyond the Frontier: A History of St. Louis to 1821* (Tucson, Arizona, 2004), pp. 128-32.

615. Jackson, *Letters*, Document 268, pp. 411-12, (Clark to Dearborn, May 18, 1807); Foley, *Wilderness Journey*, p. 163.

616. Potter, *Sheheke*, pp. 138-39.

617. Jackson, *Letters*, Document 270, p. 414 (Clark to Dearborn, June 1, 1807).

618. Elliott Coues, "Letters of William Clark and Nathaniel Pryor," *Annals of Iowa* (Des Moines, State Historical Society of Iowa), Vol 1, No. 8, (1895) pp. 615-619. Available as a pdf download at Iowa Research Online, ir.uiowa.edu. It is also found in Jackson's *Letters*, Document 279, pp. 432-38 (Nathaniel Pryor to Clark, October 16, 1807).

619. James J. Holmberg, "Seaman's Fate," *We Proceeded On* (Lewis and Clark Trail

Heritage Foundation), Vol. 26, No. 1 (February, 2000), pp. 7-9. Available on both www.lewisandclark.org (foundation website) and www.lewis-clark.org (Discovering Lewis & Clark website).

620. Jacques Nicolas Léger, *Haiti, Her History and Her Detractors* (Classic Reprint first published 1907), p. 52. Available at www.archive.org.

621. Donald Jackson, "On the Death of Meriwether Lewis's Servant," *The William and Mary Quarterly* (Omuohundro Institute of Early American History and Culture), 3rd Serl, Vol. 21, No. 3 (July, 1964), pp. 445-448.

622. Potter, *Sheheke,* pp. 125-36.

623. Jackson, *Letters,* Document 262a, pp. 394-97 (Conrad Prospectus, c. April 1, 1807).

624. Jackson, "A Notebook Sketch Among the Lewis & Clark Papers" *We Proceeded On* (Lewis and Clark Trail Heritage Foundation), Vol. 12, No. 3 (August, 1986), p. 15. Jackson discovered a drawing on the back cover of Lewis's astronomical notebook at the University of Missouri Library that looks exactly like the woodcuts that appeared in Gass's published journal. The drawing was published in *WPO* with a brief note. See archives at www.lewisandclark.org and search by issue.

625. Cutright, *History of the Lewis and Clark Journals,* pp. 25-26.

626. Jackson, *Letters,* Document 257, p. 389 (Jefferson to William Hamilton, March 22, 1807).

627. *Ibid,* Document 253, pp. 385-86 (Lewis to the Public, March 13, 1807).

628. *Ibid,* Document 295, pp. 462-63 (Lewis and Clark: Settlement of Account, August 21, 1809).

629. Jones, *William Clark,* pp. 208, 216, 238. Jackson, *Letters,* Document 388, pp. 621-23 (Thomas Ashley to Charles Chauncey, September 18, 1816).

630. Moulton, *Definitive Edition,* Vol. 2, p. 38. The *Definitive Edition,* referenced in these footnotes, is the paperback version of the original hard cover publication, *The Journals of the Lewis and Clark Expedition.* The journals may be read online at www.lewisandclarkjournals.unl.edu

631. Jackson, *Letters,* Document 431, pp. 681-684 (Mahlon Dickerson Diary, 1802-09).

632. Dillon, *Meriwether Lewis,* p. 278. Dillon is mistaken about the letter being addressed to William Clark. The letter, dated February 11, 1807, was written to Auguste Chouteau, Sr. and hand carried to him as a letter of introduction for Frederick Bates. A typed transcript is found in the Grace Lewis Miller papers.

633. The next known correspondence by Lewis is a letter dated October 28, 1807, at Ivy Creek, his mother's home in Albemarle County. Jefferson wrote a letter to him in July, 1808 saying that he had not heard from him since he parted with him in Albemarle in September. Jackson, *Letters,* Document 275, p. 418 (Lewis to Jefferson,June 27, 1807); Documents 276, p. 419 (Simmons to Lewis, July 31, 1807); Document 277, pp. 419-431 (Financial Records of the Expedition, August

5,1807); Document 280, pp. 438-39 (Lewis's receipt to William Woods, October 28, 1807); Document 286, pp. 444-46 (Jefferson to Lewis, July 17, 1808). Danisi, *Uncovering the Truth*, footnote 12, p. 374 regarding the ledgers.

634. R. Kent Newmyer, *The Treason Trial of Aaron Burr: Law, Politics and the Character Wars of the New Nation* (Cambridge, Cambridge University Press, 2012), pp. 62-71. Stewart, *American Emperor,* pp. 257-258.

635. Stewart, *American Emperor,* 259-73.

636. Jackson, *Letters,* Document 443, pp. 696-719 (Observations and Reflections of Lewis, August, 1807). None of the historians who write about Lewis's "Observations and Reflections" have made the connection between this and his planned publication. It reflects the lack of interest and understanding shown in his publication plans. For a discussion of his plans regarding the fur trade and Indians, see pp. 270-73 in Chapter 10 of this book. The compete text with annotations is also found in Danisi and Jackson's book, *Meriwether Lewis,* pp. 349-373.

637. General Wilkinson wrote to Thomas Jefferson on September 15, 1807. See Thomas Jefferson Papers, American Memory website, Library of Congress.

638. Danisi, *Uncovering the Truth,* p. 321. Lewis's account book for dates of October 8th and 20th, 1807. The accounts book is in the archives of the Missouri History Museum.

639. Jackson, *Letters,* Document 44, pp. 719-21 (Lewis to Mahlon Dickerson, November 3, 1807). This is the only known letter of Lewis to Dickerson.

640. William Clark Papers, Clark to Lewis, April 7, 1807, Missouri History Museum; James R. Bentley, "Two Letters of Meriwether Lewis to Major William Preston," *Filson History Quarterly,* Vol. 44 (April, 1970), pp. 170-75. Ambrose, *Undaunted Courage,* p. 440. Historians mention Letitia's sister as another possible candidate for marriage, Elizabeth Brackenridge, but she was 14 years old at this time. See Crotty, *The Visits of Lewis & Clark to Fincastle, Virginia,* pp. 54-55.

641. Meriwether Lewis Papers, Lewis to Lucy Marks, February 15, 1808. Missouri History Museum, Meriwether Lewis Papers, Folder 10.

642. Grace Lewis Miller, *His Excellency Meriwether Lewis and the First Publication West of the Mississippi River,* (Master's Thesis, University of Texas-Austin, August, 1948. (St. Louis, Grace Lewis Miller Papers, Jefferson National Expansion Memorial), pp. 42-49. Miller proposes that Lewis bought the press, and I agree.

643. *Ibid,* 76-78. David Kaser, *Joseph Charless: Printer in the Western Country* (Philadelphia, University of Pennsylvania Press, 1963), pp. 64-69. Library of Congress, "About Louisiana Gazette (St. Louis [Mo.]) 1809-1812." Kaser notes that the first extant issue of the *Gazette,* dated July 26th, states Charless was going to Kentucky.

644. Kaser, *Joseph Charless,* p. 69.

645. *Ibid,* pp. 70-71.

646. Grace Lewis (Miller), "The First Home of Governor Lewis in Louisiana Territory" ". . . the house in which General Clark resides." (St. Louis, *Missouri Historical*

Society Bulletin), Vol XIV (July, 1958), pp. 357-68.

647. Holmberg, *Dear Brother*, p.144.

648. *Ibid*, p. 154, 158. http://familytreemaker,geneaology.com, search for Richard Clough Anderson.

649. *Ibid*, p. 160.

650. *Ibid*, pp. 172-73.

651. *Ibid*, pp. 183-84.

652. *Ibid*, p. 212. Betts, *In Search of York* with an *Epilogue* by Holmberg, pp. 119, 163-70.

653. Grace Lewis Miller speculates Lewis moved out because he paid $125 rent on "General Clark's house" on August 31, 1808. Clark repaid him a year later. There are no entries in Lewis's account book for rent other than this. Miller was unaware of Clark's letter to his brother establishing that Lewis was living with them in November, 1808. "Life as Governor til Death," Box 48, Folder 11.

654. Danisi, *Uncovering the Truth,* pp. 221-22 (Lewis to Mrs. Lucy Marks, December 1, 1808). Grace Lewis Miller, "The Landed Estate of Meriwether Lewis in Louisiana Territory (GLM Collection, JNEM), June, 1970, pp. 1-4, 16-26, 34. The property of 3,000 acres extended approximately from North Broadway on the east; Page Blvd and Dr. Martin Luther King Drive on the south; Lucas-Hunt Road on the west; and St. Cyr Road on the north. The property intended for his mother was in the northeast portion, south of Bellefontaine Park. The Great Mound was located between Second and Broadway, and Mound and Brooklyn streets.

655. Danisi, *Uncovering the Truth*, pp. 128-29, 151.

656. Grace Lewis Miller, "Life as Governor Til Death," Box 48, Folder 11, JNEM, pp. 40-48.

657. Carter, *Territorial Papers*, Vol. XIV, pp. 196-203 (Lewis to Dearborn, July 1, 1808); p. 204 (Dearborn to Lewis, July 2, 1808); pp.212-216 (Lewis to Dearborn, August 20, 1808); pp. 221--222 (Jefferson to Lewis, August 24, 1808).

658. Danisi, *Uncovering the Truth*, p. 138, and footnote 61, p. 378. Carter, *Territorial Papers*, Vol. XIV, pp. 293-312 (Lewis to Secretary of War, August 27, 1809). Danisi comments that Donald Jackson, the editor of the *Letters of Lewis and Clark*, believed Lewis did not write to Jefferson after June 27, 1807, but didn't realize the two very long letters to the Secretary of War were also intended for Jefferson and Madison. Both letters also contained additional enclosures. These are the letters dated December 15, 1808 and August 27, 1809. 659. Carter, Territorial Papers, Vol. XIV, footnote, p. 221.

660. *Ibid*, p. 260 (Clark to Secretary of War, April 5, 1809); pp. 264-271 (Clark to Secretary of War, April 29, 1809).

661. Buckley, *William Clark, Indian Diplomat*, pp. 75-77. Marshall, *Life and Papers of Frederick Bates*, pp. 344-345 (Bates to John B. Treat, May 26, 1808).

662. Oglesby, *Manuel Lisa,* pp. 65,

663. GLM papers typed transcripts of letters of 1809. Danisi, *Meriwether Lewis,* p. 241 and note 47, p. 250.

664. Carter, *Territorial Papers*, Vol. XIV, pp. 334-345 (Judge Lucas to the Secretary of the Treasury, October 19, 1809).

665. *Ibid,* pp. 353-356 (Judge Lucas to the Secretary of War, December, 1809).

666. *Ibid,* pp. 328-330, ("A Land Claimant" to the *Missouri Gazette,* October 16, 1809). The petition is on pp. 323-327, dated October 10, 1809 and signed by a Select Committee of John Smith T, Edward Hempstead, and George C. C. Harbison. It was received by the Committee on Public Lands on January 3, 1810.

667. Marshall, *Life and Papers of Frederick Bates,* Vol. II, pp.77-79, Moses Austin to Bates, August 27, 1809. Letter of Frederick Bates to his brother, Richard Bates, July 14, 1809, pp. 67-73 Regarding Lucas, "He was never my friend." p. 70.

668. *Ibid*, pp. 108-112.

669. Danisi, *Uncovering the Truth,* Chapter 9, pp. 108-123, "Was Governor Lewis's Correspondence Intentionally Delayed?" This letter was also not included in Jackson's *Letters.*

670. Regarding the letterbook, see Gale, *The Death of Meriwether Lewis,* Document 5, Memoradum of Lewis's Personal Effects, pp. 236-238 and Document 6, p. 239. The memorandum and Coles's receipt are also found in Jackson, *Letters,* Documents 303 & 304, pp. 470-474.

671. Danisi, *Uncovering the Truth,* Appendix A. #15 "Meriwether Lewis to William Simmons, August 18, 1809, Saint Louis," pp. 225-27. The sentence is on page 227. Regarding Jackson and his omission, see p. 148, and footnote 100, p. 381 in Danisi. The full text is also found in Carter's *Territorial Papers,* Vol XIV, pp. 290-293.
The original letter is found in RG 107, L-328 (4). Received, September 8th, 1809. In examining the original handwriting, there is no doubt Jackson's omission was deliberate. On p. 6 a separate paragraph at the end of the page reads "I have reason to believe that sundry of my Letters have been lost as there re" and continues on p. 7, "main several important Subjests on which I have received no Answer." The word "remain" is split in two parts. Jackson writes the sentence with no capital at the beginning: "as there remain several important Subjects on which I have not yet received an Answer." Since the word remain is in his text, he would have seen the 13 words preceding it, "I have reason to believe that sundry of my Letters have been lost." Jackson's version of the August 18, 1809 letter is found in his *Letters,* Document 294, pp. 459-461. The footnote, " Lewis well deserved a scolding," is found on p. 445. The letter is Document 286, pp. 444-446. (Jefferson to Lewis, July 17, 1808).

Danisi has his own distortion in publishing the August 18, 1809 letter. He states it was written by Lewis to the government accountant William Simmons. Instead it was written to William Eustis, Secretary of War, in response to a letter that Lewis received from Eustis dated July 15, 1809. Danisi says the July 15th letter was written by Simmons. He accuses William Simmons of petty and malicious behavior, and sa exerting financial control over the Treasury and War Departments. His biased thinking is revealed in crediting the letters as being sent and received by Simmons rather than by Eustis. Carter in the *Territorial Papers* correctly states they were written to and from the Secretary of War. (See pp. 285-86 for the July 15, 1809 letter from Eustis.) Again, this is a deliberate alteration of an important document by a historian.

672. Grace Lewis Miller typed countless transcripts of handwritten documents in her research on Meriwether Lewis.Typed transcripts of Lewis's communications for the year 1808 fill two very fat file folders of mine. The letters cited are found there. Otherwise, a researcher should consult the microfilm in the National Archives for Lewis's correspondence.

673. Danisi, *Uncovering the Truth*, p. 123. Jackson, *Letters*, Document 210, pp.335-343 (Lewis to an Unknown Correspondent, September 29, 1806).

674. Ralph Ketcham, *James Madison: A Biography* (Charlottesville and London, University of Virginia Press, 1990), pp. 460-66; Adams, Henry, *Documents Relating to New England Federalism, 1800–1815* (Kessinger Publishing's Rare Reprints), pp. 24-25. 236-237. Brown, *The Northern Confederacy,* pp. 65-72. Isenberg, *Fallen Founder*, p. 371.

675. Ketcham, *James Madison,* pp. 481, 483, 484.

676. Côte, *Theodosia*, pp. 93-94, 105, 112-113, 162, 173, 248-250, 276.

677. Jacobs, *Tarnished Warrior*, pp. 67-68; Danisi, *Uncovering the Truth,* p.144.

678. Kennerly, *Persimmon Hill*, pp. 16-18: Jackson, *Letters,* pp. 526 (Nicholas Biddle Notes, circa April 1810).

679. Carter, *Territorial Papers*, pp. 285-86, The Secretary of War to Governor Lewis, July 15, 1809. Meriwether Lewis to Secretary of War, May 13, 1809 in Grace Lewis Miller papers, JNEM. Either in "missing files, 1808-09" which were returned by an anonymous party and found on a desk in 1980, or in the files for 1809 documents. I keep the missing files separate, as a curiosity.

680. Oglesby, *Manuel Lisa*, pp. 74-84; Potter, *Sheheke,* 154-63. Jackson, *Letters,* Document 291, pp. 451-456 (Lewis to Pierre Chouteau, June 8, 1809).

681. Susan Colby, *Sacagawea's Child: The Life and Times of Jean-Baptiste (Pomp) Charbonneau* (Spokane, WA, Arthur H. Clark Company, 2005), pp. 71-99. Jones, *William Clark*, p. 194.

682. Danisi, *Uncovering the Truth*, Appendix D, Notes on Meriwether Lewis's Real Estate Transactions and Personal Debts, p. 289.

683. Carter, *Territorial Papers*, pp. 289-90, The Secretary of War to William Clark, August 7. 1809. Danisi, *Uncovering the Truth*, p. 145, and note p. 380.

684. Wheeler, *The Trail of Lewis and Clark,* p. 70. has a photographic image of Robert Smith's handwritten. signed, refusal to pay the bill, submitted by Lewis to James Madison, on February 6th, 1809, and noted by Smith as being refused on July 8, 1809. Marshall. *Life and Papers of Frederick Bates,* pp.73-74, Bates to Albert Gallatin, July 17, 1809. Danisi, *Uncovering the Truth*, pp. 150-51.

685. Danisi, *Uncovering the Truth,* pp.150-51.

686. It is not difficult to associate Dr. Eustis to Aaron Burr and, by extension, to the Northern Confederacy. See Côte, *Theodosia,* pp. 120, 126. 141. Wilkinson and Samuel Smith were longtime friends. See James Wilkinson Papers, for correspondence between him and Smith, University of Pittsburgh Library. The papers are digitized and available for pdf download. http://digital.library.pitt.edu
It is my speculation they were aware of conspiracy plans to invade Mexico that were underway in 1809, and wanted Lewis gone from office.

687. *Ibid*, pp. 225-27. Lewis's letter of August 18, 1809, mistakenly labeled as being addressed to William Simmons, rather than to William Eustis, Secretary of War. Carter identifies the recipient correctly. Carter, *Territorial Papers,* pp. 290-293.

688. *Ibid*, Appendix D, Deed Book Transactions, Meriwether Lewis as Grantor and Grantee, pp. 292-95 and pp. 288-89. Appendix A, Will Carr to Charles Carr, August 25, 1809, pp. 227-28.

689. Holmberg, *Dear Brother*, pp. 233-36.

690. Danisi, *Uncovering the Truth*, p. 401, note 84. Lewis's handwriting may be examined in the microfilm, National Archives, RG 107, M 222, roll 38, L 101, frames 4907-31.

691. Wilkinson Papers, James Wilkinson to Samuel H. Smith, April 25, 1806, June 4, 1806, June 10 1806, regarding John B. C. Lucas.

692. Marshall, *Life and Papers of Frederick Bates*, pp. 86-92 (Bates to William Eustis, Secretary of War, September 28, 1809). Holmberg, *Dear Brother*. pp. 209-214, William Clark to Jonathan Clark, August 26, 1809.

693. Frederick Bates Collection, Missouri History Museum, James House to Frederick Bates, September 28, 1809.

694. Gilbert C. Russell to Thomas Jefferson, January 4 and January 31, 1809. American Memory, Library of Congress, *Thomas Jefferson Papers*. Holmberg, *Dear Brother*, pp. 228-233, William Clark to Jonathan Clark, November 26, 1809. Jackson, *Letters,* Document 346, Statement of Gilbert C. Russell, November 26, 1811. Oddly enough, Jackson titles the document using Russell's middle initial, C. for Christian. Russell always signed his name with the C included, and in fact the C joined his first and last names together as a single unbroken signature. The statement was signed without the C. middle initial. Jonathan Williams signature as a witness was also forged, and signed simply J. Williams, when he always signed his name Jon. Williams. The two experts who testified at the Coroner's Inquest were Gerald B. Richards and Duayne Dillon. See *The Death of Meriwether Lewis* for their

testimonies. Richards is the former head of the FBI's Documents Division. Dillon, is a California criminologist.

695. Meriwether Lewis Collection, Missouri History Museum. The letter is not found in the papers of James Madison. This copy was either never sent, or it appears to be a draft.

696. James E. Starrs and Kira Gale, *The Death of Meriwether Lewis: A Historic Crime Scene Investigation* (River Junction Press LLC, 2009, 2012), pp. 246-47. Also available on Library of Congress American Memory website, *Thomas Jefferson Papers.*

697. Holmberg, *Dear Brother*, pp. 228-233, William Clark to Jonathan Clark, November 26, 1809. See pp. 42-43 in this book for the George Rogers Clark incident.

698. Starrs & Gale, *The Death of Meriwether Lewis,* pp. 247-48, or TJ Papers.

699. As this book is going to press, Earl Weidner has turned up the name of an army officer who fits the profile of "Mr. X." His name is Major Ambrose Whitlock, and he was Paymaster for the District. He was associated with Wilkinson since the Indian Wars, having served as a sergeant at Fort Washington in Cincinnati for five years. He was a 2nd lieutenant of the First Infantry Regiment with Benjamin Wilkinson in 1801. He had testified under oath supporting General Wilkinson's version of his encounter with Major Bruff in 1805. (Wilkinson's *Memoirs*, Vol. 2, pp. 297, 720.) Governor Harrison of Indiana reported that Whitlock had just returned from Fort Madison to Vincennes on July 8, 1809. ("Tecumseh's Travels Revisited, *Indiana Magazine of History,* Vol. 96, No 2, pp. 150-68, available online). He was known as an incorruptible paymaster, who never lost or misappropriated any money. He served under General Wilkinson in building the Natchez Trace as a military road in 1801-03. Whether or not he was "Mr. X," his record demonstrates there was an officer serving on the frontier who easily fit the profile of the conspirator "Mr. X."

700. Tony Turnbow, "The Man Who Abandoned Meriwether Lewis," pp. 20-31, *We Proceeded On*, 38:2, May, 2012. Available as download, www.lewisandclark.org, Archives of the Lewis and Clark Trail Heritage Foundation.

701. Starrs & Gale, *Death of Meriwether Lewis*, pp. 248-49, or Thomas Jefferson Papers on the Library of Congress American Memories website.

702. Lewis-Marks Collection, University of Virginia Special Collections, for unsigned note regarding Neelly having Lewis's pistols and gold watch. Meriwether Lewis Collection, Missouri History Museum, for John Marks to Reuben Lewis, January 22, 1812.

703. Both letters by John Brahan may be examined on the Library of Cogress American Memory website, *Thomas Jefferson Papers.* Search for Brahan and for Neelly, October 18, 1809. Starrs & Gale, *Death of Meriwether Lewis*, 2nd edition (2012), pp. 17-22.

704. Tony Turnbow, the author of several articles in WPO on the death of Lewis, did the research regarding the court records. He found them in U. S. Federal Records, Mississippi Territory, R M1315, Atlanta NARA office, the Nashville archives in Atlanta. A descendant of Joseph Van Meter, Jim Van Meter, supplied the infor-

mation, which may be found on a genealogy site online, by searching for Colonel Joseph Van Meter.

705. Dr. William Lindsey McDonald, *Lore of the River: The Shoals of Long Ago* (Bluewater Publications, USA, expanded 3rd edition, 2007), pp. 112-114.

706. Martin Cooper Avery, "Tragedy at Grinder's Stand: The Death of Meriwether Lewis," a Master's thesis at Middle Tennessee State University, 1978, pp. 49-50.

707. Starrs & Gale, The Death of Meriwether Lewis: A Historic Crime Scene Investigation, pp. 234-36.

708. *Ibid*, pp. 251-53.

709. *Ibid*, pp. 260-262.

710. http://www.tngennet.org/lewis/meriwether.lewis.htm. Story of Meriwether Lewis as told by the Cooper Family.

711. Starrs & Gale, *Death of Meriwether Lewis*, pp. 262-65.

712. Avery, "Tragedy at Grinder's Stand, pp. 65-67.

713. Tony Turnbow, the lawyer who discovered Neelly's court record in Franklin County court, was the organizer of the event. Turnbow's family has lived in Hohenwald, Tennessee since the early 1800's. He also introduced me at my presentation at the Southern Book Festival in Nashville in October, 2009. The book talk was shown on C-Span Book TV, and both of us also appeared on Brad Meltzer's *History Decoded* series on the History Channel in 2010.

714. Jay H. Buckley, "A Postmortem Trial concerning Meriwether Lewis's Controversial Death," a chapter in John Guice's book, *By His Own Hand?,* pp. 137-38.

715. Jackson, *Letters,* Document 303, pp. 470-74 (Memorandum of Lewis's Personal Effects, November 23, 1809). Original in Lewis-Marks Manuscripts, University of Viginia Special Collections, Memorandum of items contained in two trunks, and receipt of personal effects by Issac Cole, January 10, 1810. The inventory and receipt from Isaac J. Coles are in Starrs & Gale, *The Death of Meriwether Lewis*, pp. 236-39. Contrary to Jackson's assertion, (and my own book) the William Anderson who signed the inventory was William P. Anderson, not William C. Anderson. The old fashioned writing of P. looked much like a C. It should be noted that this letter from Isaac Coles is only found in the Lewis-Marks manuscripts with the original inventory document. There is another letter by Isaac Coles dated January 5, 1810 found in the *Papers of Thomas Jefferson, Retirement Series,* Vol. 2, 16 November 1809 to 11 August 1810 (Princeton and Oxford, Princeton University Press, 2005), pp. 123-24. This letter, on the same subject, does not mention that the materials arrived "so badly assorted that no idea could be given of them. . ." The letters by Coles contain conflicting dates. The original in the Lewis-Marks collection is clearly dated January 10th. The printed transcript of the letter in the *Papers of Thomas Jefferson* is January 5th.

716. Daniel S. Dupre, *Transforming the Cotton Frontier: Madison County, Alabama 1800–1840* (Baton Rouge and London, Louisiana State University Press, 1997), p. 27.

717. *Ibid*, pp. 27, 82-87.

718. Jefferson, *Papers, Retirement Series*, Vol. 2, pp. 33-35.

719. *Ibid*, pp. 42-44.

720. Jackson, *Letters*, Document 449. (Excerpts from Clark's 1809 Journal), pp. 724-26.

721. Tony Turnbow, "The Man Who Abandoned Meriwether Lewis," *We Proceeded On*, 38:2, May, 2012. Available as PDF at the Trail Heritage Foundation archives on www.lewisandclark.org.

722. John Bakeless, *Lewis and Clark: Partners in Discovery* (New York, William Morrow and Co, 1947), p. 421, 484.

723. Jefferson, *Papers, Retirement Series,* Vol 2, pp. 208-09, 364, 672, 673n. Donald Jackson, "On the Death of Meriwether Lewis's Servant," *The William and Mary Quarterly*, Vol. 21, No. 3 (July, 1964), pp. 445-448. Jackson says that Julian Boyd, editor of the Thomas Jefferson Papers suggested the publication of the letters which Boyd had discovered. Jackson chose not to include the letters pertaining to Pernier's death in the second edition of the Letters published in 1978.

724. Jacobs, *Tarnished Warrior,* pp. 262-63.

725. Starrs & Gale, *Death of Meriwether Lewis*, p. 266.

726. James Wilkinson, *Burr's Conspiracy Exposed and General Wilkinson Vindicated Against the Slander of His Enemies* (1811), p. 97. The full version, with appendices, is available as a download at Internet Archive, www.archive.org. Years ago I did an extensive search attempting to locate the court records by contacting Natchez archives and then the University of Texas at Austin archives, but they have disappeared.The case was ultimately heard in January, 1819 in Natchez, and Wilkinson was ordered to pay Adair $2,500. Friends came to his rescue and Congress passed a bill giving Wilkinson $3,000. See Jacobs, *Tarnished Warrior,* p. 327. The September 25, 1809 letter to Eustis is found RG 107, M221A, Roll 33, W-688. The letter dated November 15, 1809 is referenced in footnote in *Tarnished Warrior*, p. 261 as being in the collection of the Chicago Historical Society, Wilkinson Papers, Vol. 3.

727. Paul Lachance, "Repercussions of the Haitian Revolution in Louisiana," a chapter in *The Impact of the Haitian Revolution in the Atlantic World,* edited by David P. Geggus (Columbia SC, University of South Carolina Press, 2001), pp. 213-19.

728. Claiborne, *Official Letter Books Volume 4*, pp. 391-393.

729. Starrs & Gale, *The Death of Meriwether Lewis, 2nd New Evidence Edition,* pp. 62, 70-71. See also footnote 733.

730. Starrs & Gale, *The Death of Meriwether Lewis*, pp. 64-69. The letters were written in French and enclosed in correspondence found in Wilkin-

son's letter to Eustis, RG 107, M221A, roll 33, W-687, at the National Archives. I had them translated by a professional translator, and published them in their entirety in the second edition of *The Death of Meriwether Lewis.* I suspect that General Wilkinson himself placed Francis Newman in the service. When the general protests, look into it further.

731. Claiborne, *Official Letter Books, Volume 4*, letters to Secretary of State Robert Smith, dated November 5, November 12, December 31, 1809. They may have been jailed for their own safety. While they were in jail, Francis Newman was promoted.

732. To learn more about Francis Newman see www.newman-family-tree.net. The website is done by Jerry Gandolfo, a descendant of one of the oldest families in New Orleans, and owner of the Historic Voodoo Museum in the French Quarters.

733. Claiborne, *Official Letter Books, Volume 5* (University of California Libraries reprint), pp. 22-24, letter to Secretary of State Robert Smith dated December 31, 1809.

734. Jacobs, *Tarnished Warrior*, p. 249.

735. Wilkinson, *Memoirs*, Vol. 2, pp. 370-73.

736. W. A. Goff, "Reuben Smith," a chapter in *Mountain Men and the Fur Trade of the Far West,* edited by Leroy Hafen (Lincoln, University of Nebraska Bison Books, 1982), Vol. 7, pp. 261-79.

737. James McCallum *A Brief Sketch of the Settlement and Early History of Giles County, Tennessee* (Southern Historical Press, 2012, first published 1928), pp. 87-90. *Nashville Democratic Clarion & Tennessee Gazette,* May (?), 1812, photocopy of page from Tennessee State Archives, My thanks to Tony Turnbow, who provided this information to me.

738. Harrison Gaylord Warren, *The Sword Was Their Passport: A History of American Filibustering in the Mexican Revolution* (Port Washington NY/London, Kennikat Press, 1943), pp. 24-25.

739. Jacobs, *Tarnished Warrior*, pp. 337-340.

740. Fisher, *Suicide or Murder?,* pp. 143-45.

741. Starrs & Gale, *The Death of Meriwether Lewis*, "Testimony of Gerald B. (Jerry) Richards," pp. 126-140, "Testimony of Duayne Dillon," pp. 164-175.

742. Jackson, *Letters*, Document 347, Gilbert C. Russell Statement, November 26, 1811, pp. 573-75. Starrs & Gale, *The Death of Meriwether Lewis,* pp. 253-256.

743. Starrs and Gale, *The Death of Meriwether Lewis, "New Evidence Edition"* published in 2012, pp. 23-28, 31-32. Thomas Danisi's book—although it has valuable information concerning the delays in Lewis's mail delivery and his finances—contains grossly inaccurate characterizations of key material. For example, Danisi writes

about the court-martial that—

> Russell . . . maintained that Wilkinson's conduct during the years he was in Mississippi and Orleans territories was exemplary and he also gave a character reference.

Major Russell actually said the general *had attempted to assassinate Seth Hunt*. Danisi knows better. He is simply building a case for accepting the Russell Statement as genuine, despite the fact he knows the two document experts testifed under oath that it was a forgery. (Danisi, *Uncovering the Truth*, p. 173).

Danisi's interpretation of other key material ignores evidence which doesn't fit his narrative. Tony Turnbow and Thomas Danisi have written articles for *We Proceeded On*. Turnbow's article, "The Man Who Abandoned Meriwether Lewis," appeared in the May, 2012 issue (Vol. 38, No. 2, pp 20-31). Danisi's article, "The Real James Neelly: Meriwether Lewis's Caretaker," appeared in the November, 2014 issue (Vol. 40, No. 4, pp. 9-26). Turnbow's letter of rebuttal will appear in the issue of May, 2015 issue of WPO. See pp. 16 and 18-19 of Danisi's article for the court case.

Danisi would like to disprove Neelly's appearance in court on October 11th, by saying that Neelly did not have to appear in court because he hired an attorney who represented him. Neelly signed a personal bond guaranteeing his appearance in court, and a jury was seated—both of which Danisi admits—which proves that he was in court on that day.

Danisi admits the Neelly letter was written by John Brahan, but claims it was signed by James Neelly, despite the evidence of a forged signature. He refuses to deal with the paid newspaper announcement created by Neelly at the Indian Agency on October 3rd, because it doesn't fit in with his scenario, which has a different travel route and a different timeline. He says the standard travel time on horseback was 10–15 miles a day, and that they did not have enough time to go to the Indian Agency. Travelers on horseback routinely traveled 30 miles a day or more. Postal riders made 50 miles a day. Danisi wants to paint a picture of Neelly as a kind and compassionate man. He writes:

> There was no else at the fort who could have guided Lewis as ably to Nashville. Neelly's willingness to take on the obviously ill and perhaps troubled Lewis at this point reveals a level of compassion and altruism consistent with a responsible caretaker, rather than an inconsiderate man who would abandon Lewis in his time of need.

Danisi is defending the Neelly letter and the story of Lewis's suicide as genuine. He never discusses the letter written by John Marks about his unsuccessful attempt to retrieve Lewis's pistols, dirk and gold watch from Neelly in 1811. Marks, however, did manage to retrieve the rifle and horse from Mrs. Neelly. "Compassion and altruism" are hardly the words for Neelly's behavior." ("The Real James Neelly," p. 12).

The court martial charges come from the trial transcript.

744. Jackson, *Letters*, Document 362, Jefferson to Paul Allen, August 18, 1813. pp. 586-593. Starrs & Gale, *Death of Meriwether Lewis,* pp. 257-59.

Bibliography

Books

Abernethy, Thomas Perkins, *The South in the New Nation, 1789–1819* (Baton Rouge, Louisiana State University Press, 1961, 1989).

———*Three Virginia Frontiers* (Gloucester MA, Peter Smith, 1962).

Adams, Henry, *Documents Relating to New England Federalism, 1800–1815* (Kessinger Publishing's Rare Reprints).

Ambrose, Stephen, *Undaunted Courage: Meriwether Lewis, Thomas Jefferson, and the Opening of the American West* (New York, Touchstone Press, Simon & Schuster, 1996).

Ammon, Harry, *The Genet Mission* (New York, Norton & Co, 1973).

Anderson, Sarah Travers Lewis (Scott), *Lewises, Meriwethers and Their Kin* (Baltimore, Genealogical Publishing Co Inc, 1984).

Bakeless, John, *Background to Glory: The Life of George Rogers Clark* (Philadelphia and New York, J. B. Lippincott co, 1957).
———*Lewis and Clark: Partners in Discovery* (New York, William Morrow & Co, 1947).

Baldwin, Leland D., *Whiskey Rebels: The Story of a Frontier Uprising* (Pittsburgh, University of Pittsburgh Press, 1939).

Betts, Robert B., with a new epilogue by James J. Holmberg, *In Search of York: The Slave Who Went to the Pacific with Lewis and Clark, Revised Edition* (Boulder, Co, Colorado Associated University Press, originally published 1985, revised edition, 2000).

Bodley, Temple, *George Rogers Clark: His Life and Public Services* (Boston and New York, Houghton Miflin Co., 1926).

Brackenridge, Henry Marie, *Journal of a Voyage Up the Missouri River in 1811* (Massachusetts, Applewood Press, reprint of second edition, 1816).

———*Recollections of Persons and Places in the West* (Bibliolife, 2009, originally published 1868),

Bradbury, John, *Travels in the Interior of American in the Years 1809, 1810, and 1811* (Google Books and University of Nebraska Press, 1986).

Brown, Charles Raymond, *The Northern Confederacy: According to the Plans of the Essex Junto, 1796-1814* (Kessinger Publishing's Rare Reprints, first published 1915).

Brown, John Mason, *The Political Beginnings of Kentucky* (Google eBooks 1890).

Buckley, Jay H., *William Clark: Indian Diplomat* (Norman and London, University of Oklahoma Press, 2008).

Burr, Aaron, *The Political Correspondence and Public Papers of Aaron Burr,* edited by Mary-Jo Kline (Princeton, Princeton University Press, 2 volumes, 1983).

Burton, Clarence, *The City of Detroit Michigan, 1701–1922,* vol 1 (Google eBook, 1922).

Capon, Lester J., editor, *The Adams-Jefferson Letters: The Complete Correspondence between Thomas Jefferson & Abigail & John Adams,* (University of North Carolina Press, 1959, renewed 1987 by Stanley B. Cappon).

Carstens, Kenneth C. and Nancy Son Carstens, editors, *The Life of George Rogers Clark, 1752-–1816: Triumph and Tragedies* (Wesport, CT and London, Praeger, 2004).

Carter, Clarence Edwin, *The Territorial Papers of the United States, Vol. XIV: The Territory of Louisiana-Missouri, 1806–1814* (Washington, U. S. Gov't Printing Office, 1949).

Chandler, David Leon, *The Jefferson Conspiracies: A President's Role in the Assassination of Meriwether Lewis* (New York, William Morrow and Company, 1994).

Cheney, Lynne, *James Madison: A Life Reconsidered* (New York, Viking, 2014).

Chernow, Ron, *Alexander Hamilton* (New York, The Penguin Press, 2004).

Childs, Frances Sergeant, *French Refugee Life in the United States, 1790–1800: An American Chapter of the French Revolution* (Baltimore, Johns Hopkins Press, 1940, Google Book reprint).

Chuinard, Eldon G., M.D., *Only One Man Died: The Medical Aspects of the Lewis and Clark Expedition* (Fairfield, WA, Ye Galleon Press,1979).

Claiborne, William, *Official Letter Books of W. C. C. Claiborne, 1801–1816,* edited by Dunbar Rowland (Nabu Public Domain Reprints, 2012, first published, 1917), Volume 4 (Sept 5, 1806–November 5, 1809).

———*Official Letter Books of W. C. C. Claiborne, 1801–1816* (Reprint from the collections of the University of California Libraries), first published 1917) Volume 5.

Clark, Daniel, *Proofs of the Corruption of Gen. James Wilkinson, and of His Connexion with Aaron Burr: A Full Refutation of His Slanderous Allegations in Relation to the Character of the Principal Witness Against Him* (Honolulu, University Press of the Pacific, 2005/1809); see also Google eBooks.

Cook, Warren L., *Flood Tide of Empire: Spain and the Pacific Northwest, 1543-1819* (New Haven and London, Yale University Press, 1973).

Cooper, Jean L., *A Guide to Historic Charlottesville & Albemarle County* (Charleston & London, History Press, 2007).

Cooper, Linda &Alice Roker with the Town of Yorktown, *Images of America: Yorktown* (Charleston SC, Arcadia Publishing, 2003).

Côte, Richard N.*Theodosia: Theodosia Burr Alston: Portrait of a Prodigy* (Mt. Pleasant, SC, Corinthian Books, 2003).

Coues, Elliott, editor, *The Expeditions of Zebulon Montgomery Pike* (2 volumes, Dover Publication, orginal publication, 1895).

Cowan, Helen I., *Charles Williamson: Genesee Promoter* (Clifton NJ, Augustus M. Kelley Publishers, 1973).

Crackel, Theodore, *Mr. Jefferson's Army: Political and Social Reform of the Military Establishment, 1801-1809* (NY & London, New York University Press, 1987).

Crotty, Gene, *The Visits of Lewis & Clark to Fincastle, Virginia* (History Museum and Historical Society of Western Virginia, 2003).

Cutright, Paul Russell, *A History of the Lewis and Clark Journals* (Norman, University of Oklahoma Press, 1976).

———*Lewis & Clark Pioneering Naturalists* (Lincoln and London, University of Nebraska Press, 1969).

Danisi, Thomas, *Uncovering the Truth About Meriwether Lewis* (Amherst, NY, Prometheus Books, 2012).

——— and John C. Jackson, *Meriwether Lewis* (Amherst, NY, Prometheus Books, 2009).

Darby, John F., *Personal Recollections of John F. Darby: Mayor of St. Louis, 1835* (St. Louis, Hawthorne Publishing Company, 1880).

Davis, Richard Beale, editor, *Jeffersonian America: Notes on the United States of America Collected in the Years 1805-6-7 and 11-12 by Sir Augustus John Foster, Bart.* (Westport CT, Greenwood Press, 1954).

Deacon, Richard, *Madoc and the Discovery of America: Some New Light on an Old Controversy* (New York, George Braziller, 1966).

De Conde, Alexander, *The Quasi-War: The Politics and Diplomacy of the Undeclared War with France, 1797-1801* (New York, Charles Scribner's Sons, 1966).

Dillon, Richard, *Meriwether Lewis: A Biography* (Lafayette CA, Great Western Books, 2003).

Dolan, Eric Jay, *Fur, Fortune, and Empire: The Epic History of the Fur Trade in America* (New York and London, W.W. Norton & Company, 2010).

Douglas, Walter B., annotated and edited by Abraham Nasatir, *Manuel Lisa* (New York, Argosy-Antiquarian, 1964).

Dupre, Daniel S., *Transforming the Cotton Frontier: Madison County, Alabama 1800–1840* (Baton Rouge and London, Louisiana State University Press, 1997).

Egerton, Douglas R. *Gabriel's Rebellion: The Virginia Slave Conspiracies of 1800 & 1802* (Chapel Hill, University of North Carolina Press, 1993).

Elkins, Stanley and Eric McKitrick, *The Age of Federalism: The Early American Republic, 1788-1800* (New York and Oxford, Oxford University Press, 1993).

Ellicott, Andrew, *The Journal of Andrew Ellicott* (Google eBook, 1803).

Ewers, John C., *The Blackfeet: Raiders on the Northwestern Plains* (Norman, University of Oklahoma Press, 1958).

Fisher, Vardis, *Suicide or Murder?:The Strange Death of Governor Meriwether Lewis* (Athens, Ohio, Swallow Press, Ohio University Press, 1962).

Fischer, David Hackett, *Albion's Seed: Four British Folkways in America* (New York and Oxford, Oxford University Press, 1989).

———and James C. Kelly, *Bound Away: Virginia and the Westward Movement* (Charlottesville and London, University Press of Virginia, 2000).

Fleming, Thomas, *Duel: Alexander Hamilton, Aaron Burr and the Future of America* (Basic Books, A Member of the Perseus Books Group, 1999).

Flores, Dan L., *Southern Counterpart to Lewis and Clark: The Freeman and Custis Expedition of 1806* (Norman, University of Oklahoma Press, 1984).

Foley, William E., *The Genesis of Missouri: From Wilderness Outpost to Statehood* (Columbia and London, University of Missouri Press,1989.

——— *Wilderness Journey: The Life of William Clark* (Columbia, MO, University of Missouri Press, 2004).

Fortier, Alcée ,*A History of Louisiana, The Spanish Domination and the Cession to the United States, 1769-1803* (Google eBooks, 1904) Volume 2.

Frick, Ruth Colter, *Courageous Colter and Companions* (Washington, Missouri, 1997).

Furstenberg, François, *When the United States Spoke French: Five Refuges Who Shaped a Nation* (New York, The Penguin Press, 2014).

Gale, Kira and James E. Starrs, *The Death of Meriwether Lewis: A Historic Crime Scene Investigation* (Omaha, Nebraska River Junction Press, 2009).

——— *The Death of Meriwether Lewis: A Historic Crime Scene Investigation* (Omaha Nebraska River Junction Press, 2012 New Evidence Edition).

Gaffney, Richard M., *Bon Appetit! and the The Life & Times of George P. Drouillard. Lewis & Clark's Shawnee Hunter and Interpreter* (Bloomington, Indiana, Author House, 2006).

Gilmer, George, *Sketches of Some of the First Settlers of Upper Georgia* (Baltimore, Genealogical

Publishing Co. Indexed Edition, 1855/1965).

Gitlin, Jay, *Bourgeois Frontier: French Towns, French Traders & American Expansion* (New Haven & London, Yale University Press, 2010).

Graf, Alan D., *Bayonets in the Wilderness: Anthony Wayne's Legion in the Old Northwest* (Norman, University of Oklahoma Press, 2004).

Green, Thomas Marshall, *The Spanish Conspiracy: A Review of Early Spanish Movements in the South-West* (Gloucester MA, Peter Smith, 1967).

Guice, John D. W., *By His Own Hand? The Mysterious Death of Meriwether Lewis* (Norman, University of Oklahoma Press, 2006).

Hall, Ted S., *The Trail Between The Rivers: The Travels of Meriwether Lewis and William Clark; August 24 thru September 26, 1805,* (Stevensville, MT, Stoneydale Press Publishing Company, 2000), Volume 1.

Hartley, Robert E., *Lewis and Clark in the Illinois Country: The Little Told Story* (Westminster CO, Sniktau Publications, 2002).

Hay, Thomas Robson and M.R. Werner, *The Admirable Trumpeteer: A Biography of General James Wilkinson* (Garden City NY, Doubleday, Doran & Company, 1941).

Haynes, Robert V., *The Mississippi Territory and the Southwest Frontier, 1795–1817,* (Lexington, University Press of Kentucky, 2010).

History of Erie County, Pennsylvania (Google eBook, 1884)

Hodes, Frederick A., *Beyond the Frontier: A History of St. Louis to 1821* (Tucson, Arizona, 2004).

Holland, Leandra Zim, *Feasting and Fasting with Lewis & Clark: A Food and Social History of the Early 1800s* (Emigrant, Montana, Old Yellowstone Publishing, 2003).

Holmberg, James, editor, *Dear Brother: Letters of William Clark to Jonathan Clark* (New Haven & London, Yale University Press, and the Filson Historical Society, 2002).

——— editor, *Exploring with Lewis and Clark: The 1804 Journal of Charles Floyd* (Norman, University of Oklahoma Press, 2004).

Holmes, Jack D., *Gayoso: The Life of a Spanish Governor in the Mississippi Valley, 1789–1799* (Baton Rouge, Lousiana State University Press, 1965).

Hurt, R. Douglas, *The Ohio Frontier: Crucible of the Old Northwest, 1720-1830* (Bloomington & Indianapolis, Indiana University Press, 1996).

Isenberg, Nancy, *Fallen Founder: The Life of Aaron Burr (NY, Viking, 2007).*

Jackson, Donald, *The Journals of Zebulon Montgomery Pike with Letters and Related Documents* (Norman, University of Oklahoma Press, 1966, 2 volumes),

———*Letters of the Lewis and Clark Expedition with Related Documents, 1783–1854* (Urbana and Chicago, University of Illinois Press, 1978, 2 volume second edition).

——— *Thomas Jefferson & the Stony Mountains: Exploring the West from Monticello* (Urbana, University of Illinois Press, 1981 and Norman, University of Oklahoma Press, 1993).

Jackson, John C., *The Piikani Blackfeet: A Culture Under Seige* (Missoula MT, Mountain Press Publishing Company, 2000).

———and Thomas Danisi, *Meriwether Lewis* (Amherst, NY, Prometheus Books, 2009).

Jacobs, James Ripley, *The Beginning of the U. S. Army, 1783–1812, (*Princeton NJ, Princeton University Press, 1947)

———*Tarnished Warrior: Major-General James Wilkinson* (New York, The Macmillan Company, 1938).

James, James Alton, *The Life of George Rogers Clark* (Chicago IL, University of Chicago Press, 1928/1978).

James, Thomas and Walter B. Douglas (editor, with notes & biographical sketches), *Three Years Among the Indians and Mexicans* (St. Louis, Missouri Historical Society, 1916). The reprint edition by BiblioLife contains the complete version of James's acount, with appendices.

Jefferson, Thomas, *The Family Letters of Thomas Jefferson* (Charlottesville, University of Virginia Press, 1986).

———*Notes on the State of Virginia* (original publication, 1787, Google books).

———*Papers of Thomas Jefferson, Retirement Series*, Vol. 2, 16 November 1809 to 11 August 1810 (Princeton and Oxford, Princeton University Press, 2005).

Jones, Landon Y., *William Clark and the Shaping of the West* (New York, Hill and Wang, Division of Farrar, Straus and Giroux, 2004).

Josephy, Alvin M., Jr., *The Nez Perce Indians and the Opening of the Northwest* (New Haven and London, Yale University Press, 1965).

Kaser, David, *Joseph Charless: Printer in the Western Country* (Philadelphia, University of Pennsylvania Press, 1963).

Kennedy, Roger G., *Burr, Hamilton and Jefferson: A Study in Character* (Oxford and New York, Oxford University Press, 2000).

———*Cotton and Conquest: How the Plantation System Acquired Texas* (Norman, University of Oklahoma Press, 2013).

Kennerly, William Clark, as told to Elizabeth Russell, *Persimmon Hill: A Narrative of Old St. Louis and the Far West* (Norman, University of Oklahoma Press, 1948).

Ketcham, Ralph, *James Madison: A Biography* (Charlottesville and London, University of Virginia Press, 1990).

Kline, Mary-Jo, *Political Correspondence and Public Papers of Aaron Burr* (Princeton NJ, Princeton University Press, 1983), two volumes.

Knopf, Richard C., *Anthony Wayne: A Name in Arms* (Pittsburgh, University of Pittsburgh Press, 1960). This is a collection of letters edited and transcribed by Knopf.

Koch, Adrian, *Jefferson and Madison: The Great Collaboration* (Old Saybrooke, CT, Konecky & Konecky, no date).

Kohn, Richard H., *Eagle and Sword: The Beginnings of the Military Establishment in America* (New York & London, The Free Press, A Division of MacMillan Publishing Co., 1975).

Kukla, Jon, *A Wilderness So Immense: The Louisiana Purchase and the Destiny of America* (New York, Alfred A. Knopf, 2003).

Larson, Edward J., *A Magnficent Catastrophe: The Tumultuous Election of 1800, America's First Presidential Campaign* (NY, Free Press, A Division of Simon & Schuster, 2007).

Léger, Jacques Nicolas, *Haiti, Her History and Her Detractors* (Classic Reprint first published 1907). Available at www.archive.org.

Lewis, Meriwether, *The Definitive Journals of Lewis & Clark*, ed. by Gary Moulton (University of Nebraska Press, 2002), 13 volumes, paperback edition.

Linklater, Andro, *An Artist in Treason: The Extraordinary Double Life of General James Wilkinson* (New York, Walker Publishing Co. 2009).

Liss, Peggy, *Atlantic Empires: The Network of Trade and Revolution, 1713-1826* (Baltimore, Johns Hopkins University Press, 1983).

Lomask, Milton, *Aaron Burr: The Conspiracy and Years of Exile, 1805-1836* (New York, Farrar, Straus, Giroux, 1982).

Loomis, Noel M. and Abraham P. Nasatir, *Pedro Vial and the Raods to Sante Fe* (Norman, University of Oklahoma Press, 1967).

Lowry, Thomas P., *Venereal Disease and the Lewis and Clark Expedition* (Lincoln and London, University of Nebraska Press, 2004).

Lytle, Richard M., *The Soldiers of America's First Army, 1791* (Lanham MD, Toronto, Oxford, The Scarecrow Press, 2004).

Malone, Dumas, *Jefferson and the Ordeal of Liberty* (Boston, Little Brown and Company, 1962).

———*Jefferson The President: First Term, 1801-1805* (Boston, Little Brown & Company, 1970).

Mann, John W. W., *Sacagawea's People: The Lemhi Shoshones and the Salmon River Country* (Lincoln and London, University of Nebraska Press, 2004).

Marichal, Carlos, *Bankruptcy of Empire: Mexican Silver and the Wars Between Spain, Britain and France, 1760–1810* (New York, Cambridge University Press, 2007).

Marshall, Humphrey, *The History of Kentucky Exhibiting an Account of the Modern Discovery; Settlement; Progressive Improvement; Civil and Military Transactions; and the Present State of the Country.* (Google eBooks, 1824).

Marshall, Thomas Maitland, *The Life and Papers of Frederick Bates* (St. Louis, Missouri Historical Society, 1926). Two volumes.

Matthews, Catharine Van Cortlandt, *Andrew Ellicott: His Life and Letters* (New York, The Grafton Press, 1908; Google eBook).

McCallum, James, *A Brief Sketch of the Settlement and Early History of Giles County, Tennessee* (Greenville, SC, Southern Historical Press, original publication date 1876).

McCullough, David, *John Adams* (New York, Simon & Schuster, 2001).

McDonald, Dr. William Lindsey, *Lore of the River: The Shoals of Long Ago* (Bluewater Publications, USA, expanded 3rd edition, 2007).

Melton, Buckner F. , Jr. *The First Impeachment: The Constitution's Framers and the Case of Senator William Blount* (Macon, GA, Mercer University Press, 1998).

Michaux, André, *Journal of André Michaux*, 1793–1796 (Nabu Public Domain Reprints, no date).

Miller, John C., *Alexander Hamilton, Portrait in Paradox* (Konecky & Koneky, by special arrangement with HarperCollins Publishers, 1959).

Miller, Robert J., *Native America, Discovered and Conquired: Thomas Jefferson, Lewis and Clark, and Manifest Destiny* (Westport CT and London, Praeger, 2006).

Moulton, Gary E., editor, *The Definitive Journals of Lewis & Clark* (University of Nebraska Press, 2002), 13 volumes, paperback edition.

Nasatir, Abraham P., editor, *Before Lewis and Clark: Documents Illustrating the History of the Missouri, 1785-1804* (Lincoln and London, University of Nebraska Press, 1990). Two volumes.

——— *Borderland in Retreat: From Spanish Louisiana to the Far Southwest* (Albuquerque, Uni versity of New Mexico Press, 1976).

Nelson, Paul David, *Anthony Wayne: Soldier of the Early Republic* (Bloomington, Indiana University Press, 1985).

Newmyer, R. Kent, *The Treason Trial of Aaron Burr: Law, Politics and the Character Wars of the New Nation* (Cambridge, Cambridge University Press, 2012).

Oberg, Barbara, editor, *The Papers of Thomas Jefferson Retirement Series*, Vol. 2, 16 November 1809 to 11 August 1810 (Princeton and Oxford, Princeton University Press, 2005).

Oglesby, Richard E., *Manuel Lisa and the Opening of the Missouri Fur Trade* (Norman and London, University of Oklahoma Press, 1963.)

Osgood, Ernest Staple, editor, *The Field Notes of Captain William Clark, 1803-1805* (New Haven and London, Yale University Press, 1964).

Quisenberry, A. C. *The Life and Times of Hon. Humphrey Marshall* (Google eBooks, 1892).

Peck, David J., D.O., *Or Perish in the Attempt: Wilderness Medicine in the Lewis & Clark Expedition* (Helena, MT, Far Country Press, 2002).

Peterson, Merrill D., *Thomas Jefferson and the New Nation : A Biography* (London, Oxford, New York, Oxford University Press, 1970).

Pike, Zebulon Montgomery, *The Expeditions of Zebulon Montgomery Pike*, edited by Elliott Coues (2 volumes, Dover Publication, orginal publication, 1895).

Pinkham, Allen V. and Steven R. Evans, *Lewis and Clark Among the Nez Perce: Strangers in the Land of the Nimiipuu* (Washburn, North Dakota, The Dakota Institute Press of the Lewis & Clark Fort Mandan Association, 2013).

Potter, Tracy, *Sheheke, Mandan Indian Diplomat: The Story of White Coyote, Thomas Jefferson and Lewis and Clark* (Far Country Press and Fort Mandan Press, 2003).

Potts, Gwynne Tuell & Samuel W. Thomas, *George Rogers Clark and Locust Grove: Military Leader in the Pioneer West and The Croghan Homestead Honoring Him* (Louisville, Historic Locust Grove, 2006).

Ravenswaay, Charles van, *St Louis: An Informal History of the City and Its People, 1764-1865* (St. Louis, Missouri Historical Society Press, 1991).

Rhyss, Isaac, *The Transformation of Virginia, 1740–1790* (Williamsburg VA, Omohundro Institute of Early American History and Culture and Chapel Hill NC, The University of North Carolina Press, 1982).

Robinson, James Alexander and Paul Alliot, *Louisiana Under the Rule of Spain, France, and the United States, 1785-1807; Social, Economic, and Political Conditions of the Territory Represented in the Louisiana Purchase* (ULAN Press sold through Amazon, 2011, original publication date, 1923).

Rodney, Thomas, *A journey Through the West: Thomas Rodney's 1803 Journal from Delaware to the Mississippi Territory*, edited by Dwight L. Smith and Ray Swick (Ohio University Press, 1997).

Rodriguez, Junius P., editor, *The Louisiana Purchase: A Historical and Geographical Encyclopedia* (Santa Barbara, ABC CLIO, 2002).

Ronda, James P., *Lewis and Clark Among the Indians* (Lincoln and London, University of Nebraska Press, 1984).

Salish-Pend d'Oreille Culture Committee and Elders Cultural Advisory Council Confederated Salish and Kootenai Tribes, *The Salish People and the Lewis and Clark Expedition* (Lincoln & London, University of Nebraska Press, 2005).

Scharf, John Thomas, *History of St Louis City, From the Earliest Periods to the Present Day* (1883, Google eBook).

Schroeder, Walter A., *Opening the Ozarks: A Historical Geography of Missouri's Ste. Genevieve District, 1760–1830* (Columbia and London, University of Missouri Press, 2002).

Seale, William, *The President's House: A History* (White House Historical Association and National Geographic, 1986) two volumes.

Sellers, Charles Coleman, *Charles Willson Peale: A Biography*, (New York, Charles Scribner's Sons, 1969).

Semonin, Paul, *American Monster: How the Nation's First Prehistoric Creature Became a Symbol of*

National Identity (New York & Londoln, New York University Press, 2000),

Sharp, James Roger, *American Politics in the Early Republic: The New Nation in Crisis* (New Haven, Yale University Press, 1993).

Shreve, Royal Ornan, *The Finished Scoundrel: General James Wilkinson Commander-in-Chief of the Army of the United States, who made intrigue a trade and treason a profession* (Indianapolis, Bobbs-Merrill Company, 1933).

Skarsten, M. O., *George Drouillard: Hunter and Interpreter for Lewis & Clark and the Fur Trade, 1807-1810* (Spokane WA, The Arthur H. Clark Company, 2003).

Slaughter, Thomas P., *The Whiskey Rebellion: Frontier Eipilogue to the American Revolution* (New York and Oxford, Oxford University Press, 1986).

Smith, Margaret Bayard, *First Forty Years of Washington Society: Portrayed by the Family Letters of Mrs. Samuel Harrison Smith (Margaret Bayard) From the Collection of Her Grandson J. Henley Smith* (Nabu Public Domain Reprints, no date, original date of publication, 1906).

Squier, Ephraim G. and Edwin H. Davis, edited and with an introduction by David J. Meltzer, *Ancient Monuments of the Mississippi Valley* (Washington and London, Smithsonian Press, 1998; original date of publication, 1848).

Starrs, James E. and Kira Gale, *The Death of Meriwether Lewis: A Historic Crime Scene Investigation* (Omaha, Nebraska River Junction Press, 2009).

——— *The Death of Meriwether Lewis: A Historic Crime Scene Investigation* (Omaha, Nebraska River Junction Press, 2012 New Evidence Edition).

Steward, Dick, *Frontier Swashbuckler: The Life and Legend of John Smith T* (Columbia, Missouri, University of Missouri Press, 2000).

Stewart, David O., *American Emperor: Aaron Burr's Challenge to Jefferson's America* (New York, Simon & Schuster, 2011).

Sugden, John, *Tecumseh's Last Stand* (Norman and London, University of Oklahoma Press, 1985).

Swagerty, William R., *The Indianization of Lewis and Clark* (Norman, OK, The Arthur H. Clark Company, a division of the University of Oklahoma Press, 2012,). Two volumes.

Swayne, Zoa L., *Do Them No Harm!: Lewis and Clark Among the Nez Perce* (Caldwell, Idaho, Caxton Press, 2003).

Syrett, Harold C., *The Papers of Alexander Hamilton* (NY, Columbia University Press, 1977).

Thompson, David, *David Thompson's Narrative of His Explorations in Western America, 1784-1812* (Google Books).

Turner, Frederick Jackson, *The Policy of France Toward the Mississippi Valley in the Period of Washington and Adams* (Google eBook, 1905).

Unger, Harlow Giles, *The French War Against America: How a Trusted Ally Betrayed Washington and the Founding Fathers* (Hoboken NY, John Wiley & Sons, Inc., 2005).

Warren, Harrison Gaylord, *The Sword Was Their Passport: A History of American Filibustering in the Mexican Revolution* (Port Washington NY/London, Kennikat Press, 1943).

Wheeler, Olin D., *The Trail of Lewis and Clark* (Scituate MA, DSI Digital Scanning, Inc. 2002, reprint of 1904 edition, two volumes).

Whitcomb, John and Claire Whitcomb, *Real Life at the White House: Two Hundred Years of Daily Life at America's Most Famous Residence* (New York and London, Routledge, 2000).

Whitaker, Arthur Preston, *The Mississippi Question, 1795–1803: A Study in Trade, Politics and Diplomacy* (New York and London, D. Appleton–Century Company Inc. and the American

Historical Association, 1934).

Wilkinson, James, *Memoirs of My Own Times* (Philadelphia, 1816, three volumes, reprint). Available on internet.

———*Burr's Conspiracy Exposed and General Wilkinson Vindicated Against the Slander of His Enemies* (1811). Available as reprint without appendices from various sources. The complete book may be found at the University of Pittsburgh Darlington Memorial Library, and on Internet Archive at www.archive.org, where it is available for download. The book is 99 pages, and the appendix is an additional 136 pages.

Williams, Gwyn A., *Madoc: The Legend of the Welsh Discovery of America* (Oxford & New York, Oxford University Press, 1987).

Wilson, Maurine T. and Jack Jackson, *Philip Nolan and Texas: Expeditions to the Unknown Land, 1791-1801* (Waco TX, Texian Press, 1987).

Winkler, John F., *Fallen Timbers 1794: The US Army's First Victory* (Oxford and Long Island City NY, Osprey Publishing, 2013).

Witthaus, R. A. and Tracy Becker, *Medical Jurisprudence Forensic Medicine and Toxicology,* vol. 4 (Google eBook, 1896).

Wolferman, Kristie C., *The Osage in Missouri* (Columbia and London, University of Missouri Press, 1997).

Wood, W. Raymond, *Prologue to Lewis and Clark: The Mackay and Evans Expedition* (Norman, University of Oklahoma Press, 2003).

Woodger, Elin and Brandon Toropov, *Encyclopedia of the Lewis & Clark Expedition* (NY, Checkmark Books, imprint of Facts on File, 2004).

Woods, Edgar, *History of Albemarle County* (Albemarle VA, Michie Co., 1901; reprint, Google Books).

Wright, Gen. Marcus J., *Some Account of the Life and Services of William Blount* (Google eBooks, 1884)

Ziak, Rex, *Down and Up the Columbia River: A unique fold-out guide, mapping day-by-day Lewis and Clark's journey from the Rocky Mountains to the Pacific Ocean—and back* (Astoria, OR, Moffitt House Press, 2005).

Articles, Manuscripts, Records, Chapters & Papers

Adams, Mary P., "Jefferson's Reaction to the Treaty of San Ildefonso," *The Journal of Southern History*, Vol. 21, No. 2, May, 1955, pp. 173-188.

Avery, Martin Cooper, "Tragedy at Grinder's Stand: The Death of Meriwether Lewis," a Master's thesis at Middle Tennessee State University, 1978.

Bellesiles, Michael A. "The Soil Will Be Soaked with Blood," a chapter in *The Revolution of 1800: Democracy, Race & the New Republic,* edited by James Horn, Jan Ellen Lewis, & Peter S. Onuf (Charlottesville, University of Virginia Press, 2002)

Bissell, Daniel. Bissell's statement concerning Burr's activities in the vicinity of Fort Massac is found at *Letters Received by the Secretary of War*, Registered Series, 1801-1870, Microfilm M221, Roll 4 (Vol.3, May 1, 1806–Dec. 31, 1808), National Archives and Records Administration. Bissell was the fort commander. A transcript is on the internet at chribbs.wordpress.com. Look under documents/Chribbs/1807-Daniel Bissell's Account of Aaron Burr.

Buckley, Jay H., "William Clark: Superintendent of Indian Affairs, 1813-38," a dissertation, University of Nebraska-Lincoln, www.digitalcommons.unl.edu.

——— "A Postmortem Trial concerning Meriwether Lewis's Controversial Death," a chapter in John Guice's book, *By His Own Hand?*

Burke, Henry Rosecrans, Papers (Missouri History Museum Archives, St. Louis). "Notes on John Smith T." (Folder 6); "Moses Austin and the Lead for Aaron Burr" (Folder 6); "Design for Treason"(Folder 8).

Caldwell, Norman. "Cantonment Wilkinsonville," *Mid-America: An Historical Review,* 31:1, Jan, 1949. pp. 3-26.

Clark, George Rogers, "Letter from George Rogers Clark to George Mason" *First American West: The Ohio River Valley, 1750-1820,* American Memory, The Library of Congress, http://memory.loc.gov

Clark, William, "Journal of General Anthony Wayne's Campaign Against the Shawnee Indians in Ohio, 1794-1795," edited by R. C. McGrane (*Mississippi Valley Historical Review,* vol. 1, Dec. 1914; Google eBooks)

———William Clark Papers, Missouri History Museum.

Coleman, C. W., "Geneaology of the Smith Family of Essex County, Virginia," *William and Mary Quarterly* Vol. 25, No. 3 (January, 1917).

Coues, Elliott, "Letters of William Clark and Nathaniel Pryor," *The Annals of Iowa* (Des Moines, State Historical Society of Iowa), Vol 1, No. 8, 1895. Available as a pdf download at Iowa Research Online, ir.uiowa.edu.

Cova, de la, Rafael, "Filibusters and Freemasons: The Sworn Obligation," *Journal of the Early Republic,* Vol. 17, No. 1 (Spring, 1997), pp. 95-120.

Cox, Isaac Joslin, editor, "Documents Relating to Zachariah Cox," *Quarterly Publication of the Historical and Philosophical Society of Ohio*, vol. VIII, Nos. 2 &3, April-June, July-Sept.,1913.

———"General Wilkinson and His Later Intrigues with the Spaniards," *American Historical Review,* Vol XIX, No. 4 (July, 1914), pp. 794-812.

——— "Opening the Sante Fe Trail," *Missouri Historical Review*, Vol. 25 (Oct. 1930), pp. 30-66.

Dabney, Virginus, "Captain Jack Jouett, Junior: American Revolutionary War Hero (1754-1822)," Jack Jouett Chapter, Daughters of the American Revolution reprint, 1966. Courtesy of the Albemarle Charlottesville Historical Society.

Danisi, Thomas C. and W. Raymond Wood, "Lewis and Clark's Route Map: James MacKay's Map of the Missouri River," *The Western Historical Quarterly,* Vol. 35, No. 1 (Spring, 2004), pp. 53-72 (The Western History Association).

Daviess, J. H., "View of the President's Conduct Concerning the Conspiracy of 1806," edited by Isaac J. Cox and Helen Swineford, *Quarterly Publication of the Historical and Philosophical Society of Ohio,* vol XII: nos. 2 & 3, April-June, July -September, 1917.

Degges, Dan, "The Mexican Association: The Secret Society Behind the Aaron Burr Conspiracy," paper by student at University of Arkansas at Monticello, Alpa-Nu-Zeta Chapter, www.academia.edu.

Earl, John L. III, "Talleyrand in Philadelphia, 1794–1796," *Pennsylvania Magazine of History and Biography,* 91:3(July, 1967).

Earman, Cynthia D., "Remembering the Laides: Women, Etiquette and Diversions In Washington City, 1800-1814," *Washington History*, Vol. 12, No. 1 (Spring/Summer, 2000), pp. 102-117

Fabel, Robin, "Bernard Lintot: A Connecticut Yankee on the Mississippi, 1775-1805," *The Florida Historical Quarterly,* vol. 60, No. 1 (July, 1981).

Gale, Kira, *Defending the Western Frontier: Manuel Lisa and the War of 1812 in the Council Bluffs Area* (Omaha, River Junction Press, 2000), booklet.

Gandolfo, Jerry, www.newman-family-tree-net (Francis Newman research).

Gilmer, George, "Papers of George Gilmer of Pen Park, 1775-1778," *Miscellaneous Papers, 1672-1865* (Virginia Historical Society,1887; Google Books).

Goff, W. A., "Reuben Smith," in *Mountain Men and the Fur Trade of the Far West* , edited by Leroy Hafen, (Lincoln, University of Nebraska Bison Books, 1982) Vol. 7, pp. 261-79.

Hay, Thomas Robson, "Charles Williamson and the Burr Conspiracy," *The Journal of Southern History,* Vol. 2, No. 2 (May. 1936), pp. 175-210.

Heckewelder, John, "Narrative of John Heckewelder's Journey to the Wabash in 1792," *Pennsylvania Magazine of History and Biography* (Google eBooks, 1888) vol. 12.

Hickey, Don, "America's Response to the Slave Revolt in Haiti, 1791–1806," *Journal of the Early Republic,* vol. 2, no. 4 (Winter, 1982), pp. 361-379.

Holden, Robert J., "Clark's Attack on Vincennes, February 5-25, 1779" in *The Life of George Rogers Clark, 1752-1816: Triumphs and Tragedies,* edited by Kenneth C. Carstens and Nancy Son Carstens (Westport CT and London, Praeger, 2004).

Holmberg, James J., "Seaman's Fate," *We Proceeded On (*Lewis and Clark Trail Heritage Foundation), Vol. 26, No. 1 (February, 2000), pp. 7-9. Available on both www.lewisandclark.org (foundation website) and www.lewis-clark.org (Discovering Lewis & Clark website).

Hunt, Robert R., "Crime and punishment on the Lewis and Clark Expedition," *We Proceeded On,* Vol. 15: 2&3, May and August, 1989. www.lewisandclark.org

Jackson, Donald, "Jefferson, Meriwether Lewis, and the Reduction of the United States Army, *Proceedings of the American Philosophical Society,* Vol. 124, No. 2 (April 29, 1980), pp. 91-96.

———"On the Death of Meriwether Lewis's Servant," *The William and Mary Quarterly* (Omuohundro Institute of Early American History and Culture), Vol. 21, No. 3 (July, 1964), pp. 445-448.

———"A Notebook Sketch Among the Lewis & Clark Papers" *We Proceeded On* (Lewis and Clark Trail Heritage Foundation), Vol. 12, No. 3 (August, 1986)

Jarman, Paul, "Madoc, 1795: Robert Southey's Misdated Manuscript," *The Review of English Studies,* New Series, Vod. 55, No. 220 (June, 2004), pp. 355-373 (Oxford University Press).

King, Grace, "The Real Philip Nolan," *Publication of the Louisiana Historical Society, Proceedings & Reports,* vol. X (1917).

Knopf, Richard C., "The Fort Adams Incident," (Columbus, OH, The Anthony Wayne Parkway Board, Ohio State Museum no date) pp. 2-3.

———"Two Journals of the Kentucky Volunteers, 1793 and 1794," *The Filson Club History Quarterly*, vol. XXVII, no. 3, pp. 247-281.

———*Anthony Wayne and the Founding of the United States Army* (Columbus, Ohio, Anthony Wayne Parkway Board, 1961; a typescript of 75 copies).

Koch, Adrienne and Harry Ammons, "The Virginia and Kentucky Resolutions: An Episode in Jefferson's and Madison's Defense of Civil Liberties," *The William and Mary Quarterly,"* Third Series, Vol. 5, No. 2 (April, 1948), pp. 145-176.

Lachance, Paul, "Repercussions of the Haitian Revolution in Louisiana," a chapter in *The Impact of the Haitian Revolution in the Atlantic World,* edited by David P. Geggus (Columbia SC, University of South Carolina Press, 2001).

Lange, Robert E., "Private George Shannon: The Expedition's Youngest Member—1785 or 1787–1836, *We Proceeded On,* Vol 8:3, July, 1982, pp. 10-15. WPO archives at www.lewisandclark.org

Large, Arlen, "Additions to the Party: How An Expedition Grew and Grew," *We Proceeded On* (Lewis and Clark Trail Heritage Foundation) Vol. 16, No.1, Feb, 1990. WPO archives at www.lewisandclark.org.

Lewis, Meriwether Papers, Missouri History Museum.

Lewis-Marks Manuscripts, University of Virginia Small Special Collections Library.

Long, Stephen Meriwether, "British Lieutenant Colonel Banastre Tarleton and the American Revolution: Drama on the Plantations of Charlottesville," *Meriwether Connections,* vol 24, no. 2, Apr-June, 2005. http://homepages.rootsweb.ancestry.com/~meriweth/ (accessed 1/7/13).

Loomis, Noel M. "Philip Nolan's Entry into Texas in 1800," *The Spanish in the Mississippi Valley,* edited by Francis McDermott (Urbana, University of Illinois Press, 1974).

May, Robert E., "Young American Males and Filibustering in the Age of Manifest Destiny: The United States Army as a Cultural Mirror," *The Journal of American History,* Vol. 78, No. 3 (Dec.,1991), pp. 857-886.

Mayer, David P., "Cantonment Wilkinsonville 1801: Burr, Wilkinson, the U. S. Army and Treason," unpublished paper, dated March 7, 1985, Erie, Pennsylvania. The paper is available at the Cairo Public Library, Cairo, Illinois.

McGhee, Peter, "Meriwether Lewis" transcript, University of Virginia Special Collections, MSS 4730-a.

Miller, Grace Lewis, "Grace Lewis Miller Papers," Jefferson National Expansion Memorial, St Louis.

———*His Excellency Meriwether Lewis and the First Publication West of the Mississippi River,* Miller's Master's Thesis, University of Texas-Austin, August, 1948.

——— "The First Home of Governor Lewis in Louisiana Territory" ". . . the house in which General Clark resides." (St. Louis, *Missouri Historical Society Bulletin*), Vol XIV (July, 1958), pp. 357-68. Miller says the Clark family would have lived here at least through early 1809.

Moore, Robert, "The Making of a Myth: Did the Corps of Discovery Actually Eat Candles," *We Proceeded On,* Vol. 25, No. 2, May, 1999,

Moulton, Gary E., "The Missing Journals of Meriwether Lewis," *The Montana Magazine of Western History,* Vol. 35, No. 3 (Summer, 1985), pp. 28-39.

Moyers, William Nelson "A Story of Southern Illinois, the Soldiers Reservations, including the Indians, French Traders, and Some Early Americans," *Journal of the Illinois State Historical Society,* (1908-1984), Vol. 24: No. 1 (Apr, 1931), pp. 26-104.

Musselman, Joseph, founding producer and editor, *Discovering Lewis & Clark* website, www.lewis-clark.org.

Narrett, David E., "Liberation and Conquest: John Hamilton Robinson and U. S. Adventurism toward Mexico, 1806-1819," Western Historical Quarterly, (Utah State University and Western History Association), Vol. 40, No. 1, (Spring 2009), pp.23-50.

Nasatir, Abraham P. "John Evans, Explorer and Surveyor," *Missouri Historical Review,* Vol. 25, Issues 2 , 3, 4, 1931, pp. 219-239, 432-460, 585-608.

Nichols, Roger, "The Arikara Indians and the Missouri River Trade," *Great Plains Quarterly,* Paper 1656 (http://digitalcommons.unl.edu/greatplainsquarterly/1656).

Robinson, Doane, editor, "A Sioux Indians View of the Last War with England," *South Dakota State Historical Collections,* Vol. 5 (1910).

Rusche, Timothy M., "Treachery Within The United States Army," *Pennsylvania History,* vol. 65,

Autumn, 1998, pp. 478-491.

Second Auditors Journals of the Accountants for the War Dept (National Archives, Washington DC, Record Group 217, Entry 366, Volume #8 ("H"), pp. 4302-4305, Account of Meriwether Lewis.

Scofield, Merry Ellen "The Fatigues of His Table: The Politics of Presidential Dining During the Jefferson Administration," *Journal of the Early Republic* (University of Pennsylvania Press), Vol. 26, No. 3 (Fall, 2006), pp. 449-469.

Simmons, David A., *The Forts of Anthony Wayne* (Historic Fort Wayne pamphlet publication, 1977).

Smith, Dwight, ed., "The Ohio River in 1801—Letters of Jonathan Williams, Junior, *The Filson Club Historical Quarterly*, Vol. 27, No. 3, (July, 1953)

Smith, Ronald D. , "Napoleon and Louisiana: Failure of the Proposed Expedition to Occupy and Defend Louisiana," *Louisiana History: The Journal of the Louisiana Historical Association,* Vol. 12, No. 1, Winter, 1971, pp. 21-40.

Turnbow, Tony, "The Man Who Abandoned Meriwether Lewis," *We Proceeded On*, 38:2, May, 2012, pp. 20-31.

Turner, Frederick Jackson, "Documents on the Blount Conspiracy," *American Historical Review,* vol. X, No. 3, April, 1905.

———"Correspondence of Clark and Genet: Selections from the Draper Manuscripts," *Annual Report of the American Historical Association, 1896."* Internet Archive, https://archive.org.

Wagner, Mark, "Officer Country: Clothing and Personal Items from Cantonment Wilkinson (11PU282) in Southern Illinois" *Midcontinental Journal of Archeology*, 37:2, Fall, 2012.

Wagner, Mark, "Searching for Cantonment Wilkinson: An 1801-1802 U.S. Army Post in Pulaski County, Illinois," 2003 [November], Paper Presented At the Illinois Archaeological Survey Annual Meetings, Southern Illinois University, Carbondale, IL.

Wayne, Anthony, Letters Relating to the Death of Maj.-Gen. Anthony Wayne, *The Pennsylvania Magazine of History and Biography,* vol. 19. no. 1 (1895) pp. 112-115.

———"Notes and Queries," *The Pennsylvania Magazine of History and Biography,* vol. 15, no. 2 (1891) pp. 247-248.

Whitaker, Arthur Preston, "James Wilkinson's First Descent to New Orleans in 1787," *The Hispanic American Historical Review,* vol 8: I (February, 1926)

Wilkinson Papers, for correspondence between James Wilkinson and Samuel Harrison Smith, University of Pittsburgh Library. The papers are digitized and available for pdf download. http://digital.library.pitt.edu,

Williams, David, "John Evans' Strange Journey: Part I. The Welsh Indians," *The American Historical Review,* Vol. 54, No. 2 & 3 (Jan. & Apr., 1949) (Oxford University Press on behalf of the American Historical Association), pp. 277-295 and 508-529.

Williams, Gwynn A., "Welsh Indians: The Madoc Legend and the First Welsh Radicalism," *History Workshop*, No. 1 (Spring, 1976) (Oxford, *Oxford Journals*, Oxford University Press).

Williams, Jonathan Archives, Manuscripts Dept., The Lilly Library, University of Indiana, Bloomington.

Witte, Stephen, "Republicanizing the Army: The Military Peace Establishment Act of 1802" in *Discovering Lewis & Clark* at www.lewis-clark.org.

Zontine, Patricia L., "Lucy Meriwether Lewis Marks (1752-1837): Her Life and Her World." (Robert H. Smith International Center for Jefferson Studies, 2009). www.monticello.org.

Index

Search for special category lists to find—assassination targets; boundary lines; campsites; expedition members; forts; Indian tribes; Indian chiefs/leaders; historians and biographers; rivers; traders and trappers; treaties